Key Technologies for Tunnel Construction under Complex Geological and Environmental Conditions

Series Editor

Qihu Qian, Nanjing, China

This series is written on the basis of several difficult projects, aiming at complex geological conditions such as complex topography and landform types, complex geological structure and poor geotechnical engineering geology, systematically describes the Key Technologies for Tunnel Construction under Complex Geological and Environmental Conditions like mountains, underwater, soft and uneven strata, from the aspects of geological exploration, early warning and risk control, design, construction and digital application. It's a monograph that introduces innovative achievements and core technologies formed in the construction of complex geological tunnels in recent years. The series contains both texts and illustrations. It can be used as a reference for engineers, technicians, scientific researchers, teachers and students of related specialties in tunnels and underground engineering.

This series systematically describes the Key Technologies for Tunnel Construction under Complex Geological and Environmental Conditions like mountains, underwater, soft and uneven strata, from the aspects of geological exploration, early warning and risk control, design, construction and digital application.

Contents of the series are mainly from a number of key projects, such as Yuelongmen Tunnel of Chenglan Railway in Sichuan Province, Jinan Yellow River Tunnel Project, Qiyueshan Tunnel of West Hubei Expressway, Jiaozhou Bay Tunnel of Qingdao, Nanjing Yangtze River Tunnel and Wuhan River Crossing Tunnel, as well as the 973 Projects of the State-Mechanism, Prediction, Early Warning and Control of Major Water Inrush Disasters in Deep and Long Tunnels, Science and Technology Project by the Ministry of Transport "Study on Key Technologies of Underwater Large Section Shield Tunnel Construction under Complex Geological Conditions", National Defense Research Project "Three-Dimensional Advance Prediction Technology for National Defense Underground Engineering".

More information about this series at https://link.springer.com/bookseries/16378

Guohou Cao · Hao Liu

Three-Dimensional Exploration Technology of Tunnel Geology

SSTP

Springer

Guohou Cao
Chinese Army Research Institute for Engineering Design
Kunming, Yunnan, China

Hao Liu
Yunnan Aerospace Engineering Geophysical Detecting Co., Ltd.
Kunming, Yunnan, China

Translated by
Xi Yang
Kunming, China

ISSN 2662-2904 ISSN 2662-2912 (electronic)
Key Technologies for Tunnel Construction under Complex Geological and Environmental Conditions
ISBN 978-981-16-9224-6 ISBN 978-981-16-9225-3 (eBook)
https://doi.org/10.1007/978-981-16-9225-3

Jointly published with Shanghai Scientific and Technical Publishers
The print edition is not for sale in China (Mainland). Customers from China (Mainland) please order the print book from: Shanghai Scientific and Technical Publishers.

This Springer imprint is published by the registered company Springer Nature Singapore Pte Ltd.
The registered company address is: 152 Beach Road, #21-01/04 Gateway East, Singapore 189721, Singapore

Foreword by Jun Sun

In recent years, the development of infrastructure construction in China has been developing vigorously. The construction of "The Belt and Road" has been continuously promoted. Many major projects have mushroomed vigorously, such as the Three Gorges Project, the Qinghai-Tibet Railway, the South to North Water Diversion, the three vertical and four horizontal high-speed rail network, the Hong Kong-Zhuhai-Macao Bridge, Shanghai Center Tower, and the construction with China's aid of the Jakarta-Bandung high-speed railway, the China-Laos railway, the China-Thailand railway, the Gwadar Port, the Piraeus port, and so on. There is no doubt that China has become one of the countries with the largest number of major projects in the world. In terms of construction scale, technical difficulty and capital investment, these major projects are among the top not only in China, but also around the world, and even in the first place. A series of major key technical problems emerged in the construction process of these projects, many of which have been well optimized and solved through analysis, exploration, and innovation. Some even created new technical means and methods on the basis of the original theory and technology, and applied for a large number of technical patents. For example, the Shanghai Center Tower, 632 m, as the world's tallest green building, has achieved many scientific achievements in many aspects such as super high-rise design, green construction, construction supervision, and building information modeling (BIM) technology. It has applied for eight invention patents and authorized 12 practical new technologies. Only in the aspect of structural engineering, it has been applied to many innovative technological innovations, such as the support technology of super deep foundation pit, the technology of super high pumping concrete, the installation technology of complex steel structure and the technology of structural crack control, some of which have reached the world level. These optimizations, breakthroughs, and innovations will be very valuable references for Chinese engineers and technicians.

During the plenary session of the National People's Congress held in early March 2016, many representatives said that a great deal of technological innovation and development is a key driving factor for realizing China's macroeconomic strategic adjustment during the 13th Five Year Plan period, and a fundamental support and key driving force for realizing the overall development of the country under the overall layout.

At the same time, in the face of the opportunity of a new round of scientific and technological revolution, only by making innovations and breakthroughs in key core technologies one by one can we realize the overall leap of social productivity, make the level and ability of our scientific research achievements and engineering technology control enter the ranks of developed countries as early as possible and as quickly as possible, so as to continuously improve the technological competitiveness in the world, and the national strength will be stronger! At present, many engineering technology innovation has been widely recognized, but there are still many problems in the promotion and application of innovation results. In the field of major engineering construction, after the key engineering technical problems have been broken through and solved in practice, it is necessary to further sort out and summarize the new theories or methods, and then widely apply them to production practice again, which in turn will promote further innovation and development of technology, which is a huge driving force for the sustainable development of technology. One of the most effective aspects is to systematically summarize the innovative achievements and publish a set of technical monographs. This is also the significance of publishing the series of "Research on Key Technologies of Major Engineering Construction". With the promotion of academic innovation as the main goal, the series of "Research on Key Technologies of Major Engineering Construction" has the following characteristics:

1. Focus on major projects and key projects. At present, China's infrastructure construction is booming in various fields, and various kinds of engineering projects continue to mount. From the perspective of project volume and technical difficulty, we have selected a number of major projects and key projects, based on which, we have summarized the professional theory and technology to make them into books. Due to the large number of fields and disciplines involved in various projects, and the intersection and integration of disciplines. It is difficult to set a series of books with a single discipline. Therefore, based on the main line of Engineering categories, a series of books have been written in eight fields, i.e., tunnel and underground engineering, bridge engineering, railway engineering, highway engineering, super high-rise and large-scale public buildings, water conservancy engineering, port engineering, urban planning and architecture, basically covering the main fields of engineering construction in China, with a view to providing professional and technical reference guidance for major engineering construction in the future. As there are many fields and specialties involved, there are similarities and differences between technologies. In the field of cross technology, they are dealt with according to the specific situation to avoid content duplication and disjunction.
2. Highlight common technology and innovation achievements, and focus on application technology theorization. Focusing on a series of key technical problems in major projects in recent years, and based on the innovation achievements and technological breakthroughs made by the project, the series of books sorts

out pertinently the technical experience and scientific research achievements of common, key or significant promotion value in each series, and makes in-depth, systematic and detailed analysis and elaboration from the perspective of technical methods and engineering practice experience, and provides practical theoretical basis and application reference for the solution of similar problems and the improvement of technology. In the "Series of *Key Technologies for Tunnel Construction under Complex Geological and Environmental Conditions*" (Academician Qihu Qian is the director of the editorial board), the key problems in the current tunnel and underground engineering construction are systematically described and the corresponding professional technical theory system is formed, including the prediction, warning and risk control of major water inrush disaster in deep and long tunnels, the treatment of soft and hard uneven stratum and extremely soft stratum in shield engineering, a sort of rectangular shield method, underwater shield tunnel, ground access shield method tunnel, extra-long highway tunnel, three-dimensional geological detection of tunnel, rapid detection of shield tunnel disease, digitalization of tunnel and underground engineering, new anchorage material for large deformation tunnel in soft rock, and so on, and the key problems are solved in different degrees in the research and effectively implemented in the follow-up project.

3. Paying attention to the practical value of the project. The technical achievements involved in the series have been adopted many times in China, which have been proved to be reliable and effective in practice, and some of them have obtained technical patents. The series emphasizes the theory as the guide, the application as the key, and the case as the illustration. All the technical achievements are required to take the engineering project as the background and the production practice as the support, so that the series not only has the academic connotation, but also has the important engineering application value. For example, the "Series of *Key Technologies for Long-span Bridge Construction and Maintenance*" (academician Jielian Zheng is the director of the editorial board, academician Zhengqing Chen is the deputy director), focusing on major bridge projects such as long-span suspension bridges, long-span bridges across the sea, multi-tower cable-stayed bridges, long-span concrete-filled steel tube arch bridges, long-span pedestrian bridges, large-scale spatial suspension bridges with variable width, etc., focuses on the design innovation theory, construction innovation technology, technical breakthrough of construction difficulties, bridge structure health monitoring and state assessment, maintenance and maintenance during operation period, etc. The main contents include the vacuum-assisted pouring technology of large-scale concrete-filled steel tube structure, the spatial cable plane suspension bridge system with large proportion and variable width, the new eddy current damping vibration reduction technology, the cable hoisting and cable-stayed buckle construction of long and large bridges, the super large-scale deep-water foundation super high composite bridge tower, intelligent deformation monitoring, and the integration of construction and maintenance based on BIM, etc. These technologies are based on major construction projects, including the first bridge of Zhejiang

Yangtze River, the second bridge of Hejiang Yangtze River, Wushan Yangtze River Bridge, Nanpanjiang bridge of Guangxi Guangzhou railway, Zhangjiajie Grand Canyon bridge, Xihoumen Bridge, Jiashao bridge, Hong Kong-Zhuhai-Macao Bridge, Humen second bridge, etc. in the book, the relevant contents of specific project cases are analyzed in detail, with good application reference value.

4. Focus on the hot spot, focus on risk analysis, disaster prevention and mitigation, health detection, engineering digitization, and other emerging branches in recent years. Under the guidance of the principles of green and sustainable development, in recent years, technological innovation in the field of infrastructure construction has many achievements and examples in energy conservation and emission reduction, low-carbon environmental protection, green civil engineering, risk analysis, disaster prevention and mitigation, health detection (remote wireless video monitoring), safety and economy in the life cycle of the project, reliability and durability, construction technology organization and management, digitization, etc. Series of books are also reflected in these aspects, in order to give full play to the role of series books in promoting the long-term, green, and sustainable development of major project construction.
5. Establish an open framework. As a result of the above characteristics, the progress of each volume of the series is different, so the open framework is adopted, and flexible way of publishing in stages is adopted in the setting of each volume of the subsequent series.
6. The editor in chief has first-class academic level, which lays a solid foundation for the academic quality of the series. The chief editor of each series is the academic authority in this field and has an important academic status and influence in this field, for example, Prof. Zhengqing Chen, academician of Chinese Academy of engineering, chief scientist of Project 985, bridge structure and wind engineering expert; Prof. Jielian Zheng, academician of Chinese Academy of engineering, bridge design and construction expert; Prof. Qihu Qian, academician of Chinese Academy of engineering, protection and underground engineering expert; Prof. Zhiqiang Wu, academician of Chinese Academy of engineering, urban planning and construction expert, etc. The main writers who participated in the writing were all the front-line personnel active in the research, education, and engineering of infrastructure construction in China, who had undertaken major engineering construction projects or national major scientific research projects, mainly from China Railway Tunnel Bureau Group Co., Ltd., CCCC Tunnel Engineering Bureau Co., Ltd., China Railway Construction Co., Ltd., CCCC highway planning and Design Institute Co., Ltd., and the army Institute of engineering design, Qingdao Metro Group Co., Ltd., Shanghai Urban Construction Group, China Merchants Chongqing Communications Research and Design Institute Co., Ltd, Tianjin Urban Construction Group Co., Ltd., Zhejiang Communications Planning and Design Institute, Tongji University, Hehai University, Southwest Jiaotong University, Hunan University, Shandong University, etc. In addition to undertaking the heavy tasks of engineering construction and scientific research and teaching, all experts have

contributed their wisdom, knowledge, and sweat to the progress of engineering technology in China. On behalf of the General Editorial Committee of the series, I would like to express my heartfelt thanks and respect for your hard work.

At present, not only the domestic infrastructure projects are in the ascendant, but also under the initiative of "The Belt and Road", China's major overseas projects are booming, and the demand for high-level engineering technology is becoming increasingly urgent. It is believed that the publication of the series of books can provide certain assistance for the development of major engineering construction and the progress of innovative technology in China.

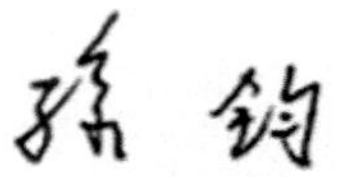

Shanghai, China
December 2017

Jun Sun

Mr. Jun Sun first-class Honorary Professor of Tongji University, senior academician of the Chinese Academy of Sciences, well-known expert in geotechnical mechanics and engineering at home and abroad. Chief Editor of the series *Research on Key Technologies of Major Engineering Construction.*

Foreword by Qihu Qian

Since entering the twenty-first century, with the continuous development of the economy, the continuous improvement of comprehensive national strength and the continuous application of high and new technology, China's tunnel and underground engineering have achieved unprecedented rapid development. China has become the country with the largest scale, the largest number, the most complex geological conditions and structural forms, and the fastest development of construction technology in the world. At the same time, with the continuous increase of urban subway construction, the number of cross river and cross sea tunnel projects continues to increase. National key construction projects such as long-distance water supply, underwater transportation, West to East Gas Transmission and other projects will involve the problem of crossing rivers. The construction of railway, highway, municipal administration, water supply, gas supply, flood control, hydropower, and other tunnel projects will greatly increase the number of tunnels.

In the aspect of tunnel construction technology, the technical system of high-speed railway tunnel has been basically formed; the construction technology of long large tunnels with complex geological conditions in difficult and dangerous mountainous areas has made continuous progress; the construction technology of large section soft surrounding rock tunnels has made great progress; the construction technology of urban large-span shallow buried tunnels and river crossing underwater tunnels has made breakthroughs; the research and development and manufacture of tunnel boring machines have made great progress, these marks the tunnel construction technology in China which has reached a new level of development. In particular, China has a vast territory, complex geological conditions, and extremely complex geological conditions are the main factors restricting the safe and efficient construction of tunnels, which are recognized as the difficulties of tunnel construction. Relying on a large number of key and difficult projects, such as Guanjiao tunnel of Qinghai-Tibet railway, West Qinling Tunnel of Lanzhou-Chongqing railway, immersed tunnel of Hong Kong-Zhuhai-Macao Bridge, Gaoligongshan tunnel of Dali-Ruili railway, Yangtze River Tunnel of Sanyang road in Wuhan, etc., China has made great achievements and more innovative achievements in tunnel and underground engineering, especially in tunnel construction under complex geological and environmental conditions. Therefore, in view of the complex geological conditions such as complex terrain and landform type, complex geological structure and poor engineering geology of rock and soil mass, based on the breakthrough and innovation achievements of key and difficult projects, the editorial board of the series and Shanghai Science and Technology Press jointly planned the series.

From the perspective of geological exploration, early warning and risk control, design, construction and digital application, the series systematically sorts out the key technologies of tunnel construction under complex geological and environmental conditions, such as mountains, underwater, soft and hard uneven strata. In the form of academic monographs, it introduces the innovative achievements and core technologies formed in the process of complex geological tunnel construction in recent years. The innovative achievements and technologies involved and introduced in the series are at the leading level in China, some of which have formed core technologies with independent intellectual property rights, and have been applied in major projects, making the series have the characteristics of cutting-edge, original, innovative, and leading. For example, Kairong Hong's ***Shield Tunneling Technology in Hard-Soft Uneven Strata and Extremely-Soft Strata*** expounds shield treatment technology of soft and hard uneven stratum and extremely soft stratum of underwater tunnel from theory, technology, and engineering cases, which represents the latest theory and practice of soft and hard uneven stratum tunnel and underground engineering. Based on the scientific research project "National Defense Engineering Geological Prediction and Quality Nondestructive Testing System" presided over by the author, and combined with the research and application results of the team for many years, Guohou Cao and Hao Liu's ***Three-Dimensional Exploration Technology of Tunnel Geology*** comprehensively introduced the new theory, new method, and new technology of tunnel complex geological geophysical exploration. Shucai Li's ***Monitoring Method and Early Warning Technology of Tunnel Water Inrush*** relying on the national major scientific research instrument and equipment development project "Comprehensive Geophysical Exploration Instrument for Quantitative Advance Prediction of Tunnel Unfavorable Geology Used in the Roadheader Construction", the National Natural Science Foundation for outstanding young scientists project "Mechanism and Disaster Control of Tunnel Water Inrush and Mud Inrush" and other major scientific research projects, introduces the advanced prediction technology of water inrush and mud inrush disaster source in complex geological tunnels and its application in large tunnel engineering, etc. The application cases in the series also include yuelongmen tunnel of Chengdu-Lanzhou Railway in Sichuan, Yellow River Tunnel Project in Jinan, Qiyueshan Tunnel of West Hubei expressway, Jiaozhou Bay Tunnel in Qingdao, Yangtze River Tunnel in Nanjing, and River Crossing Tunnel in Wuhan.

Under the strategic background of vigorously promoting the "One Belt and One Road" construction, implementing the innovation driven development strategy and building a powerful transportation nation, we hope that the publication of the series will not only better summarize the above technological achievements and promote the popularization and application of innovative technologies, but also play a catalytic role in the breakthrough of basic theoretical research and common key technologies, and can play a positive role in the cultivation of technological innovation mode and professional talents. In the process of discussion, planning, organization, compilation, and review, the series has been greatly supported by relevant large enterprises, colleges and universities, research institutions, societies, and associations. Many experts have put forward many very good suggestions and ideas for the series in their busy schedule. Thank you all.

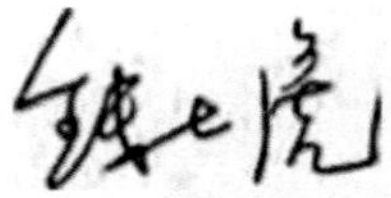

Beijing, China
August 2018

Qihu Qian

Mr. Qihu Qian Mr. Qihu Qian professor at the Army Engineering University of the Chinese People's Liberation Army, academician of the Chinese Academy of Engineering, and the director of the editorial board of Series of *Key Technologies for Tunnel Construction under Complex Geological and Environmental Conditions.*

Preface

As one of the "Series of *Key Technologies for Tunnel Construction under Complex Geological and Environmental*", this book aims to help technicians engaged in geotechnical engineering understand the latest achievements and applications of tunnel geological three-dimensional detection technology, and provide reference and guidance for engineering construction and related research. The guidance is mainly for engineering and technical personnel with certain practical experience and can also be used for postgraduate study and advanced college students to expand reading.

By the end of 2017, 16229 highway tunnels with a total length of 15285 km and 14547 railway tunnels with a total length of 15326 km have been built in China. China has become a major tunnel country in the world. Especially under the guidance of the national "The Belt and Road" initiative, tunnel engineering, as a key node in traffic construction, will usher in a new construction peak.

However, under the circumstance of more and more complex geological conditions, construction environment and structural form, safety control risk of tunnel construction is getting higher and higher. The biggest safety risk of tunnel construction stems from the unclear geological conditions, and it is difficult to formulate preventive measures to avoid major safety accidents such as mudslides and landslides. The majority of tunnel engineering builders look forward to a scientific detection method that can intuitively and accurately predict the geological conditions in the tunnel, facilitate the formulation of reasonable and feasible construction plans, reduce project construction risks, improve construction efficiency, and save construction investment.

Geological prediction is an important guarantee for tunnel construction safety. The geophysical method has a wide range of detection, high efficiency, and little impact on construction. It has been widely used in geological prediction at home and abroad. The geophysical methods used in tunnel geological prediction mainly include seismic wave method and electrical method. The existing methods play an active role in tunnel construction, but they also have certain limitations. For example, one-dimensional or two-dimensional observation methods are often used, which has low detection accuracy and poor intuitiveness. Based on the existing advanced forecasting technology, the author and his research team have developed a complete set of three-dimensional tunneling equipment and technology, which has improved the detection accuracy, realized the three-dimensional visualization of the detection results, and has carried out a lot of success applications in engineering practice.

This book is a summary of the author and his research team's many years of theoretical results and engineering applications. It is based on the scientific research project "National Defense Engineering Geology Prediction and Quality Nondestructive Testing System" hosted by the author and combines part of the research and application results of Yunnan Aerospace Engineering Geophysical Testing Co., Ltd. From theory to practice, with illustrations and texts, supplemented by a large number of engineering examples, this book comprehensively introduces the new theories, new methods, and new technologies of tunnel complex geological and geophysical exploration, focusing on three-dimensional detection of seismic wave method, three-dimensional detection of high-power electric sounding, three-dimensional exploration of ground penetrating radar, and other innovative technologies.

The book is divided into seven chapters. Chapter 1 introduces the theory, the current status and development trend of tunnel geological exploration. Chapter 2 combs and analyzes the complex geological conditions and the current detection technology. Chapter 3 systematically expounds the three-dimensional detection theory of seismic wave method, signal acquisition and analysis methods, signal processing and imaging technology, and introduces the complete set of equipment and related software developed by the author as well as application examples. Chapter 4 explains the principle of three-dimensional detection of water content in the surrounding rock of the tunnel and introduces the three-dimensional detection technology of high-power electric sounding and the three-dimensional detection technology of ground penetrating radar and application examples. Chapter 5 introduces the three-dimensional detection technology of the tunnel side slope and analyzes the factors affecting the stability of the side slope and detection application examples. Chapter 6 introduces the three-dimensional detection technology of tunnel shield and its application. Chapter 7 introduces in detail the application and effect of the three-dimensional detection technology and method after the sudden geological disaster in the tunnel combined with the typical engineering cases.

In the process of writing this book, great support and guidance was received from Xiao Li (Researcher), Prof. Yunsheng Wang, Prof. Weiping Zhang (Senior Engineer), and Yonggui Zhao (Researcher). Heartfelt thanks to them!

The new technology of three-dimensional exploration of tunnel geology is still in the stage of active exploration and development in China, and it belongs to the interdisciplinary category. Due to the rush of time and the limitations of the author's understanding, the omissions and inadequacies of this book are inevitable. Suggestions and comments are welcome.

Kunming, China

Guohou Cao
Hao Liu

About This Book

This book comprehensively introduces the new theories, new methods, and new technologies of current tunnel complex geological and geophysical detection, focusing on the theories, methods, and application of innovative technologies such as seismic wave three-dimensional detection methods, high-power electrical sounding three-dimensional detection, ground penetrating radar three-dimensional detection, etc.

The book is divided into seven chapters. Chapter 1 introduces the theory, the current status and development trend of tunnel geological exploration. Chapter 2 combs and analyzes the complex geological conditions and the current detection technology. Chapters 3–7 respectively describe the seismic wave method three-dimensional detection technology, tunnel circumference three-dimensional detection technology of rock water content, three-dimensional detection technology of tunnel side slope, three-dimensional detection technology of tunnel shield, three-dimensional detection technology after tunnel geological disaster, and its application in engineering.

This book is a summary of the author and his research team's many years of theoretical results and engineering applications. It aims to help technicians engaged in geotechnical engineering understand the latest achievements and applications of tunnel geological three-dimensional detection technology, and provide reference and guidance for engineering construction and related research.

Contents

Chapter 1
Introduction

1.1 Overview of Engineering Geophysical Exploration

1.1.1 Engineering Geophysical Exploration Concept

Engineering geophysical exploration is the exploration and testing method of engineering geology and hydrogeology survey based on the principle of geophysics. It is a branch of geophysical exploration, called engineering geophysical exploration for short. Due to the differences in physical properties such as density, magnetism, electrical conductivity, elasticity, and radioactivity of various rocks or geological bodies, people use different methods and different instruments to measure the natural or artificial geophysical field, analyze and study the variation of the physical field caused by these differences in physical properties, and then infer and interpret to understand the underground geological conditions, or people use instruments to directly measure the physical properties of the rock mass to provide the parameters required for engineering design. The extensive and correct application of engineering geophysical exploration in engineering geological survey can speed up the surveying speed and reduce the cost, and also obtain the physical parameters of the rock mass in situ, which can promote the quantitative evaluation of engineering geological conditions.

1.1.2 Engineering Geophysical Exploration Method and Application

The application of engineering geophysical exploration in engineering geological surveys in China began in the early 1950s. Engineering geophysical exploration methods are mainly divided into two categories: elastic wave method and electrical (electromagnetic wave) method. Basically, the gravimetric method and the magnetic

G. Cao and H. Liu, *Three-Dimensional Exploration Technology of Tunnel Geology*,
Key Technologies for Tunnel Construction under Complex Geological
and Environmental Conditions, https://doi.org/10.1007/978-981-16-9225-3_1

method are not used. In the "Regulations for Geophysical Exploration of Water Conservancy and Hydropower Engineering" edited by Professor Baixun Xiao in 2005, these two methods have not been mentioned at all. Radiological prospecting is also rarely used. Natural radioactivity logging has not been carried out, and density logging has also been carried out rarely. The traditional methods of engineering geophysical exploration include seismic refractive wave method and emission wave method, direct current method, acoustic wave and electromagnetic wave perspective, integrated logging and underwater acoustic exploration. New engineering geophysical exploration methods include ground penetrating radar (GPR), Rayleigh surface wave (RSW), vertical acoustic reflection (VAR), high-density resistivity (HDR), transient electromagnetic method (TEM), controlled source audio-frequency magnetotelluric (CSAMT), underground television, crosshole electromagnetic wave and acoustic tomography, etc. In addition, there are some new special methods for special geological tasks, such as anchor quality detection method, tunnel advance prediction system, flow field method for detecting leakage, additional mass method for detecting rockfill density and foundation bearing capacity, underground pipeline detection system, etc.

When the artificially excited seismic wave encounters the interface between two media with different density and elasticity when propagating underground, reflection, transmission, and refraction of the wave will occur. The reflected wave returns directly to the ground, and the transmitted wave enters the lower stratum through the interface. Only the part of the seismic wave whose incidence angle θ_1(the angle to the interface normal when the interface is incident) is equal to the critical angle i (the value is determined by the ratio of the seismic wave propagation velocity v_1 to v_2 of the upper and lower strata) will produce a crease along the plane angle of the incidence angle after reaching the interface, slide forward on the interface at the interface speed, and form a new wave at the place where it arrives. The new wave shoots to the ground at the critical angle with the normal of the interface, which is called refraction wave. Reflected waves and refracted waves return to the ground to be accepted by the preset geophone, and the seismic prospecting instrument records the propagation time and vibration characteristics from the seismic source to the geophone. The area around the source where refracted waves cannot be received is called a blind zone. The propagation time is determined by the travel of these waves and the seismic wave velocity of the medium along the way. After the distance between the seismic source and the detection point is selected, the travel of the wave depends on the depth of the interface, so it can be used for geological exploration. The use of reflected waves is called the reflected wave method, and the use of refracted waves is called the refracted wave method.

Due to the shallower exploration depth of engineering survey, the refraction wave method has less interference than the reflected wave method. It is easier to identify and can measure the interface velocity, so as to understand the lithology of the lower layer and detect faults, so the application of the refraction wave method is more common. However, the refracted wave method requires a strong seismic source. There are also the limits of the blind zone, and the lower wave speed must be higher than the upper. In order to avoid these shortcomings, in recent years, the engineering survey department has also strengthened research and experiments on the shallow reflected wave method. Seismic exploration is mainly used in engineering surveys to determine the depth and shape of geological interfaces, such as the thickness of overburden, weathered layer, landslide body, and groundwater level, as well as to detect faults, fracture zones, etc. (see color map). The reflected wave method can also explore karst caves.

Elastic waves are excited in rock mass or soil layer. The velocity of these elastic waves (including P-wave and S-wave) and the attenuation of energy in the process of propagation are measured by instruments, so as to obtain the dynamic elastic coefficient of rock mass and evaluate the mechanical properties of rock mass. This method is different from seismic exploration in that it mostly uses direct waves. Elastic wave testing can provide a basis for engineering geological classification of rock masses, determine the slack range of surrounding rock in underground caverns, check the effect of grouting, and perform non-destructive detection of the quality of dam foundations and pile foundations. It can also provide parameters for the seismic design of engineering buildings and the study of sand liquefaction.

Various geophysical detection techniques are also used in drilling. The comprehensive application of the logging method can be divided into drilling geological section in detail, detecting the weak interlayer, determining the location and occurrence of faults and fissures, fracture zones, determining the resistivity, elastic wave velocity, porosity, density, mud content of the formation, water content, determining the location of the aquifer, and determining the salinity, velocity and flow direction of groundwater.

Various geophysical methods have their own specialties, and they all have their limitations determined by the principle of the method. In many cases, the data of a single method is more solvable. In order to obtain ideal survey results, several effective comprehensive geophysical prospecting methods must be selected according to local conditions for specific tasks.

1.1.3 Engineering Geophysical Exploration Mission

At present, China's highway, railway construction, hydropower station construction, urban construction, and other infrastructure construction are in a period of rapid development, which provide a broad space for development of engineering geophysical exploration technology. The status of engineering geophysical exploration technology in China's engineering construction is increasing day by day, but its problems are more and more complicated. Engineering builders have higher and higher requirements for engineering geophysical exploration technology, and engineering geophysical exploration is facing unprecedented opportunities and severe challenges. It can be said that engineering geophysical prospecting is indeed an old and young profession.

At present, the main tasks of engineering geophysical exploration in China include: detection of overburden layering and thickness, detection of hidden structures and fracture zones, detection of karst and groundwater activity laws, detection of engineering rock weathering shapes and zones, detection of artificial blasting damage and unloading zones depth, detection of weak interlayer, detection of landslide, detection of dyke hidden danger, quality inspection of dike cutoff walls and other concealed works, advanced prediction of artificial tunnel construction, groundwater detection, dam foundation rock mass quality detection, environmental radioactivity detection, concrete surface microcracks and internal defects detection, consolidation grouting and curtain grouting quality detection, cavern concrete lining quality detection, artificial slope anchor rod and anchor cable anchoring quality detection, building foundation and foundation pile detection, rockfill (soil) body density and foundation bearing capacity inspection, leakage inspection of embankments and reservoirs, quality inspection of rockfill dam panels, inspection of steel lining and concrete contact status, detection of hydrogeological parameters, testing of geophysical and mechanical parameters, quality inspection of airport runways and highways, urban underground pipeline detection, coal mine goaf detection, artificial slope shotcrete thickness detection, embankment protection slope (bank) riprap distribution and thickness detection, concrete internal steel reinforcement equipment detection, etc.

1.1.4 Main Features of Engineering Geophysical Exploration

(1) The detection object is shallow complex medium, and the detection target is small scale. Engineering geophysical exploration is usually explored to a depth of less than 200 m, with a maximum depth rarely exceeding 500 m. The medium is often non-uniform, viscoelastic, anisotropic rock mass or a complex of various materials. The geometric scale of the target body is usually meter or even centimeter.

(2) Detection problems are complex and changeable, and its precision is required to be high. For example, the detection of loose layers of artificial slopes, anchor rods and cables, faults, weak interlayers, complex karsts, concrete cracks, impervious walls, and advanced tunnel predictions are not only complicated by the types of boundary conditions and problems, but also has very high requirements for the resolution and spatial positioning of structural and environmental geotechnical defects. Sometimes it is even required to distinguish weak interlayers and cracks as small as a few millimeters and accurately locate them. In addition, the physical detection of water conservancy and hydropower projects is also restricted by some special factors, such as limited site, large terrain influence, strong noise background, etc.

(3) The results are required to be obtained quickly, and the results can be tested immediately.

(4) The team is scattered with large quantities and wide areas, unlike the relatively concentrated exploration team for oil and metal mines.

1.1.5 Development History of Engineering Geophysical Exploration in China

China's engineering geophysical exploration began in 1951. The first work is that Mr. Gu Gongxu led the electrical method group in Beijing Shijingshan area to explore water sources. In the 1950s, engineering geophysical exploration teams were established only for systems such as geology, hydropower, and railways. In 1954, the railway system applied electrical methods to carry out geological exploration in Baocheng Railway. In 1955, the hydropower system applied electrical exploration to the Mokou Hydropower Station of Yongding River to understand the engineering geological problems of the dam foundation. In 1956, during the site selection of the Yangtze River Three Gorges Project, the electrical method was used to investigate the engineering geology of the dam area. The 1960s and 1980s was the first stage of development of China's engineering geophysics. Since the 1960s, China's construction, metallurgy, coal, petroleum, nuclear industry, machinery and electronics, weapons, aerospace, shipping, agriculture, commerce, textile, chemical industry, culture (archeology) and other systems have successively established their own engineering geophysical exploration teams. In the late 1970s, various related universities and research institutes also carried out engineering geophysical exploration work. After 1980, engineering geophysical exploration began to develop rapidly, and it is still in a period of rapid development until today.

The publication of the Journal of Engineering Geophysics in 2004 ended the history of engineering geophysical exploration without its own publications.

1.2 Analysis of Current Status of Engineering Geophysical Exploration at Home and Abroad

1.2.1 Status of Domestic Engineering Geophysical Exploration and Detection Technology

China's engineering geophysical exploration technology and equipment level has been among the world's advanced ranks. With the large-scale economic construction of the country, the domestic geophysical exploration industry began to develop from its own professional fields to engineering geophysical exploration and detection in the 1980s. In the early 1990s, it was expanded in a large area. Various methods were also emerged one after another, which were applied in many engineering fields in different aspects.

(1) **Current status of engineering geophysical exploration applications**

A series of engineering geophysical exploration and detection methods have been developed in response to the need for comprehensive exploration of the foundations of large-scale projects such as railways, highways, subways, construction sites, ports, and airports. The main geological tasks of the engineering geophysical exploration in these fields are to determine the thickness of the surface loose layer, the thickness of the Quaternary system, the depth of bedrock, the location of the fault and fracture, the underground cavity and the water bearing capacity of the stratum. Common methods include electrical method (conventional electrical method, high-density resistivity method), microgravity, high-precision magnetic force, ground penetrating radar (GPR), shallow earthquake, acoustic wave, audio frequency electromagnetic method (AMT), and controlled source audio frequency magnetotelluric method (CSAMT), transient electromagnetic method (TEM), radioactivity, crosswell CT, well logging, etc. In the geophysical exploration of railway engineering, the application of high-density electrical method and seismic refraction tomography has achieved good results in detecting unfavorable geological phenomena such as small faults and caves in the front. For the exploration of deep tunnels, controlled source audio magnetotelluric method (CSAMT) uses artificial field sources to enhance the signal strength, and the detection effect is also obvious.

(2) **Exploration and detection technology**

In terms of railway subgrade detection, many units are also trying to varying degrees to use transient Rayleigh wave technology, ground penetrating radar technology, transient electromagnetic technology, and high-density resistivity technology for detection research. Dynamic resistivity imaging system (CCR) and multi-frequency electromagnetic system (GEM300) introduced by Eastern Geophysical Company are also used in some methods and experiments. However, due to the complexity of the shallow situation, the working conditions required for the method, and the work efficiency, further studies are needed before it can play a role in engineering detection.

In the aspect of highway system detection, ground penetrating radar has achieved good results in monitoring road quality and detect the quality of steel and concrete. Now it has formed a standardized market area. In the advance prediction of tunnels, the main application technologies include ground penetrating radar, tunnel seismic prediction (TSP) advance prediction, negative-vision velocity method, ultrasonic method, infrared water exploration method, etc. These technologies have different effects in different geological conditions. Regarding the detection of tunnel lining, it is now common practice to use ground penetrating radar. This method is efficient, and the detection effect is good. Pile foundation inspection is also a major application area for engineering geophysical exploration. There are high-strain (large-strain) and low-strain (small-stain) technologies for pile inspection when the pile soil is not damaged. Through detection, rich pile-soil information can be obtained such as pile integrity (integrity, broken pile, segregation, mud, etc.), pile geometry (diameter expansion, etc.), and single pile bearing capacity (vertical and horizontal).

1.2.2 Status of Engineering Geophysical Exploration and Detection Equipment at Home and Abroad

With the continuous application of engineering geophysical exploration technology in various countries around the world and the rapid development of digital circuits, a series of targeted shallow detection instruments have been developed. In the depth of the detection target, there are many types of instruments ranging from radar equipment with a resolution of a few centimeters to low-frequency electromagnetic instruments with a resolution of thousands of meters. According to the method, there are mainly elastic wave instruments, electrical instruments, and ultrasonic instruments.

Due to the many characteristics of shallow seismic exploration instruments in stimulating source, instrument performance, structure, etc., as well as the advantages of low development and manufacturing costs, quick results, and wide range of uses, many manufacturers have been developing shallow seismic exploration instruments in recent years. Chongqing Geological Instrument Factory, Changchun Institute of Geology, Beijing Institute of Hydroelectric Geophysical Exploration, Xi'an Petroleum Instrument Factory, and other units have developed and produced instruments for shallow exploration. In foreign countries, in addition to OYO, GEOMETRIC, BISON, EG&G, EDA, etc., continuously improving the original instruments, SCINTREX has also begun to develop and produce shallow seismic instruments. At present, foreign instruments are mostly 12/24 channels, and there are also single and four channels. Domestic instruments are available in single, three, six, twelve, and 24 channels. The seismic instruments used for shallow exploration abroad mainly include R 24 shallow seismometers, NZ 48, MARK 6, RAS 24, SUMMIT, and other seismographs. Domestically, there are DZQ24 from Chongqing Geological Instrument Factory, GDZ24 from Xi'an Petroleum Instrument Factory, SWS multifunctional seismograph from Beijing Hydropower and Geophysical Exploration Institute, and RS2416C multifunctional seismometer from Wuhan, all of which have the functions of refraction, reflection, surface wave, micromotion, pile foundation detection, etc. Tianjin Haitai Optoelectronics Technology Co., Ltd. can produce all kinds of seismic instruments with low power consumption and large dynamic range.

Many domestic manufacturers produce DC-based instruments, especially high-density resistivity instruments, whose performance has reached the international advanced level. The performance of domestic transient electromagnetic instruments is also constantly improving. Tianjin Haitai Optoelectronics Technology Co., Ltd. has been able to produce CDMA electrical instruments with high signal-to-noise ratio and low power consumption. In the field of audio magnetotelluric, there is still a gap in China. The Xi'an Geophysical Equipment Branch of the Eastern Geophysical Company (formerly Xi'an Petroleum Instrument Factory) has established a project to apply for the development of such instruments. In foreign countries, electromagnetic instruments are developing rapidly, mainly including V5 magnetotelluric instrument, V5 2000 magnetotelluric instrument, V6 multi-function electric instrument, V8 multi-function electric instrument, MTU 5A high-frequency electromagnetic instrument, MT1 magnetotelluric instrument, MT24 magnetotelluric instrument, EH4 high frequency electromagnetic instrument, GDP32 multi-function electromagnetic instrument, PROTEM transient electromagnetic instrument, ground penetrating radar, GEM300 multi-frequency electromagnetic instrument, etc.

1.2.3 Development Trend of Engineering Geophysical Exploration and Detection at Home and Abroad

(1) **Technical development characteristics**

① Three high—high sensitivity, high resolution, high precision.

② Three more—multi-parameter, multi-function, multi-series. In order to improve the accuracy of the test results, a comprehensive analysis of multiple parameters is required. For the accurate detection and description of the formation, not only the resistivity, magnetic susceptibility, polarizability, density, longitudinal wave velocity, but also the shear wave velocity and other parameters are used. Only in this way can it be described comprehensively. At the same time, during the propagation of various waves, the degree of attenuation of the amplitude and the change of the spectrum can also reflect the change of a physical property of the object. Therefore, the use of these parameters for comprehensive analysis will be one of the future development directions of engineering geophysical exploration technology.

③ Three optimizations—optimize information, optimize data interpretation (pretreatment), optimize display. Just as CT technology, nuclear magnetic resonance and other technologies have caused a revolution in diagnostic technology in the medical community, the use of computer digital imaging technology for detection will inevitably greatly improve the detection level. Although CT imaging technology has been initially applied to the field of non-destructive detection, its mapping accuracy method needs to be improved, and the scope of application needs to be expanded. For example, the current two-dimensional plan view develops to the three-dimensional map, and the image interpretation of the engineering geophysical prospecting section develops in the direction of more understandable three-dimensional visualization.

(2) **Application development characteristics**

① Deepening—on the basis of ensuring accuracy, deepen the depth of detection. In recent years, large-depth tunnels have continuously appeared in railway and highway systems, and the detection accuracy of deep unfavorable geological bodies has become increasingly demanding.

② Developing—expanding the scope. For example, non-destructive detection is mainly applied to the quality inspection of cement concrete, but engineering construction materials are various. Even for concrete materials, some structures are difficult to apply, such as the strength of hollow bricks, the strength and durability of hollow floors. Therefore, with the advancement of non-destructive detection theory, methods and technology, the application of detection methods will be developed in various fields of engineering construction quality detection.

③ Refining—improve accuracy to discover small geology, such as small karst, coal mine small structure, concrete component flaw detection, etc., which requires the acquisition instrument to have high sampling, high number of passes, high transmission rate, and other characteristics.

In short, engineering geophysical exploration is a non-destructive detection technology, featuring high sampling density, high speed, low cost, high-technology content, and wide service fields. Compared with drilling, engineering geophysical exploration can achieve a continuous profile for the subterranean formation, rather than "one hole view" or "number of holes view", which overcomes the presumed characteristics of the formation between the wells. This is the advantage of engineering geophysical exploration. Some experts pointed out that engineering geophysical exploration technology will have a big leap development, which will be an indispensable exploration tool in engineering construction, and will play a major role in the future national economic construction. In recent years, Eastern Geophysical Corporation has made high-density three-dimensional exploration for coal mining, with an area of 2.5 km^2. The goal is to understand the information clearly of the coal seam of 1–2 m and the stratum of 1–2 m, providing reliable geological data for the development of coalfields.

Engineering geophysical exploration and detection technology are advancing with the development of science and technology. In the future, the development of geophysical technology will be from geometric seismology to physical earthquake (actually, physical seismology has been developed at present). Instruments and equipment will develop from electromechanical to photoelectric age. Due to the development and advancement of technology, the accuracy of the problems to be solved will continue to increase. Therefore, engineering geophysical exploration and detection will play a greater role in various engineering fields such as railway engineering, highway construction, coal mining, and water conservancy facilities construction. Vigorously developing engineering geophysical exploration and detection technology, improving the effect of engineering geophysical exploration and detection, comprehensively promoting the application of engineering geophysical exploration in engineering construction, and increasing the investment in engineering geophysical exploration and detection in engineering construction will greatly improve the quality of engineering construction. It will play a greater role in reducing the project risk and the total cost of project construction.

1.3 Prospects for the Development of Tunnel Detection Technology Under Complex Geological Conditions

At present, China is carrying out large-scale infrastructure construction, and it is necessary to build a large number of railways and highway tunnels. With the influence of the national "One Belt, One Road" initiative, and with the advancement of tunnel construction technology and the shift of construction scale, the focus of tunnel construction will shift to the western region with complex terrain and geological conditions, and a large number of deep-buried long tunnel projects will appear. Therefore, higher requirements will be put forward for tunnel engineering survey, and the requirements will be more strict and precise, so the detection technology of tunnel construction under complex geological conditions will be a huge challenge.

To this end, the author and related scholars have conducted in-depth research on tunnel detection technology. For example, the induced polarization method, transient electromagnetic technology, and complex frequency conductivity technology are applied to the advanced detection of the geological structure of the tunnel to infer the surrounding rock structure and water content of the tunnel. Based on the theory of stationary random process, the wave velocity structure of underground medium is obtained by using micro motion detection technology, so as to carry out advanced geological detection of tunnel.

At the same time, with the development of information technology, the traditional interval tunnel detection cannot meet the requirements of engineering refinement construction, especially the tunnel construction often has great uncertainty. With the development of information technology, the real-time fine detection of tunnel detection, combined with the information system platform to display the detection results in real time, will be the development trend of future tunnel construction.

In summary, regarding the development of tunnel detection technology under complex geological conditions, the author believes that on the one hand, the application of some new detection methods will be developed, and the existing methods will be improved to increase the scope of application. On the other hand, it will combine the development of information technology to make the existing detection technology develop in the direction of three-dimensional, refined and real-time, and promote the progress of traditional tunnel engineering detection through the development of information technology.

Chapter 2
Complex Geological Conditions and Detection Technology of Tunnel

Since the reform and opening up, China's highway construction industry has developed rapidly. In 2016, the mileage of the new expressway in China exceeded 6000 km, with the total mileage exceeding 130,000 km. In 2017, China will also add more than 5000 km of expressways. According to the highway network development plan announced by the State Council, China is accelerating the construction of the main skeleton of the national expressway network. The new road network consists of 7 capital radiation lines, 9 north–south vertical lines and 18 east–west horizontal lines, referred to as "7918 network." It can be basically completed during the "12th Five-Year Plan" and "13th Five-Year Plan" period. It is estimated that the total mileage of highways in China in 2018 will reach 140,000 km, far exceeding the total mileage of highways in the USA (the total mileage of highways in the USA is about 100,000 km), ranking first in the world. The total mileage of the highway network will exceed 4 million kilometers, and there will be 658,100 highway bridges with a total length of 30.4831 million meters. Among them, there are 2051 extra-large bridges with a total length of 3.4698 million m, and 49,489 bridges with a total length of 11.6704 million m. There are 7384 highway tunnels nationwide, with a total length of 5.1226 million m. Among them, there are 265 extra-long tunnels with a total length of 1.1388 million m, and 1218 long tunnels with a total length of 2.0208 million m. With the in-depth development of China's highway construction, especially the rise of highway projects in the central and western regions, the geological conditions it faces will become more and more complicated. Whether it is the engineering design process or the construction process, the requirements for geological exploration technology will be more and more higher.

China is a large tectonic geological country with complex regional geological conditions and large spatial differences. In combination with the theory of plate tectonics, Yuqing Huang divided the geotectonic structure of China into the Paleo-Asian tectonic domain, the Pacific Ocean tectonic domain and the Tethys Himalayan tectonic domain. According to the regular evolution relationship between the secondary tectonic units formed by tectonic movements in different periods of

G. Cao and H. Liu, *Three-Dimensional Exploration Technology of Tunnel Geology*, Key Technologies for Tunnel Construction under Complex Geological and Environmental Conditions, https://doi.org/10.1007/978-981-16-9225-3_2

geological structure, Dekun Sang divided China's geotectonic structure into five tectonic domains, namely paleo Asian tectonic domain, Tarim North China tectonic domain, Kunlun Qinqi tectonic domain, paleo Cathaysian tectonic domain, and Tethys tectonic domain. Regardless of the geological tectonic school, the consensus is that the geological conditions in southwestern China are particularly complex. Taking Yunnan Province as an example, the strata in the province are well-developed and the sedimentary types are diverse. The magmatic activity is intense and often continues to be active in the same area, forming a large-scale tectonic magma belt. Metamorphic rocks are widely distributed and have various metamorphic effects. The crustal activity is generally strong. The Jinshajiang-Ailaoshan fault zone that obliquely traverses the entire region in the northwest-southeast direction roughly divides Yunnan Province into east and west parts. The eastern part belongs to the western margin of the Yangtze quasi-platform or the Paleozoic accretion of the quasi-platform. The western part belongs to the Tanggula-Qamdo-Lanping-Simao fold system and the Gangdise-Nyainqentanglha fold system. Several major faults, including the Nujiang fault, Changning-Lancang fault, Lancangjiang fault, Muli-Lijiang fault, Honghe fault, Panzhihua-Chuxiong concealed fault and Kaiyuan-Qiubei-Guangnan concealed fault, etc. subdivide the area into multiple secondary tectonic units.

With the deepening of the development of the western region, especially under the guidance of the "One Belt, One Road" initiative, the focus of the national expressway construction will gradually shift to the central and western regions. Therefore, the complex geological conditions in the southwest region will bring new challenges to road construction. The key to road construction lies in bridges and tunnels, especially the latter. Bridges and tunnels for highway construction in Southwest China account for up to 50% of the route length, and most of them are long and large tunnels. If the geological conditions of the tunnel are unclear during the excavation process, there will often be serious sudden disasters, so the geological prediction is very important. How to ensure the safety of tunnel construction under complex geological conditions has become an important task for researchers and plays a vital role in the economic development and social stability of underdeveloped areas in China.

2.1 Tunnel Complex Geological Conditions

According to the current geotechnical engineering survey standards in China, the main features of complex geological conditions are: strong development of adverse geological effects and geological disasters, complex topography and landform types, complex geological structures, with obvious changes in lithology and lithofacies, and poor engineering geological properties of rock and soil, and poor geotechnical engineering geological properties, complex conditions of karst fissure water or other hydrogeological conditions and human engineering activities that damage the geological environment are intense. At present, the geological environment faced by the tunnel construction is at least one or more of the above, especially for tunnels with larger buried depths (e.g., buried depths of 200 to 300 m or more). Geological exploration before design is impossible to find out the complete geological conditions on the tunnel excavation elevation. Therefore, landslides, karst caves, underground rivers, water gushing, and mud gushing occurred from time to time during construction, resulting in heavy losses of persons, materials, property, construction period, cost and other aspects. Based on the engineering inspection practice of complex geological conditions of tunnels for many years, the author summarizes the complex geological conditions of tunnels into the following ten items.

(1) **Underground karst**

Karst is the result of water's chemical dissolution of carbonate rock, gypsum, salt rock, and other soluble rocks, as well as the combined action of erosion, latent erosion and collapse. It often forms crevices, karst pipes, karst troughs, karst caves, or underground rivers in rocks, resulting in the destruction and change of rock structure, and resulting in a special topography and landscape. Since the process of dissolution is very slow, the collapsed karst is already present, not formed later. Due to the beaded development characteristics of karst, underground karst is not an isolated individual, so it is mainly manifested as the relative stability of the filling when it is in a stable state. Once the filling is carried away, collapse will occur under the repeated action of the karst groundwater system. Based on the above geological origin, it is considered that the karst filling is generally water, mud or miscellaneous fill. The main effects of karst on tunnel construction are: A large amount of karst water rushes into the tunnel, which is easy to cause gushing and mud outburst in the tunnel. Deformation and collapse cause ground subsidence, drop the groundwater level, and affect the surrounding environment.

(2) **Karst fissure water**

Fissure water refers to groundwater stored in hard rock fissures, and its damage to tunnel surrounding rock has become one of the hot research issues in tunnel construction in karst areas. The rock fissure space is the place where the fissure water is stored and moved, so the type, nature, and development degree of the fissure directly affect the burial, distribution, and movement of the fissure water. The development and distribution of fissures in rock formations are extremely uneven. The spatial distribution of fissures is uneven and directional, resulting in uneven distribution of fissure water and hydraulic connection anisotropy. Therefore, in different sections, the water conductivity and water storage capacity of the rock formation are very different. Even in the same rock formation in the same section, the water output can vary by several tens of times or even hundreds of times.

The distribution of fissure water can be distributed in layers, veins or bands. In the rock formation with uniform crack development, good openness and connectivity, and less filling, the fissure water is layered, with good hydraulic connection and unified groundwater surface. In the section with uneven fissure development and poor connectivity, especially in the section with local fissure distribution, the fissure water is distributed in a vein, forming a water-bearing fissure system. The hydraulic connections between the various aquifer systems in the same rock formation are poor, and there is often no uniform groundwater surface. The hydrodynamic properties of fissure water are complex and exhibit significant anisotropy during flow. In rock formations, the fissures along a certain direction are well developed, with good openings, and strong water conductivity, while along other directions, the fissures are poorly developed or underdeveloped. The fissures with strong water conductivity often become channels for concentrated flow of groundwater. The velocity of fissure water in the fissures is generally small, and the water flow is mostly laminar. The flow of groundwater follows Darcy's law. However, in some wide fissures, when the hydraulic gradient is large, it can be in a turbulent state.

(3) **Fault fracture zone**

The fault fracture zone refers to the significant relative displacement of the rock formations on both sides after the rock formed fractures of joint structure. Due to the tectonic activity, the stratum is fractured in tension or compression, and the fractured strips composed of non-single fractures with a certain width and considerable extension length in the rock mass lose its continuity and integrity. The fracture zone associated with the fault is filled with rock fragments, gravel, and clay material formed by the fault that are torn off by the fault wall, and some are reconsolidated to form fault breccia, cataclastic rock, and mylonite or fault gouge and so on. In addition, due to the uneven distribution of the stress bands, derived cracks may also occur near the fault plane. Outcrop research shows that the fault fracture zone is mainly composed of section fillings and derived fractures. However, in different types of strata, the faulted zones have different developmental structures and different development scales, and the width of fault fracture zones is also different. The section fillings and the derived cracks near the fault plane are distributed symmetrically or asymmetrically. The boundary of the section fill is the main fracture surface, and the boundary of the fracture zone is determined by the specific conditions of the study area. The nature of the fault is different, so the degree of development of the section fill and the derived crack is also different. In the tunnel construction in such a rock mass, the surrounding rock stress, deformation and failure mode will be very complicated. Once the construction method or process is unreasonable, serious squatting, roofing, water inrush, or even the sliding of the mountain may occur. Long-term remediation is not good and should be avoided.

(4) **Underground river**

Karst groundwater is one of the leading factors inducing water inrush (water gushing) geological disasters in tunnels. The karst water is not only a kind of groundwater with special properties but also an active geological force. In the process of its movement, it continuously interacts with the rock, transforms its own occurrence environment, and forms a unique distribution and movement characteristics. For example, limestones are dissolved to form different shapes of dissolution gaps, such as dissolution funnels, caves, etc. Some huge underground pipelines and caves can form underground dark rivers, and the karst development is spatially different, resulting in extremely uneven distribution of karst water. At the same time, the karst space is mainly developed on the basis of the fissure space. The directionality of the fissure space and the anisotropy of its permeability are inherited and aggravated in the karst medium. Therefore, the anisotropy of water permeability is another notable feature of karst media. Sometimes, in different cross sections of the same hydraulic system, the permeability coefficient, hydraulic gradient, and permeate flow rate are different, and laminar flow and turbulent flow coexist. On the other hand, it also shows that the water level and flow process of karst water show strong seasonal changes, and its water level can vary by several tens of meters, and the flow rate varies by several tens of times.

The amount of water in karst aquifers is often abundant. The degree of water enrichment in karst aquifers is closely related to the degree of karst development. Generally speaking, the rock has good solubility and the groundwater flow is smooth, while the alternately strong areas are well-developed karst areas and are also areas where karst water is enriched. There are often abundant karst waters in the following sections: the distribution area of thick pure limestone, the structural fracture part of soluble rock, the interface between soluble rock and non-soluble rock or between highly soluble rock and weakly soluble rock, the oxidation zone of sulfide deposit, the surface water body, and other karst water discharge parts.

(5) **Weak surrounding rock**

Weak surrounding rock refers to the engineering rock mass which can produce significant plastic deformation under the action of engineering force. It has the characteristics of low strength, loose structure, and easy disturbance. The engineering performance is seriously deteriorated and the bearing capacity is extremely low. The concept of soft rock varies in the geotechnical field according to industry definitions, and there are many definitions of soft rock. The International Society of Rock Mechanics (ISRM) defines rocks with uniaxial compressive strengths of 0.5 to 25 MPa as soft rocks, such as strongly broken slate, carbonaceous slate, phyllite, intercalated sandstone, coal-bearing shale, mylonite, water-rich fine siltstone and sericite soft schist. The weak surrounding rock is the main factor causing the landslide in the tunnel construction. Under the condition of certain geostress level (or buried depth), the tunnel construction of soft rock (higher classification of surrounding rock) is prone to landslide or large plastic deformation. The factors that form weak surrounding rock are diverse and complex, which puts high demands on geological exploration and subsequent construction design and disposal plans.

Rock is a kind of non-uniform geological material with complex internal structure. On a small scale, the rock contains microscopic features such as mineral particles, pores, cement and microcracks. Under the action of external load, the inoculation, nucleation, expansion and interaction of microdefects within the rock determine the macroscopic deformation and fracture characteristics of the rock. In engineering applications, rock can be simplified into a homogeneous material. However, when examining the failure process and mechanism of rock, if the microscopic non-uniformity of the rock medium is neglected, the deformation mechanism of the rock cannot be fundamentally understood. In fact, the gestation and development of large deformation of weak surrounding rock in deep-buried tunnels in high ground stress areas is a typical problem of localization and evolution of damage. Taking fine siltstone as an example, when the water content of the surrounding rock reaches 16%, the state changes from a solid state to a plastic state, and when the water content exceeds 32%, it is further converted into a fluid state. The external construction disturbance will accelerate the liquefaction process, which will cause construction difficulties, equipment difficult to move, and invert cracking and structural deformation and settlement.

(6) **Landslide body**

When the tunnel passes through the unstable landslide body, the deformation of the slope will cause a large impact on the tunnel lining structure due to the joint action of multiple factors. Therefore, when selecting the tunnel line, the survey designer should try to select the most reasonable route plan, and try not to cross the disaster-prone areas such as collapse, slip, etc., but the route is difficult to adjust due to geological, topographical or other conditions, so the entrance and exit of the tunnel will inevitably be within the range of influence of the landslide. The interaction between the tunnel and the landslide body, especially in the mountain and hilly area, when the tunnel passes through the mountain body with potential landslide body, the tunnel excavation at the entrance section of the tunnel will disturb the stability of the surrounding rock, cause the redistribution of stress and cause disturbance to the landslide body. Meanwhile, the landslide thrust of the landslide will also increase the force of the lining structure, causing deformation and cracking of the lining structure. When the tunnel crosses the mountain with more complicated geological and topographic conditions, the entrance instability is easy to occur, which causes the tunnel lining structure to be subjected to additional external force, especially the slope instability, which is easy to cause slope slip. The landslide thrust acts on the tunnel lining, causing the lining structure to be excessively stressed and cracked.

(7) **Ground stress and large deformation of surrounding rock**

Ground stress is the fundamental source of surrounding rock pressure in tunnels. High geostress is one of the main factors that endanger the safety of underground engineering. In recent years, many important projects at home and abroad have encountered high geostress problems to varying degrees. Gravity action and regional tectonic movement are the main causes of ground stress, especially the horizontal tectonic movement has the greatest influence on the formation and characteristics of ground stress. The formation of tectonic stress field is complicated, and its distribution in space is extremely uneven. It is found in the statistical analysis of the distribution law of geostress field in China that the distribution law of geostress field in small area is more meaningful for guiding the measurement of ground stress and engineering construction.

Surrounding rock is a special medium for the realization of underground engineering and supporting structure. Its basic physical and mechanical properties reflect the working ability and nature of rock. Especially under the complicated geological conditions of deep high stress, the physical and mechanical properties of surrounding rock are even more important. In the general sense, large deformation of surrounding rock refers to a series of serious engineering problems caused by large deformation of surrounding rock during construction under the condition of soft lithology of tunnel engineering.

Therefore, geostress is one of the important basic materials for geological environment and crustal stability evaluation, geological engineering design and construction. It is not only an important factor determining regional stability but also the fundamental force of deformation and destruction of various underground or ground excavation geotechnical engineering. It is a prerequisite for determining the mechanical properties of engineering rock masses, conducting rock mass stability analysis, and realizing scientific excavation design and decision-making in geotechnical engineering.

(8) **Gas**

In the process of tunnel engineering development in depth, it is often encountered that it is necessary to crosscomplex formations, such as coal-bearing strata with weak rock formations or high-pressure gas coal seams. At this time, the presence of gas in the formation will flow into the tunnel excavation space, and the tunnel is generally defined as a gas tunnel. Highway gas tunnels generally appear in the mountainous hilly areas of western China. These areas have geological conditions for the formation of gas. Once the tunnels are excavated and exposed, the gas generated may be released into the tunnel excavation space in different forms. Both the release process of gas in the stratum and the subsequent existence process in the construction space may cause gas disasters. Therefore, the construction of the highway gas tunnel has more prominent safety problems than the conventional tunnel. It is necessary to adopt a variety of specific preventive measures such as preevaluation of tunnel gas level and gas work area, advanced gas geological forecast, gas monitoring and prediction, ventilation, explosion-proof electromechanical equipment, gas treatment in landslide space, coal uncovering and prevention of outbursts, etc., which increases the complexity of the construction process.

In the tunnel coal seam, the storage state of gas is generally divided into two types: free state and adsorption state. The so-called gas content refers to the sum of the free and adsorbed gas contents. The coalification in the coal seam generates a large amount of gas. However, the actual storage of gas in the coal seam is very small, mainly because the gas content permeated through the soil layer after gas generation accounts for a certain proportion. Therefore, the gas content in the coal seam is not only affected by coalification, but also related to the surrounding geological conditions. In addition to the dangers of general tunnels during the construction of gas tunnels, there is also a special danger of gas accidents, so safety accidents often occur during the construction process. It is mainly characterized by five types of gas outburst, gas explosion, gas burning, gas suffocation, and gas poisoning.

(9) **High ground temperature**

With the development of underground engineering, the high ground temperature problem has become an important engineering problem for tunnel projects with deep embedding and complex geological environment and has attracted extensive attention from scholars at home and abroad. According to past experience, when the local temperature is greater than 30 °C, it can be judged as high ground temperature. However, in actual engineering, high-temperature conditions greater than 35 °C may be encountered in regional active fault zones, abnormal ground temperature areas, high ground stress or deep-buried (greater than 1500 m) tunnels. This environment will increase the difficulty of ventilation and cooling of the construction site.

The high ground temperature has a great impact on the construction and operation of the tunnel. Whether it is the working environment, materials, structural safety, tunnel equipment, operation management, etc., all are plagued by high ground temperature. The adverse effects mainly include the following aspects:

① In terms of impact on the construction environment. When workers are working in a high-temperature environment, the efficiency tends to decrease, and the lack of oxygen in the plateau in the western region will even pose a serious threat to the safety of construction workers.
② In terms of the impact on construction materials. When selecting materials, it is necessary to consider the reliability in high-temperature environment. For example, the material needs to be resistant to high temperature, whether it is explosive for blasting, waterstop for waterproofing, waterproof board, etc., all need special materials.
③ In terms of the impact on structure. High temperature is prone to temperature stress, which causes cracking of the concrete structure lining. In the long run, structural cracking often leads to unfavorable stress and reduced durability.
④ In terms of the impact on mechanical equipment. The tunnel not only has high temperature, but also high humidity. Mechanical equipment is seriously affected. Its conversion efficiency is reduced, and failures are gradually increasing, causing the progress of the project to be dragged down.
⑤ In terms of the impact on operational maintenance. After completion, the tunnel must be maintained. If the temperature is too high, the service life of the decoration materials will be greatly reduced, and it will have a greater impact on the working environment of maintenance personnel. The fan starts frequently, and the operating cost is much higher than that of a normal tunnel.

Therefore, it is necessary to study the temperature field distribution of tunnels to solve the problem of tunnel freezing and thawing in cold regions, high-temperature tunnel problems, tunnel fire disasters, etc.

(10) Radioactivity

In some provinces of China, granite is widely distributed, and objectively there is a possibility of higher natural radioactivity. During the excavation of the expressway tunnel, the inevitable excavation may cause additional γ external irradiation and (or) inhalation of radon to produce internal irradiation to construction workers. The drainage outside the tunnel may have high radioactivity index. If it is discharged into the environment without taking corresponding measures, it may affect the surrounding water environment. Unreasonable treatment and disposal of abandoned earth and stone with abnormal radioactive indicators in tunnel construction may also have adverse effects on the environment. Therefore, radioactive monitoring of tunnel engineering is necessary to detect anomalies in time and take effective protective measures and to ensure environmental and public safety.

2.2 Tunnel Geological Diseases

The unfavorable geological disasters in tunnel construction mainly include karst, landslide, water gushing, and mud outburst. It can be seen from past engineering accidents that bad geology is one of the main causes of geological disasters during tunnel construction. Only by strengthening the prediction and prevention of adverse geological disasters and preventing problems before, they can effectively reduce the occurrence of engineering accidents.

(1) **Karst**

Karst refers to the general term for various surface and underground dissolution phenomena caused by chemical erosion, disintegration, mechanical damage, transportation, and sedimentation of soluble rock formations (mainly carbonate rocks, some sulfate rocks, halide rocks, etc.) by surface water and groundwater. Karst cave is the most common karst phenomenon during tunnel construction, and it is mainly dominated by the dissolution of karst water.

The impact of karst geology on tunnel construction is mainly the damage of karst water and karst caves. The presence of karst water in the surrounding rock of the tunnel makes the strength of the surrounding rock low. The presence of osmotic water pressure increases the probability of engineering accidents such as sudden water inrush and mud outburst, which seriously affects the quality and safety of the tunnel during the construction period and after vehicle inspection. The existence of karst caves will affect the stiffness and stress changes of the surrounding rock of the tunnel. During the tunnel construction process, it will easily cause disasters such as falling blocks and collapse of the surrounding rock in the tunnel construction, which affects the construction progress, and may bring the loss of manpower and financial resources to the tunnel construction.

(2) **Collapse**

When constructing a tunnel in a bad geological section, there are often phenomena such as collapse of the surrounding rock and sliding of the side wall, and even serious accidents such as roof collapse. These phenomena are called landslides. The collapse accident in the tunnel construction process is ultimately caused by insufficient understanding of the engineering geological characteristics of the surrounding rock of the tunnel. The form of collapse of the tunnel during construction is rich and varied. According to the location of the collapse of the tunnel, the collapse can be divided into three types: vault collapse, sidewall collapse, and the collapse of the tunnel face. Vault collapse is the collapse of the tunnel arch. It usually occurs when the tunnel has just been excavated, and the initial branch has not been completed or just completed.

Due to the instability of the upper rock mass caused by excavation, the phenomenon of collapse occurs. Sidewall collapse occurs at the waist of the tunnel. There are two cases of the sidewall collapse. One is that the surrounding rock is relatively intact, but the vertical joint is developed. Due to various factors of the surrounding rock of the tunnel, a plastic zone appears, the pressure of the surrounding rock increases, and the intense redistribution of the stress inside the surrounding rock promotes the development of joints and fissures in the surrounding rock, and the vertical joints penetrate vertically to form a landslide. The other is that the surrounding rock of the tunnel is extremely broken and is in the form of crushed stones. Failure to timely support after excavation will cause a large area of rock fragments to collapse. The collapse of the tunnel face is caused by the poor rock mass in front of the tunnel and the instability of the tunnel face.

(3) **Mud outburst and water inrush**

Tunnel water inrush is a kind of dynamic damage phenomenon that destroys the structure of aquifer in the process of tunnel excavation, makes the hydrodynamic conditions and the mechanical equilibrium state of surrounding rock change dramatically, so that the energy stored in the underground water body is released in the form of fluid and sometimes solid matter in the form of instantaneous high-speed movement. When the water is accompanied by a large amount of solid matter (especially muddy matter), it is called tunnel mud. Water gushing is the key to mud outburst. Without groundwater as the power and carrier, mud outburst is impossible to occur.

The main sources of tunnel water are surface water bodies, karst caves and underground rivers, aquifers, and fault water. Surface water bodies mainly include rivers, lakes, ponds, etc. For most mountain tunnels, rivers and lakes are not the main source of water inrush from tunnels. The water volume in karst caves and underground rivers is often very large, and there is a large supply of water. Once the tunnel excavation blasting encounters or is close to these water source bodies, when the hydrostatic pressure exceeds the strength of the rock mass, the groundwater will surge or break through the rock mass to form a strong water surge. Aquifer water, i.e., bedrock fissure water and pore water, is the most common source of groundwater in tunnel construction. Atmospheric precipitation mainly penetrates into permeable rock formations or rocks, and then flows into tunnels through exposed aquifers. Fault water, i.e., fault fissure water, is a common source of groundwater in tunnel construction. Fault fracture zone usually is a good passage and storage place for groundwater.

During tunnel construction, tunnel water inrush often causes relaxation or concentration of tunnel surrounding rock stress, causing damage to surrounding rock, resulting in reduced stability of surrounding rock, thereby increasing the risk of tunnel water inrush. At the same time, the water inrush in the tunnel often causes changes in the topography of the tunnel, which may cause secondary disasters, such as mud outburst disasters, which cause large tunnel construction safety accidents and large economic losses.

2.3 New Detection Technology for Complex Geological Conditions

In recent years, engineering geophysical exploration has been greatly improved in survey capability and survey accuracy. International engineering geophysical exploration has become an indispensable and most effective high-tech means to solve the problem of engineering construction geology. Geophysical exploration is a method of investigation to solve geological problems by observing and studying changes in various geophysical fields. In nature, different physical functions have different physical fields. For example, there is a gravitational field (gravity exploration) in the space where gravity acts, the space where natural or artificial electricity (magnetism) acts has an electric field (magnetic field) (electrical exploration or electromagnetic exploration), and the space where waves propagate has a wave field (seismic exploration), etc. The different geotechnical media that make up the crust tend to differ in density, elasticity, electrical, magnetic, radioactive, and thermal conductivity. These differences will cause local changes in the corresponding geophysical field. For this kind of geophysical field change related to the local change of underground rock and soil medium, it is called anomalous field. Engineering geophysical exploration is to observe the distribution and variation characteristics of these geophysical fields through special instruments for specific engineering projects, and then combine the known geological data for analysis and study, infer the nature of underground geotechnical media and environmental resources, etc., so as to achieve the goal of solving the problem of geological problems.

In recent years, with the development of geophysical technology, its application in engineering geology has become more and more extensive. Because geophysical exploration technology is economical, fast, and effective, especially it does not cause damage to the object to be detected, so that it shows strong vitality. At present, with the development of computer technology and the constant innovation of various inversion methods, geophysical technology is developing in the direction of higher detection accuracy, wider detection range and more accurate interpretation, showing unprecedented broad development prospects and being widely used in the fields of engineering, environment, disaster geological survey, etc. More and more people are paying attention to it. It is believed that in the near future, geophysical techniques will play a more important and irreplaceable role in the field of engineering geology.

Combining with the development of several geophysical exploration technologies at home and abroad, the author briefly discusses the application and development of seismic wave method and electrical method in engineering geology based on the analysis of previous application examples. These classic techniques will be mentioned frequently later in this book. According to the different characteristics of these methods and the application effects in different projects, the application scope, advantages, and disadvantages of each application are summarized, and the future research directions and development prospects of various methods are discussed and predicted. Finally, the good application effect and development prospect of integrated geophysical exploration in engineering geology are analyzed.

2.3.1 *Seismic Wave Method*

The seismic wave method is divided into a reflected wave method and a transmitted wave method (i.e., seismic CT), which utilizes the density difference of the underground medium to identify the geomechanical characteristics of the underground medium in the longitudinal and lateral directions through the received seismic transmitted wave or reflected wave signal. Further, the range of geological anomalies is inferred.

2.3.1.1 Seismic Reflection Wave Method

The reflection characteristics of the underground medium strictly follow the reflection theorem, that is, the larger the wave impedance difference, the larger the reflection coefficient, the stronger the amplitude, and the weaker the opposite. Therefore, the geological anomalies such as karst and fracture zone will cause the same axis difference and amplitude variation on the seismic section. Based on this principle, the researchers developed a number of advanced geological prediction systems designed specifically for tunnels and underground projects, such as the Swiss TSP system, the Japanese HSP system, and the US TRT system. The principle is to use a high-sensitivity geophone receiver to collect seismic waves generated by multiple seismic excitation points arranged on a single sidewall of a tunnel and the reflected waves when they encounter different reflective interfaces in the surrounding rock propagation. The location of the reflective interface is processed and interpreted by the data, combined with the specific geological conditions, to predict the faults and rock fracture zones that affect the construction. The system is mainly used to predict the nature, location, and scale of the unfavorable geology in front of, above and below the tunnel face. The maximum detection distance is 150 m in front of the face, and the highest resolution is not less than 1 m.

The biggest advantage of seismic reflection wave method is that it has long detection distance, high resolution, and strong anti-interference ability. Depth migration imaging is generally used in inversion imaging to improve interpretation accuracy and accuracy of prediction. Therefore, this method has a good application prospect.

However, the advanced geological prediction system currently used has many problems in practical work. The most important problem is that the interpretation of unfavorable geological conditions lacks clear indicators and relies more on experience, especially the experience of geological experts. In addition, it is still impossible to detect and identify fault zones, saturated zones, and caverns with geometrical shapes of cylinders or cones that are almost parallel to the tunnel direction. This will also be the focus of the next research work. The problems that can be solved by the advanced geological prediction system have a certain gap with the problems that the construction unit needs to solve directly (the evaluation of the surrounding rock level and the possibility of collapse). In order to solve this problem, the technicians responsible for the aggregation should supplement some knowledge of geomechanics. It is best to track the geological work. In order to improve the accuracy of advanced geological prediction, in addition to improving the level of interpretation, it is best to apply two or more long-term forecasting methods to verify each other, so as to integrate multiple solutions into one solution as much as possible. With the gradual complication of the geological conditions in the tunnel construction process, the one-dimensional and two-dimensional data interpretation results can no longer meet the exploration needs, and more accurate three-dimensional visualization has become the current development trend of the method. The self-developed AGI T3 advanced geological prediction system has been able to realize three-dimensional visualization of data and then accurately locate underground anomalies.

2.3.1.2 Seismic CT Technology

Seismic CT is a new exploration method for engineering geophysical exploration, also known as tomography, which is an example of the application of CT technology in geophysics. It can accurately and reliably invert the visual image of the internal structure of the geological body from the external measurement data of the geological body (structure). It has the characteristics of high resolution, good reliability, and intuitive image, which is a very effective survey method in engineering exploration.

The working principle of seismic CT is similar to that of medical CT. Medical CT uses X-rays to penetrate the human body and image human tissue through observations of ray intensity attenuation. Seismic CT uses seismic wave rays to see through the geological body (structure) and image the geological body (structure) through the observation of seismic wave travel time and energy attenuation. When the seismic wave penetrates the geotechnical medium, its velocity is related to the elastic modulus, shear modulus, and density of the geotechnical medium. The rock with high density and high strength has large modulus, high wave velocity and short walking time, and vice versa. The complete hard rock mass has a high wave velocity, and the broken rock mass and the loose soil mass have a low wave velocity. The compact rock mass with small porosity has small energy attenuation, and the broken rock mass and loose soil mass have large energy attenuation. Therefore, the seismic CT image can reliably reflect the distribution boundary of various types of rock and soil mass and the degree of fragmentation and distribution of the broken rock mass. Seismic CT is especially suitable for studying the distribution boundaries and mechanical properties of various strata, structures, rock masses and soils. In engineering geological surveys, it is often used to explore the location, shape, and mechanical strength of faulty zones, dense joint zones, water-bearing zones, cavities, and weathered zones. In the exploration of engineering geological structure of tunnel surrounding rock, seismic CT is the preferred method to find out the geological reasons and extent of tunnel surrounding rock collapse and the development, extension, and mechanical properties of the fault zone in the rock mass.

With the deepening and extensive application of seismic CT technology, seismic CT has now developed into a series of methods. The imaging physical quantities include multiple types such as wave velocity, energy attenuation, Poisson's ratio,

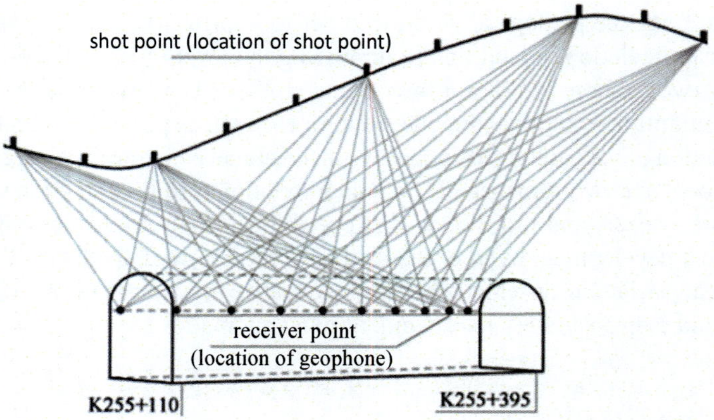

Fig. 2.1 Observation system design of seismic CT

etc. The imaging method can use a combination of direct waves, reflected waves, refracted waves, surface waves, etc., and can use various observation conditions such as boreholes, tunnels, slopes, mountains, and ground to perform two-dimensional and three-dimensional geological imaging. In the construction of railways, highways, and hydropower in the mountainous areas of western China, seismic CT can be widely used in engineering geological surveys and disaster remediation of projects such as lines, sites, tunnels, and slopes to solve complex geological conditions.

Depending on the purpose of the survey and the needs of the study, seismic CT surveys typically require the placement of two perspective sections, as shown in Fig. 2.1. In order to find out the engineering geological structure of the tunnel and the overlying rock mass, to find out the scope of tunnel collapse and roof fall, geometry, and mechanical strength of the collapsed body, and to understand in detail the development and extension of the fault and its relationship with the collapse of the surrounding rock, it is generally necessary to arrange two vertical seismic CT sections along the tunnel axis along the straight-line section of the tunnel's upline and downline.

2.3.2 Electrical Method

Electrical method refers to the use of the difference in electric field generated by artificial sources observed in the underground medium to sense the changes in the electrical parameters (apparent resistivity or permittivity) of underground geological abnormal bodies, which can then be used to infer the parameters of the underground structure, water content, and so on. The electrical method can be divided into a resistivity method and an electromagnetic wave method, wherein the high-density resistivity method is the representative of the resistivity method, and the representative of electromagnetic wave method is ground penetrating radar method.

2.3.2.1 High-Density Resistivity Method

High-density resistivity method is an array exploration method, also known as automatic resistivity system, which is the development of direct current method. Its function is equivalent to the combination of three-pole sounding and electric profile method. The artificial electric field is formed by supplying power to the underground through the electrode, and the distribution of the electric field is closely related to the distribution of the resistivity ρ of the underground rock and soil medium. Through the measurement of the artificial electric field in different parts of the surface, the distribution of the apparent resistivity ρ_s of the underground medium is understood, and the underground geological structure can be inferred and interpreted according to the distribution of the apparent resistivity of the rock and soil medium. The method is particularly sensitive to the water content of the surrounding rock. The surrounding rock is broken and contains water, and its apparent resistivity is significantly reduced. The apparent resistivity of intact and hard rock is significantly higher than the apparent resistivity of the fault zone or the fracture zone and the surrounding rock of the water-rich zone, as shown in Fig. 2.2. This method is clear in principle and intuitive in image, and it is a geophysical method with high resolution. In recent years, with the improvement of computer data acquisition technology, the exploration efficiency has been greatly improved. The coverage area and detection depth of the profile have been increased, and reliable data can be obtained in a strong interference environment, which greatly improves the signal-to-noise ratio and can accurately detects geological bodies. This method has been widely and successfully applied in engineering and hydrogeological exploration and exploration of mineral resources and water resources.

High-density resistivity measurements often use the most commonly used Winner devices for engineering surveys (Fig. 2.3). When measuring, AM = MN = NB = AB/3 is an electrode spacing, and the detection depth is AB/3. A, B, M, and N move

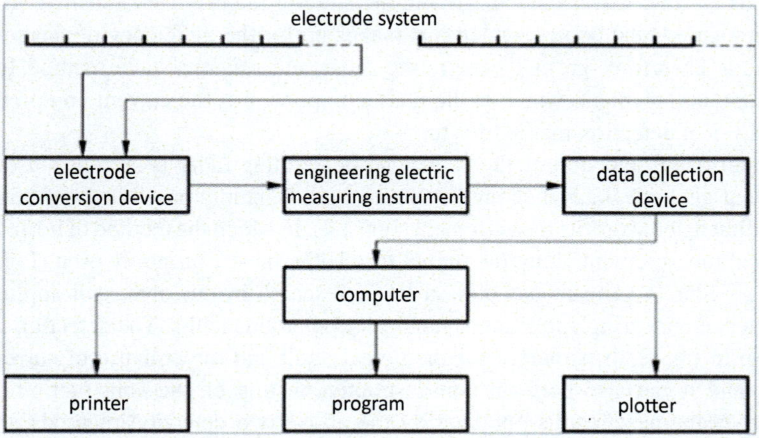

Fig. 2.2 Schematic diagram of working system of high-density resistivity method

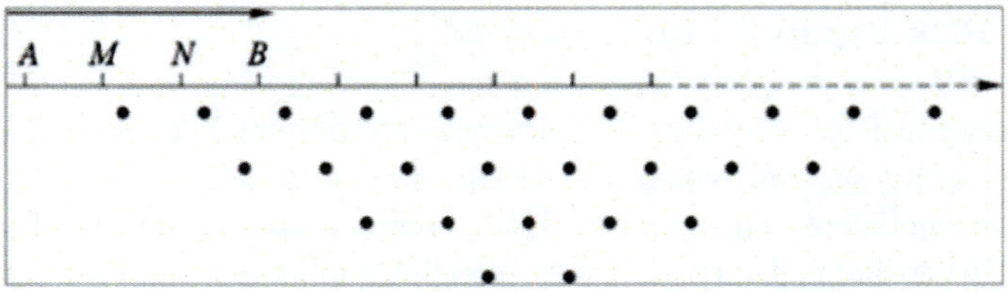

Fig. 2.3 Schematic diagram of cross-sectional scanning of Wenner device

to the right at the same time point by point to obtain the first layer section line. Then AM, MN, and NB increase one electrode spacing, and *A*, *B*, *M*, and *N* move to the right at the same time point by point to obtain another layer of section line. In this way, continuous scanning and measurement are carried out, and the inverted trapezoidal geological section is obtained.

2.3.2.2 Ground Penetrating Radar

The ground penetrating radar emits high-frequency electromagnetic waves (frequency range 100 MHz to 2 GHz). It can observe the changes of the electrical parameters (dielectric constant ε and electrical conductivity σ) of the underground geological anomalies by observing the changes of the received reflected wave waveforms, and then hydrogeological parameters such as porosity, water content, and salinity can be derived. Due to its high-frequency characteristics, ground penetrating radars are very sensitive to the changes in physical and geometric parameters and can give high-resolution imaging results of underground structures. Compared with traditional excavation geophysical methods, such as core drilling and logging, ground penetrating radar has the characteristics of high efficiency, low cost and wide detection range. Compared with other non-excavation geophysical exploration methods, such as seismic wave method, ground penetrating radar is very sensitive to water-bearing geological bodies, which just makes up for the deficiency of seismic wave method. Therefore, ground penetrating radar and seismic wave method form an effective complement, which is the main component of the current comprehensive geophysical detection method for tunnels.

In terms of detection methods, ground penetrating radar is currently used as an efficient geological advance prediction method. It mainly based on common-offset two-dimensional profile measurement (Fig. 2.4), in which the method of horizontal or vertical measurement along the tunnel face is the most commonly used (Fig. 2.4a). China's Shucai Li team has done a lot of research in this area, and applied it to practical engineering applications, and achieved good results. The detection method shown in Fig. 2.4b is mainly for the tunnel vault and the collapse of surrounding rock, and it can also perform non-destructive testing on the construction quality. The measurement results obtained by the above two detection methods are two-dimensional radar reflection profiles, and the interpretation of the results mainly

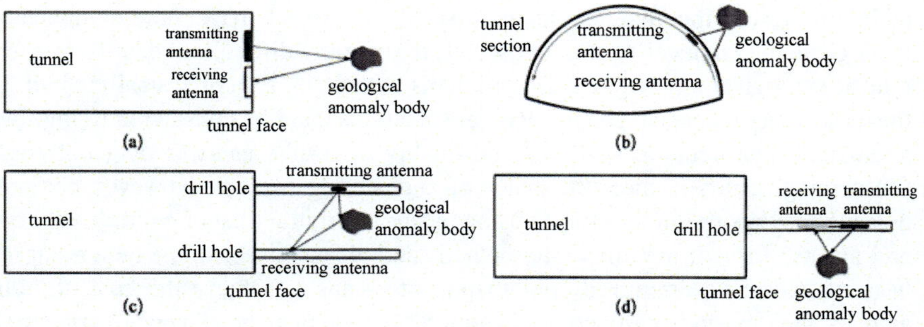

Fig. 2.4 Schematic diagram of ground penetrating radar measurement mode in tunnel

depends on signal processing techniques such as denoising, filtering, background field removal, and Fourier transform or Hilbert transform.

The detection methods shown in Fig. 2.4c, d require advanced drilling on the tunnel face. Compared with the crosshole method, although single-hole detection can also identify the depth and distance of the target geological body, it cannot give its azimuth and direction. Therefore, in view of the current ground penetrating radar measurement methods in the tunnel, the two-dimensional reflection profile is mainly used, and there is a problem that the space detection range is limited. In order not to increase the workload and reduce the efficiency of data collection as much as possible, the space detection range of the ground penetrating radar is expanded as much as possible. The author's research group proposed a detection mode based on three-dimensional scanning measurement. The specific method is shown in Fig. 2.5.

Since the application of various geophysical methods is based on certain physical premise, and geological, geophysical and boundary characteristics have a great impact on the test results, these methods have certain conditions and limitations. In

Fig. 2.5 Three-dimensional spatial scanning measurement mode of ground penetrating radar in tunnel

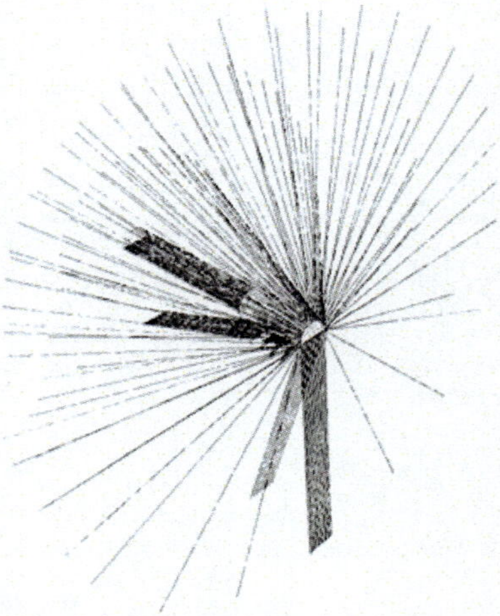

addition, most of the large and medium-sized key projects have more complicated geological and engineering problems, so it is generally difficult to identify or solve related geological and engineering problems with a single geophysical method. In this case, comprehensive geophysical prospecting should be considered to improve geological interpretation accuracy and quality of results analysis of geophysical prospecting results to meet the needs of engineering surveys. However, comprehensive geophysical prospecting is by no means an arbitrary list of multiple methods and means, nor is it that the more methods and means invested, the better, but the best method or an optimized combination of means to achieve the goal of "reliable technology and reasonable economics". Geological geophysical exploration that does not meet this requirement cannot be said to be a true integrated geophysical exploration. Different geophysical methods are applied in the same engineering project, which can often complement each other and verify each other. For example, Zaofeng Jiang applied high-density resistivity method, ground penetrating radar and seismic shallow reflection method to the leakage detection of a certain reservoir dam foundation, which has effectively identified the cause of reservoir leakage under complex geological conditions. Bin Wu et al. explored large-scale landslides on the Sichuan-Tibet Highway by means of six explored methods including reflected wave method, refracted wave method, transient Rayleigh wave exploration method, DC symmetrical quadrupole electric sounding method, combined profile method and high-density resistivity method. The application has achieved good results, which fully demonstrates the good exploration results of the comprehensive engineering geophysical exploration methods.

To sum up, the full use of geological and geophysical features, combined with a small amount of drilling, and the reasonable use of various combinations of geophysical methods are the basis for the application of geophysical methods in geological engineering, the direction and trend of the application and development of geophysical methods in the field of geological engineering, and the main way to overcome the ambiguity of geophysical methods.

Chapter 3
Three-Dimensional Advanced Prediction Technology for Tunnel Geological Structure

The tunnel is an underground construction project, and its surrounding rock geology is complex and changeable. It often encounters bad geological sections such as landslides, collapses, rock piles, caverns, high-stress and high-strength formations, loose formations, and soft soils, as well as special geological sections such as swelling formations and unsolidified water, surrounding rocks, karst caves, faults, rock bursts, quicksand, and gas overflow strata. They have different properties, shapes and scales, and the conditions of formation and variation are very complex. The geological conditions have the characteristics of suddenness. The geological data provided by the design and survey cannot be consistent with the actual situation from beginning to end, and the level of detail cannot fully meet the specific requirements of tunnel construction. If there is no reliable advanced geological prediction, potential geological hazards may become disasters. Safety and engineering accidents frequently occur during tunnel construction, ranging from extending the construction period and increasing investment, while destroying machinery and equipment, and even causing casualties, and the handling of accidents is difficult. The geological advance prediction of the tunnel during the construction process can detect the geological diseases that may be encountered in advance, and forecast the possible accidents and dangers, so as to take timely measures to prevent the accident and ensure the safety of the construction. At the same time, it also provides a basis for changes in tunnel construction methods and support forms, reduces blindness in construction, and saves tunnel investment. Therefore, advanced tunnel forecasting has been listed as a necessary part of tunnel construction.

G. Cao and H. Liu, *Three-Dimensional Exploration Technology of Tunnel Geology*, Key Technologies for Tunnel Construction under Complex Geological and Environmental Conditions, https://doi.org/10.1007/978-981-16-9225-3_3

3.1 Comment on Advanced Prediction Technology of Tunnel Geological Structure at Home and Abroad

The advanced prediction of tunnel geology has always been an international frontier subject, and all countries are studying this difficult problem. In 1980, relevant research began in Europe. After ten years, the Americas and China began their research work in this area. Because it is related to the safety, schedule, quality, and funding of the project construction, it has always been the focus of research, especially in complex geological conditions, its geological advance prediction is even more important. The world-famous Underwater tunnel, which crosses the Alps, used advanced geological forecasting technology throughout the construction process, and the effect was remarkable. The geological conditions in western China are complex, with a large number of orogenic belts and fracture zones, and abundant groundwater. Tunnel construction is prone to geological disasters such as water inrush, landslides, and mudslides. Therefore, advanced geological forecasting technology must be used to solve the above-mentioned problems in the construction of various tunnels.

At present, the advanced prediction methods for tunnel geological structures commonly used at home and abroad are divided into direct forecasting method and indirect forecasting method. The former includes advanced drilling and advanced flat guidance methods, which are one-sided, time-consuming, labor-intensive, and costly. The latter includes elastic seismic wave method, etc., which has the advantages of comprehensive, fast and practical.

The indirect prediction method of tunnel geological structure is also called physical detection method. The elastic seismic wave method is the most commonly used indirect forecasting method. Various elastic seismic wave prediction methods have been developed, such as negative apparent velocity method, horizontal sonic profile method (HSP) method, land sonar method, tunnel seismic advance prediction system (TSP), tunnel seismic tomography (TST), and true seismic reflection tomography (TRT) have been developed. This kind of method focuses on the study of the changes in the mechanical properties of rocks, can predict the structural characteristics and the degree of fragmentation of the surrounding rock in front of the tunnel or underground geological conditions equivalent to the mechanical properties, and has a better prediction effect on fault zone and broken zone.

In the early 1990s, Swiss Amberg Measurement Technology Co., Ltd. developed a set of advanced forecasting system equipment based on the principle of seismic reflection, the Tunnel Earthquake Prediction System (TSP), which uses a one-dimensional observation method (Fig. 3.1), can detect and predict the soft and hard changes of surrounding rock, faults, fracture zones, joint fissure development, water, cave and surrounding rock categories. TSP has been widely used in tunnel construction in developed countries such as Switzerland, Germany, France, Italy, Austria, and Japan. This technology was officially introduced into China in the late 1990s, and it has become one of the most widely used methods for earthquake prediction in China.

The TRT method was developed by the US NSA Engineering Company in the late twentieth and early twenty-first centuries and has been widely used in Europe

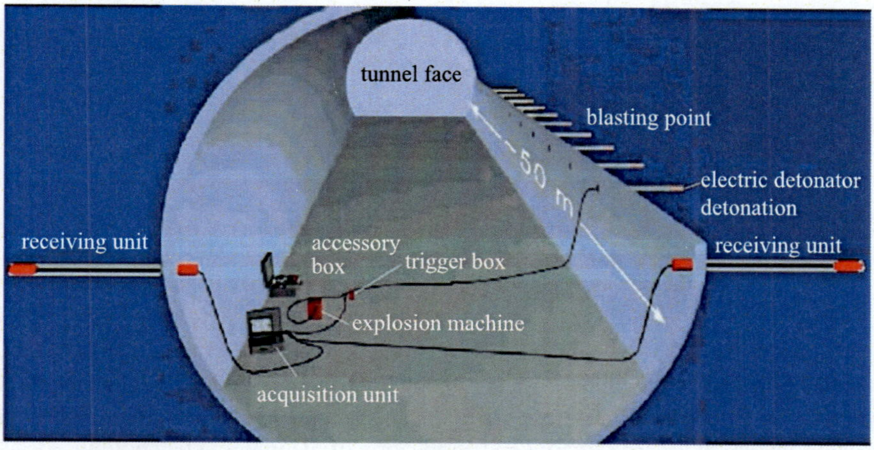

Fig. 3.1 TSP technical work diagram

and Asia. The method uses seismic migration tomography to detect the geological conditions in front of the tunnel face by analyzing the reflected seismic wave signal generated when the seismic wave encounters the interface of the wave impedance difference and timely predicts the hidden disasters of fault, fracture zone, weak zone, and water body. The TRT method is obviously different from the negative apparent velocity method and TSP in both observation mode and data processing method. In observation, it realizes spatial multi-point excitation and reception, thus obtaining rich spatial wave field information, conditionally improving the accuracy of detection and positioning. In terms of data processing, it introduces velocity scanning and migration imaging methods to make full use of kinematics and dynamics information. The resulting image is more intuitive and contains various data such as changes in rock mechanical properties and structural combinations.

The tunnel advance prediction method TRT uses a three-dimensional observation method (Fig. 3.2) and a one-dimensional imaging method. The location of geophones

Fig. 3.2 Schematic diagram of stratum geological conditions obtained by TRT

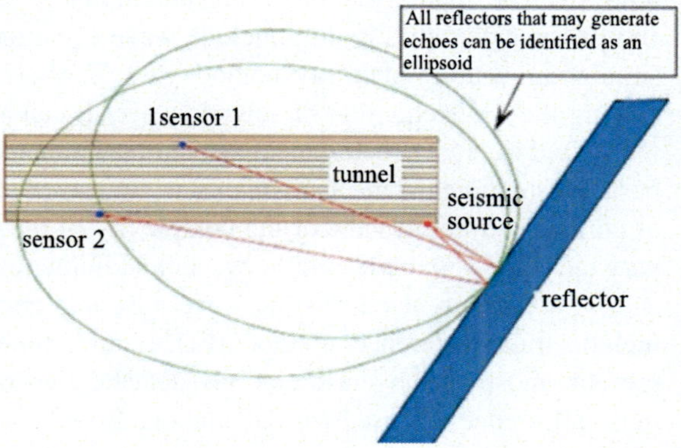

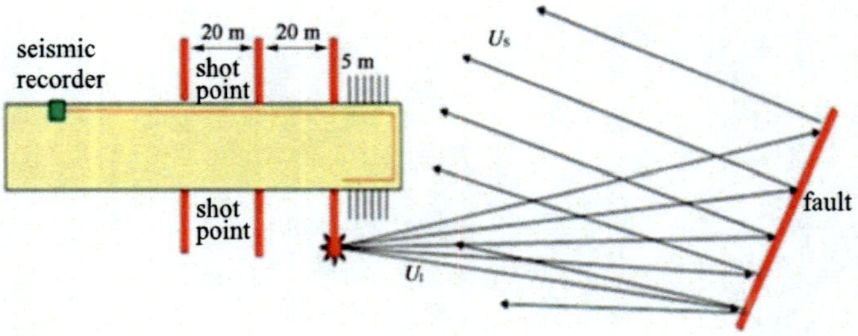

Fig. 3.3 Schematic diagram of TST advanced forecast observation system

and shot points in the tunnel adopts a spatial distribution method, which can obtain scattering gathers with different offset distances, which provides prerequisites and data foundation for the research and application of 3D imaging technology.

Tunnel seismic tomography (TST) is a forecasting system suitable for tunnel environments developed by researcher Yonggui Zhao. The system adopts two-dimensional array observation mode (Fig. 3.3). In the data processing, directional filtering is performed first, and only the front effective reflected wave is retained. On this basis, the wave field is separated, and at the same time, the scattering theory is used to improve the detection resolution.

The former Chengdu Military Region Engineering Research and Design Institute and Yunnan Aerospace Engineering Geophysical Exploration and Inspection Co., Ltd. simulated the law of seismic wave propagation in the underground three-dimensional space based on TSP and TRT technology and developed a complete set of advanced geological prediction technology for three-dimensional imaging tunnels, AGI-T3, which can be more accurate predicting the geological conditions in front of the master. The spatial positioning of bad geological bodies is more accurate, more accurate, and the results are displayed more intuitively. It plays an important role in ensuring the construction quality and safety of the supporting project. The popularization and application of this technology is of great significance in the construction of tunnel projects in China.

Tunnel geological structure advance prediction seismic wave field simulation technology can be divided into physical simulation and numerical simulation. Numerical simulation techniques mainly include wave equation method and ray method. The simulation accuracy of the ray method is low. The wave equation method is essentially solving the wave equation of seismic wave. It can accurately simulate the seismic wave field in any complex medium and provides more mathematical and physical basis for studying the mechanism of seismic wave propagation and the interpretation of complex strata. In view of the complexity of the tunnel seismic wave field, it is very meaningful to carry out the wave field numerical simulation technology.

The commonly used seismic wave field numerical simulation methods mainly include finite-difference method (FDM) and finite element method (FEM). At present, most of the elastic or viscoelastic two-way wave equations based on finite-difference are used for calculation. Boore uses the finite-difference method for CT imaging simulation of seismic waves in SH-wave inhomogeneous media. Carcione et al. proposed a simulation method for seismic wave propagation in linear viscoelastic media. Talezcr et al. studied seismic wave propagation in linear viscoelastic media. Robertsson et al. gave a finite-difference simulation method of viscoelastic wave. Graves et al. gave a simulation method for three-dimensional velocity stress equation staggered-grid finite-difference method for elastic wave propagation. Carcione et al. studied the numerical simulation of Biot longitudinal wave in porous elastic media. Ozdenvar et al. gave a staggered-grid finite-difference simulation method for seismic waves in poroelastic media. Carcione et al. proposed a staggered-grid finite-difference simulation method for seismic wave propagation in porous viscoelastic media. Qiaodeng He et al. studied seismic wave numerical simulation methods in transversely isotropic media. Yanguang Wang et al. gave a high-order difference forward simulation method suitable for strong lateral speed change. Zhenglin Pei et al. presented a staggered-grid high-order finite-difference simulation method for elastic wave propagation in three-dimensional arbitrary anisotropic media. Liangguo Dong used the staggered-grid high-order difference method to numerically simulate the propagation of elastic waves under undulating ground. Zhiyang Dai et al. used the convolutional differential operator to realize the numerical simulation of seismic wave field and verified its validity and correctness by using the model. Xiaohua Zhou et al. used the staggered-grid finite-difference method to simulate the micromotion signal, and the results show that the measured results obtained by this method are more in line with the measured results. These methods have strong seismic wave propagation simulation capabilities and can well describe the propagation characteristics of seismic waves in small-scale heterogeneous anisotropic media (pores, fractures, etc.). Bohlen et al. carried out integrated seismic imaging system (ISIS) in the tunnel to carry out tunnel prediction in advance through numerical simulation, simulated the tunnel advance prediction by three-dimensional finite-difference method, and simulated the transformation characteristics of shear wave and surface wave on the side wall of the tunnel in detail. Jingbo Liu et al. used nonlinear finite element software LS-DYNA to simulate the influence of blasting on tunnel structure and compared it with on-site blasting to prove the effectiveness of the simulation method. Guangyin Lu and others used the high-order staggered-grid difference method to carry out the full wave field numerical simulation calculation of the tunnel advance forecast. At present, in the field of engineering geophysics, it can quickly and effectively numerically simulate full-wave fields of 2D and 3D seismic, including software for simultaneously simulating P-wave, S-wave, surface wave, multi-wave multi-component wave field characteristics. However, in the field of structural mechanics, a variety of mature multi-wave multi-component finite element numerical simulation software has been used to solve elastic constitutive equations, such as ANSYS and PLAXIS.

From the perspective of the equation structure, compared with the fully elastic wave equation, the elastic constitutive equation has only one more first-order differential term with respect to time—the damping term. Therefore, the reasonable media parameter, the Rayleigh coefficient, is adjusted so that the damping term is approximately 0, and the elastic constitutive equation is degenerated into a fully elastic wave equation. And the simulated source in these software can be loaded with any custom data, so it is convenient to load the simulation with the noise source function extracted by field test and physical simulation, which is obviously closer to the actual situation compared with the simulation with Ricker wavelet in general geophysical numerical simulation. The key technology in the numerical simulation is the adjustment of the Rayleigh coefficient of the medium and the reasonable loading of the absorbing boundary conditions.

3.2 Three-Dimensional Wave Field Theory of Underground Surrounding Rock

3.2.1 Basic Theory of Seismic Wave

3.2.1.1 Generation and Classification of Seismic Wave

(1) Generation of seismic wave

When the explosion of explosives is used as the vibration source to excite seismic waves in a rock formation, the surrounding rock formations can be roughly divided into three layers for analysis. In the innermost layer, that is, around the explosive package, the strong pressure generated by the explosion will be far greater than the ultimate strength of the rock, so a cavity is exploded in the center of the rock layer to form a destruction circle (Fig. 3.4). The second layer is located outside the failure zone. As the distance from the seismic source increases, the corresponding pressure decreases, but it is still greater than the elastic limit of the rock. Although the rock will not be damaged, there will be corresponding plastic deformations, which appear as a series of ring or radial cracks. The third layer is located outside the plastic deformation layer. Because it is farther from the seismic source, the corresponding

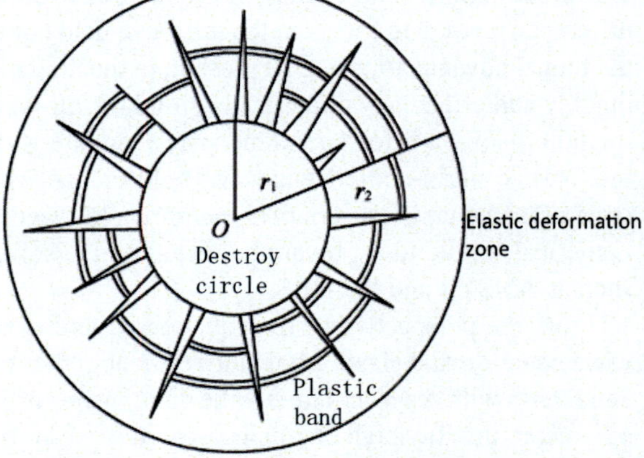

Fig. 3.4 Schematic diagram of the impact of the explosion on the surrounding rock

pressure has been reduced to within the elastic limit. Therefore, under the action of the short-term impact force generated by the explosive explosion, the rock in this layer only produces elastic deformation. Since the radii of the first two layers are relatively small relative to the receiving distance, the focus is on the properties of the third layer of elastic deformation. Therefore, the seismic waves studied here more refer to the elastic seismic waves propagating in the rock.

(2) Classification of seismic wave

According to the different strain caused by different disturbance, it can be divided into longitudinal waves and transverse waves. Under the disturbance of expansion and contraction force, the elastic medium will produce corresponding volumetric strain, and the wave formed by continuous propagation of volumetric strain form is called longitudinal wave (also called P-wave). Under the disturbance of rotational force, the elastic medium will produce corresponding shear strain, and the wave formed by the continuous propagation of the shear strain is called transverse wave (also called S-wave).

(1) Longitudinal wave. The salient feature of the longitudinal wave is that the direction of propagation of the wave during propagation is consistent with the direction of vibration of the particle, as shown in Fig. 3.5. Further, since the longitudinal wave forms a sparse zone (also called an expansion zone) and a dense zone (also called a compression zone) at intervals in the medium, the longitudinal wave is also called a sparse wave or a compression wave. Relative to other fluctuations, the longitudinal wave velocity is the fastest when propagating in the same medium, and the corresponding wave velocity is usually represented by the symbol V_p. Determined by the form of vibration, longitudinal waves can be present in any medium.

(2) Transverse wave. The characteristic of the transverse wave is that the direction of propagation of the wave in the process of propagation is perpendicular to the

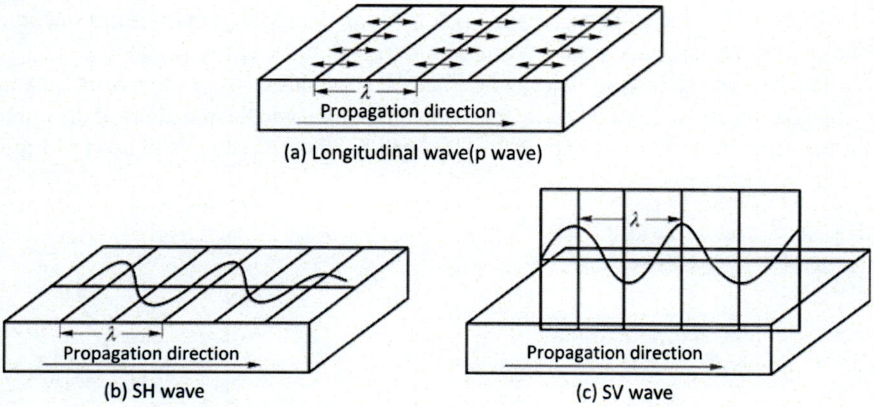

Fig. 3.5 Schematic diagram of longitudinal and transverse wave propagation

direction of vibration of the particle, and the corresponding wave velocity is usually represented by the symbol V_s. Due to the different vertical directions, the transverse wave can be divided into SH-wave and SV-wave, as shown in Fig. 3.5b and c. Specifically, the SH-wave is a waveform component of the transverse wave waveform in a horizontal plane, and the corresponding SV-wave is a waveform component of the transverse wave waveform in a vertical plane. Determined by its form of vibration, shear wave propagation requires the medium to transmit shear forces, so shear waves can only propagate in elastic solid media, but not in liquids and gases.

3.2.1.2 Propagation Principle of Seismic Wave

(1) Huygens-Fresnel principle

Assuming that the underground is a homogeneous medium, the vibration source is located at point O on the ground. After starting from point O, the seismic waves generated by the explosion propagate in all directions at the same time, as shown in Fig. 3.6. If the point where the vibration just started in the medium at a certain time t is connected into a curved surface, then the curved surface is said to be the wave front at time t. In the same way, if the points where the vibration just stopped at the same time are connected to form a curved surface, the curved surface is said to be the wave rear at this time. At the same time, the particle outside the wave front will not vibrate because the wave has not been propagated, while the particle inside the wave back has stopped to vibrate, and the corresponding area between the two curved surfaces is in the state of vibration, which is called the vibration zone. Since the shape of the wave front is related to the wave velocity of the medium, the wave front shape changes correspondingly when the wave velocity of the medium changes. So if in a homogeneous medium, the corresponding wave front is a series of concentric spheres (hemispheres only) with the source o as the center, this wave can be called a spherical wave (and when the radius of the spherical wave is very large, it will be simplified to plane waves). But if propagating in a non-uniform medium, the corresponding wave front becomes an irregular surface.

The Huygens principle mainly describes the correlation law of wave front forward propagation, and its content is: In an elastic medium, when the positions of the points of the same wave front at known time t, these points are used as the newly formed

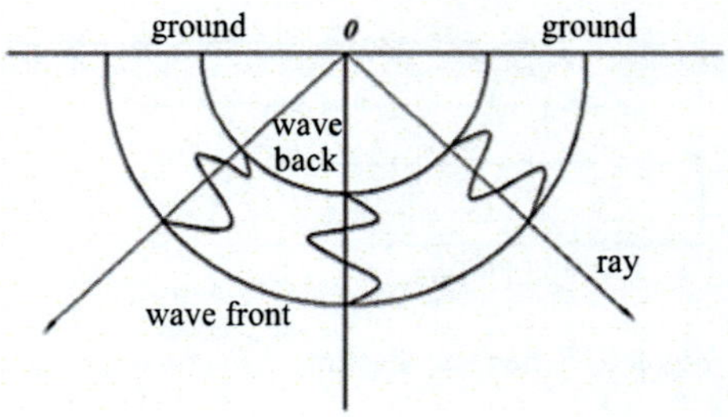

Fig. 3.6 Schematic diagram of wave front, wave back, and rays

point seismic sources at that time, and wavelets are generated to propagate outward. After any time Δt, the envelope composed of the wave front of the corresponding wavelet is the wave front of the total wave at time $t + \Delta t$. From the definition of the Huygens principle, it is known that the wave front at any time can be obtained by the method of drawing using the known wave front. That is, under the condition of uniform medium, let the wave front at the known time t be $S_{0,}$ the corresponding wave velocity be recorded as v, and the wave front at the time $t + \Delta t$ can be obtained by the following method: Taking some points on the wave front S_0 as the center of the circle, then, $\Delta r = v\Delta t$ is used as the radius to draw a circle, and then the corresponding curved surface is the wave front of each wavelet. Then the envelope surface common to all wavelet fronts is made, and the envelope surface is the new wave front S of the wave at time $t + \Delta t$. The new wave front S in Fig. 3.7 is the new wave front S of the wave at $t + \Delta t$.

It can be seen that although the Huygens principle gives the spatial geometric position of the wave propagation, it does not further describe the physical state of the wave reaching the corresponding position. In fact, Fresnel has perfected Huygens' principle very well. Correspondingly, it is proposed that the corresponding wavelet generated at each point on the wave front will produce interference superposition on the observation point, and the superposition result can be the total vibration that can be actually observed at that point. Through this perfection, the physical meaning of the Huygens principle is more clear, and the resulting Huygens–Fresnel principle (also known as the wave front principle) is applicable to both homogeneous and inhomogeneous media.

(2) **Fermat principle and time field**

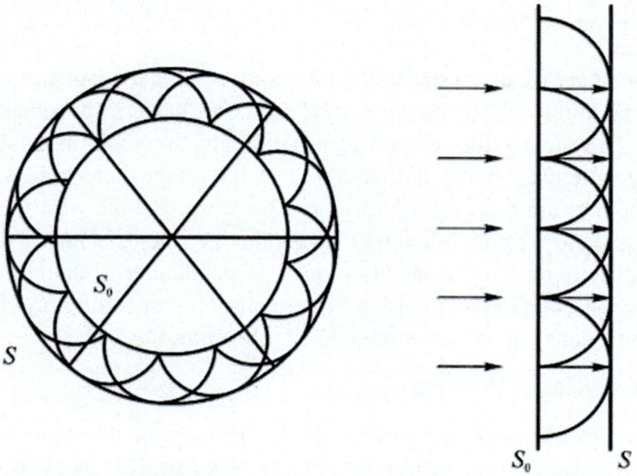

(a) Schematic diagram of the wave front of spherical wave

(b) Schematic diagram of the wave front of plane wave

Fig. 3.7 Schematic diagram of spherical wave and plane wave front

The Fermat principle is defined as follows: The travel time of the seismic wave in the direction of the ray is the smallest compared to the travel time used to travel on any other path, because the wave will travel along the path with the smallest travel time, and this least path is called the ray. In the case of uniform media, the shortest path for propagation between two points is the straight line connecting the two points directly. Therefore, under homogeneous conditions, the propagation of elastic waves generated by a point seismic source can be described by a cluster of radiating lines with the seismic source as the center. Similarly, the rays of a plane wave are a set of parallel lines perpendicular to the wave front. In the case of uneven media, the corresponding best path will become a curve, but it can always be guaranteed that the ray is perpendicular to the wave front.

According to Fermat principle, the time t used for wave front propagation of elastic waves is a function related to spatial position under elastic medium conditions. The corresponding expression is as follows:

$$t = (x, y, z) \tag{3.1}$$

By determining the above functional relations, the spatial distribution of the wave front at different times can be known. Based on its practical significance, the numerical space–time relationship determined by Eq. (3.1) is the time field.

It can be seen that the time field is only a scalar field, so it can be represented by its isosurface, and this isosurface is called an isochronous surface. The corresponding isochronous surface equation is expressed as follows:

$$t(x, y, z) = t_i \tag{3.2}$$

Although the definitions are different, at the same time, the wave fronts coincide with each other.

As mentioned above, under the action of a point source, the isochronous surface in a uniform and infinite elastic medium will become a cluster of concentric spherical surfaces with the source as the center. Correspondingly, the isochronous surface will become a curved cluster under the condition of inhomogeneous media. It is also known that the rays are perpendicular to the wave front, and the wave fronts at any time in the time field are coincident with each other in time. It is finally determined that the ray clusters on the time field are also perpendicular to the corresponding isochronous planes, and the relationship between the corresponding ray clusters and the isochronous plane clusters is orthogonal, as shown in Fig. 3.7.

(3) Apparent velocity theorem

The speed at which a seismic wave propagates in the direction of a wave ray is called the ray velocity, which is also the true velocity of the wave. This true speed *V* can be measured only if the direction of the observation is exactly the same as the direction of the wave, but this is difficult to achieve. In reality, the speed value of the wave can only be tested on the ground, and the corresponding measured speed value will

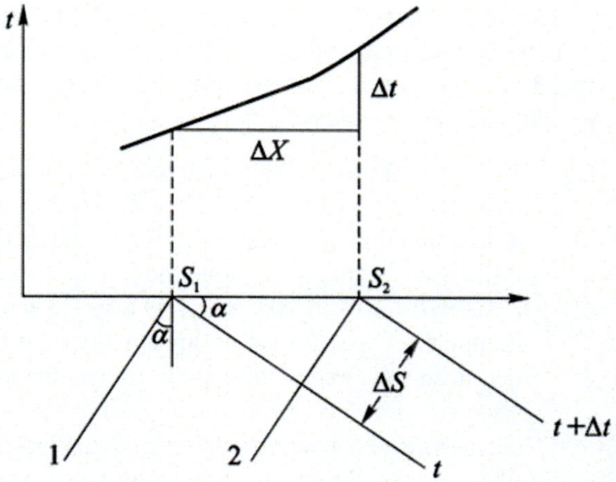

Fig. 3.8 Schematic diagram of apparent velocity principle

not be the true speed of the wave, but is called the apparent speed, which is usually expressed by V_α, as shown in Fig. 3.8.

Where S_1 and S_2 represent the positions of two detection points. The track spacing is set to ΔX. And there is a set of plane waves from the ground, respectively, the time taken to reach the point S_1 along the ray 1 is t, and the time taken to reach the point S_2 along the ray 2 is recorded as $t + \Delta t$, then the apparent speed measured on the ground is as follows:

$$V_\alpha = \frac{\Delta X}{\Delta t} \tag{3.3}$$

Also note that the ray distance between two points S_1 and S_2 is ΔS, then the true speed can be expressed as follows:

$$V = \frac{\Delta S}{\Delta t} \tag{3.4}$$

From Fig. 3.8, the corresponding geometric relationship can be obtained, $\Delta S = \Delta X \cdot \sin\alpha$, and the above two equations can be combined:

$$V = \frac{\Delta S}{\Delta t} = \frac{\Delta X}{\Delta t} \cdot \sin\alpha = V_\alpha \sin\alpha \tag{3.5}$$

It can be turned into the following expression:

$$V_\alpha = \frac{V}{\sin\alpha} \tag{3.6}$$

where α is the angle between the wave front of the plane wave and the ground (also the angle between the wave ray and the ground normal).

The relationship shown by Eq. (3.6) is called the apparent velocity theorem.

It can be seen that the theorem shows the numerical relationship between apparent speed and true speed. Further from the apparent velocity theorem, the following conclusions can be drawn:

(1) When $\alpha = 0°$, the wave propagates perpendicular to the direction of the line, corresponding to $V_\alpha \to \infty$. Then the wave front will reach the ground points at the same time. Because there is no time difference at each point, the speed of the wave propagation along the line is similar to infinity.
(2) α varies from 0° to 90°, which is also the most common case. At this time, the apparent speed must be greater than the true speed, that is, $V_\alpha > V$. In fact, in seismic exploration applications, due to the difference in distance, the reflected wave received by the near-shot track has a higher apparent velocity, and the reflected wave time difference of adjacent tracks is smaller. The received reflected wave has a lower apparent velocity and a larger time difference of reflected waves of adjacent tracks.
(3) When $\alpha = 90°$, the wave is directly incident on the observation point along the survey line direction, corresponding to $V_\alpha = V$. Obviously, when the propagation direction of the wave is consistent with the direction of the survey line, the apparent velocity is equal to the true velocity at this time.

3.2.1.3 Seismic Wave on the Elastic Interface

The above analysis is based on the premise of homogeneous media, but the actual geological section is generally composed of multiple rock layers with different geological ages and different lithology. Therefore, in order to be closer to the real situation, the rock formation with the same origin and lithology can be used as the homogeneous medium of isotropicity, and the studied geological section can be used as a layered combination of several uniform media. In order to facilitate the study, the occurrence of each layer in the layered medium is set to be horizontal, which constitutes the simplest and practical horizontal layered medium model, as shown in Fig. 3.9. By understanding the relevant propagation laws of seismic waves on the

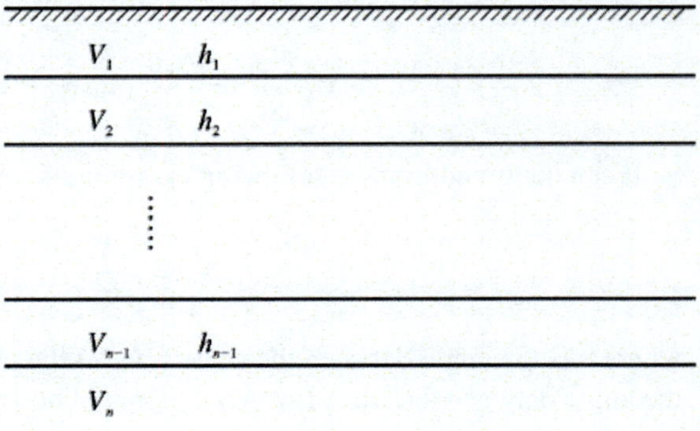

Fig. 3.9 Schematic diagram of the horizontal layered medium model

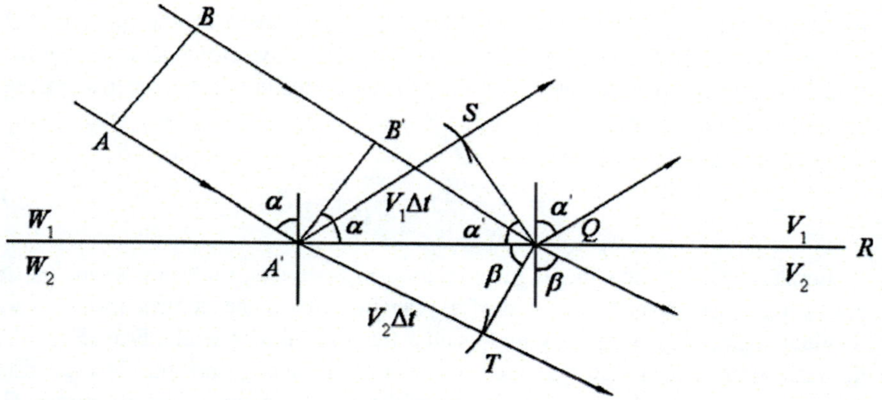

Fig. 3.10 Schematic diagram of reflection and transmission of seismic wave

interface between the dielectric layers of this simple layered medium model, it can be extended to any interface.

(1) *Reflection and transmission of seismic wave*

The following first obtains the propagation path corresponding to the back and the scattering from the angle of the wavelet generated by the new seismic source and then gives the corresponding definition. Suppose the elastic space is divided into upper and lower parts by the interface *R*, denoted as W_1 and W_2, as shown in Fig. 3.10, and the longitudinal wave velocity in the upper half space is denoted as V_1, and the longitudinal wave velocity in the lower half is denoted as V_2. A set of plane waves are projected to the elastic interface *R* at an incident angle α, and the wave front of the plane wave at this time is AB. When the plane seismic wave front continues to advance until the wave front is *A'B'*, point *A* corresponding to the previous wave front reaches the point on the interface *R* first. According to Huygens' principle, it is possible to use a point on the interface as a new seismic source and propagate around as a new disturbance in the form of wavelets generated from this point. After the Δt time, when the B' point on the wave front propagates to the corresponding *Q* point on the elastic interface *R*, the wavelet from the *A'* point has already propagated Δt time in the W_1 medium at the velocity V_1, and at the same time has propagated for Δt time in the W_2 medium, and its velocity is denoted as V_2. Now taking the point as the center of the circle, and drawing the arcs with V_1 and V_2 as the radius, and drawing the respective tangents of the two arcs from the *Q* point, the obtained tangent points are correspondingly recorded as the *S* point and the *T* point.

In the wave propagation diagram deduced by analysis, the new wave front QS connected by the *Q* point and the *S* point and the original wave front *A'B'* of the incident wave are in the same medium, then the wave represented by the wave front QS is called the reflection wave. The new wave front generated in the medium W_2 is denoted as QT, and the wave corresponding to it is referred to as the transmitted wave. The wave fronts of the three waves of incident wave, reflected wave and transmitted wave, respectively, correspond to the wave front *A'B'*, the wave front QS and the wave front QT. The angle between the three wave fronts and the interface *R* is, respectively, denoted as α, α', β, then the corresponding geometric triangle relationship is as follows:

$$\frac{\sin\alpha}{V_1} = \frac{\sin\alpha'}{V_1} = \frac{\sin\beta}{V_2} = P \tag{3.7}$$

Equation (3.7) shows the angular relationship between the three waves on the elastic interface. For convenience of explanation, it can be defined that α is the incident angle, α' the reflection angle, and β the transmission angle. From Eq. (3.7), the reflection angle is always equal to the incident angle, and the corresponding transmission angle will vary with the ratio of the upper and lower medium speeds. On the same interface, the corresponding incident wave, reflected wave, and transmitted wave all have the same ray parameter *P*, and $P = \sin\alpha_i / V_i$. The above is the famous Snell's law, also known as the reflection–transmission law.

In order to simplify the interpretation process, the relationship among incident, reflection, and transmission on the elastic interface should also conform to Snell's law, according to the nature of ray perpendicular to the wave front.

The difference lies in the meaning of the corresponding incident angle α, reflection angle and transmission angle, which are changed to the angle formed by the incident ray, reflected ray α' and transmission ray β and the normal line of the elastic interface *R*, as shown in Fig. 3.11.

(2) **Refraction of seismic wave**

The seismic wave that is being transmitted does not necessarily produce reflected waves when encountering the geological interface. If the wave velocity V_2 of the lower medium is greater than the wave velocity V_1 of the upper medium, a new wave may be generated on the interface with the change of the incident angle. Plane seismic waves (plane waves are an approximation of real spherical waves) are, respectively, projected onto the interface at different angles of incidence. It is known from Snell's law that as the incident angle α continues to increase, the transmission angle β will

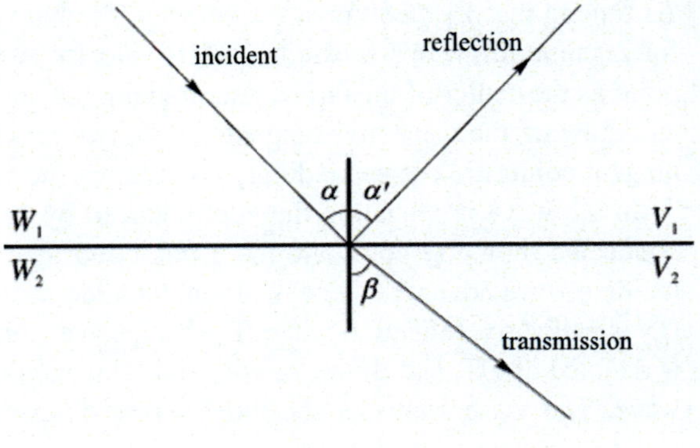

Fig. 3.11 Schematic diagram of the relationship between incidence, reflection, and transmission

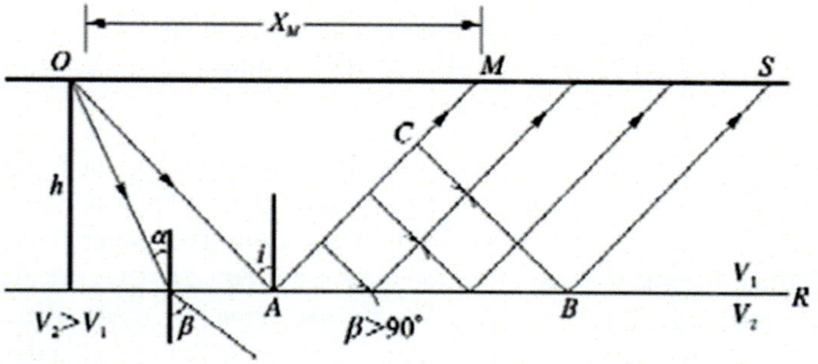

Fig. 3.12 Schematic diagram of refracted wave formation

also increase, and the corresponding transmitted wave rays will deviate from the normal and move closer to the interface. At this time, β is always greater than α, so there is an angle that α can reach. A situation of $\beta = 90°$ will occur; that is, the transmitted wave will only slide along the interface at the speed of V_2, thus becoming a sliding wave, as shown in Fig. 3.12. The incident angle at this time can be defined as the critical angle i, and the corresponding expression is as follows:

$$\sin i = \frac{V_1}{V_2}, \ i = \arcsin\left(\frac{V_1}{V_2}\right) \tag{3.8}$$

It can be seen that the critical angle can be obtained by knowing V_1 and V_2.

From the principle of wave front, any point along the path of high-speed gliding wave can be regarded as a new seismic source that produces wavelets at the beginning of the moment. Since the elastic medium in the space is continuous, the corresponding particle motion must also be continuous. Therefore, when the vibration of the lower medium particle is generated, the upper medium will be disturbed, and correspondingly, new fluctuations will occur in the upper medium. This kind of fluctuation is called refracted wave.

As shown in Fig. 3.12, the seismic wave produces a gliding wave at point A, which starts at point A and slides forward at speed V_2. After a period of time Δt, its wave front reaches point B, corresponding to the distance AB = $V_2 \Delta t$. At the same time, point A produces a semicircular wavelet propagating in the upper medium, and its radius is recorded as AC = $V_1 \Delta t$. Now the tangential line of the corresponding wave front arc from point B is recorded as BC, and BC is the wave front corresponding to the refracted wave at this time. Determined by the relevant geometric relationship, the acute angle ∠ABC formed between the interface and the interface is the same as the critical angle i. The wave front of the refracted wave is formed by drawing the corresponding envelope of the semicircle wavelet generated by each new point source on the interface. According to the condition that ΔABC is a right triangle, the following relationship can be obtained:

$$\sin\angle\mathrm{ABC}=\frac{\mathrm{AC}}{\mathrm{AB}}=\frac{V_1\Delta t}{V_2\Delta t}=\frac{V_1}{V_2}=\sin i \tag{3.9}$$

And $\angle\mathrm{ABC}=i$.

Similarly, according to the condition that the wave front and the ray are perpendicular, it can be seen that the ray of the refracted wave is a cluster of parallel lines perpendicular to the BC phase of the wave front, and the angle between the corresponding line and the normal of the interface is equal to the critical angle.

It can be seen from Fig. 3.12 that, as the first ray of the refracted wave, ray AC can only be received from point M on the ground, so point M is also called the beginning point of the refracted wave. Correspondingly, the refracted wave cannot be received within the range from the focal point O to the focal point M, so this region is called the blind area of the refracted wave.

It can be easily obtained from Fig. 3.12 that the ray AC is the first ray of the refracted wave, and only when the refracted wave is received from the point M on the ground, the point M is called the starting point of the refracted wave. Correspondingly, from the focal point O to point M, the refracted wave cannot be received, so this interval is called a blind zone of the refracted wave. Then the expression of the blind zone X_{M} is as follows:

$$X_M=2\mathrm{h}\tan i=2\mathrm{h}\tan\left[\arcsin\left(\frac{V_1}{V_2}\right)^2\right]=2\mathrm{h}\left[\left(\frac{V_2}{V_1}\right)^2-1\right]^{-\frac{1}{2}} \tag{3.10}$$

It can be seen that the blind zone X_{M} is related to the refractive interface depth h and the corresponding wave velocity ratio V_2/V_1 of the upper and lower media, and X_{M} is proportional to h, and inversely proportional to V_2/V_1. Especially when $V_2/V_1=1.4$, the corresponding X_{M} is 2 h, and this can be an application rule; that is, in order to observe the refracted wave, it is necessary to satisfy the condition that the offset is greater than 2 times the depth of the refractive interface.

From the above, it can be seen that the conditions required for the formation of refracted waves are much harsher than those for reflected waves. Not only the difference of wave velocities on both sides of the interface is required, but also the wave velocity V_2 of the lower layer medium can be greater than the wave velocity V_1 of the upper layer medium. Therefore, there are very few refraction interfaces encountered in the actual application process.

(3) **Wave conversion and energy distribution on the elastic interface**

① Wave conversion on the elastic interface. When the elastic longitudinal wave P is incident on the geological interface at an angle α, the medium near the interface is subjected to the force perpendicular and parallel to the interface, corresponding to the normal stress and the shear stress. Under the above-mentioned disturbance action, four different kinds of waves will be generated in the upper and lower media, which are correspondingly called reflected longitudinal wave RP and reflected transverse wave RS, and transmitted longitudinal wave TP and transmitted transverse wave TS, as shown in Fig. 3.13. Available from

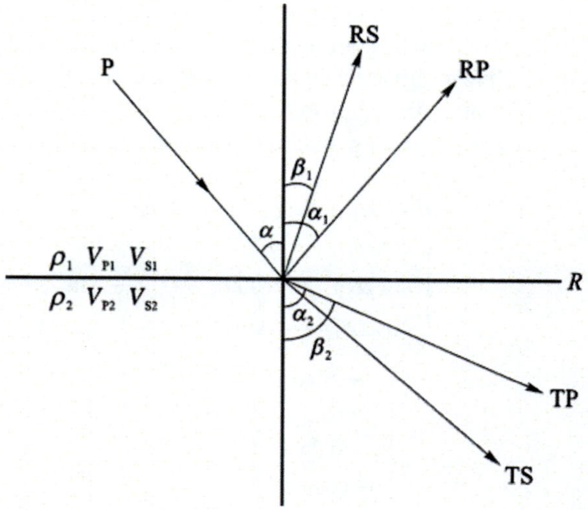

Fig. 3.13 Schematic diagram of splitting and conversion of waves generated when a longitudinal wave is incident

Snell's Law:

$$\frac{\sin\alpha}{V_{P1}} = \frac{\sin\alpha_1}{V_{P1}} = \frac{\sin\beta_1}{V_{S1}} = \frac{\sin\alpha_2}{V_{P2}} = \frac{\sin\beta_2}{V_{S2}} = P \quad (3.11)$$

where α—P-wave incident angle.

α_1, β_1—Reflection angle corresponding to transverse wave and transverse wave.

α_2, β_2—Transmission angle corresponding to transverse wave and transverse wave.

Equation (3.11) also shows that when the elastic longitudinal wave P is incident on the corresponding point A on the interface R, the incident wave will form two reflected waves and two transmitted waves, and if the incident longitudinal wave is included, there are a total of 5 kinds of fluctuations at the corresponding point A. Since the reflected longitudinal wave RP and the transmitted longitudinal wave TP are identical in waveform to the incident longitudinal wave, they are called the same type of waves. Similarly, since the other two fluctuations are inconsistent with the incident longitudinal wave on the waveform, they are called converted waves.

② Energy distribution achieved on the elastic interface. In order to study the problem of energy distribution, the proportional relationship between amplitude and energy can be used to start with the amplitude of vibration. The ratio of the reflected longitudinal wave RP and the reflected transverse wave RS to the vibration amplitude of the incident longitudinal wave P corresponds to Rpp and Rps, and the ratio of the transmitted longitudinal wave TP and the transmitted transverse wave TS to the amplitude of the incident longitudinal wave vibration corresponds to Rpp and Rps. Based on Snell's law and the corresponding boundary conditions (including the stress continuity conditions existing on the interface and the displacement continuity conditions), a linear algebraic equation system, called Zoeppritz equation, is finally derived, which describes the mathematical relations of the above various waves. The corresponding matrix expression is as follows:

$$\begin{bmatrix} \sin\alpha_1 & \cos\beta_1 & -\sin\alpha_2 & -\cos\beta_2 \\ \cos\alpha_1 & -\sin\beta_1 & \cos\alpha_2 & -\sin\beta_2 \\ \sin 2\alpha_1 & \frac{V_{\mathrm{P1}}}{V_{\mathrm{S1}}}\cos 2\beta_1 & \frac{\rho_2 V_{\mathrm{S2}}^2 V_{\mathrm{P1}}}{\rho_1 V_{\mathrm{S1}}^2 V_{\mathrm{P2}}}\sin 2\alpha_2 & \frac{\rho_2 V_{\mathrm{S2}} V_{\mathrm{P1}}}{\rho_1 V_{\mathrm{S1}}^2}\cos 2\beta_2 \\ \cos 2\beta_1 & -\frac{V_{\mathrm{S1}}}{V_{\mathrm{P1}}}\sin 2\beta_1 & -\frac{\rho_2 V_{\mathrm{P2}}}{\rho_1 V_{\mathrm{P1}}}\cos 2\beta_2 & \frac{\rho_2 V_{\mathrm{S2}}}{\rho_1 V_{\mathrm{S1}}}\sin 2\beta_2 \end{bmatrix} \begin{bmatrix} R_{\mathrm{PP}} \\ R_{\mathrm{PS}} \\ T_{\mathrm{PP}} \\ T_{\mathrm{PS}} \end{bmatrix} = \begin{bmatrix} -\sin\alpha_1 \\ \cos\alpha_1 \\ \sin 2\alpha_1 \\ -\cos 2\beta_1 \end{bmatrix} \tag{3.12}$$

In the case where the incident angle α of the incident longitudinal wave P and the density ρ_1 of the upper medium, the longitudinal wave velocity V_{P1}, the transverse wave velocity V_{S1}, and the corresponding parameters ρ_2, V_{P2}, V_{S2} of the lower medium are known, the angle α_1, α_2, β_1, β_2 can be calculated first by using Snell's law, and then the linear algebraic equations are solved, which can calculate out the four unknowns of R_{PP}, R_{PS}, T_{PP} and T_{PS}. With the amplitude, the corresponding energy distribution relationship can be determined.

It can be seen from Eq. (3.12) that the reflection coefficient and the transmission coefficient are not only related to the incident angle α, but also related to many other parameters, such as ρ_1/ρ_2, $V_{\mathrm{P2}}/V_{\mathrm{P1}}$ and $V_{\mathrm{S2}}/V_{\mathrm{S1}}$. The change of these parameters can also cause the change of the coefficients shown, which will affect the specific relationship of energy distribution. Since there are too many parameters, it is difficult to directly judge the relationship. Therefore, it first start with the special case of normal incidence, and then expand to the general incidence.

When the normal incident occurs, the incident wave is incident on the normal, and the corresponding incident angle is 0°. According to Snell's law, it can be known that the other corresponding angles α_{P}, α_{S}, β_{P} and β_{S} are also 0°, and the corresponding Zoeppritz equations are simplified as:

$$\left.\begin{aligned} R_{\mathrm{PS}} - T_{\mathrm{PS}} &= 0 \\ R_{\mathrm{PP}} - T_{\mathrm{PP}} &= 1 \\ R_{\mathrm{PS}} + T_{\mathrm{PS}}\frac{\rho_2 V_{\mathrm{S2}}}{\rho_1 V_{\mathrm{S1}}} &= 0 \\ R_{\mathrm{PP}} + T_{\mathrm{PP}}\frac{\rho_2 V_{\mathrm{P2}}}{\rho_1 V_{\mathrm{P1}}} &= -1 \end{aligned}\right\} \tag{3.13}$$

Solving Eq. (3.13) can get that

$$\left.\begin{aligned} R_{PS} &= T_{PS} = 0 \\ R_{PP} &= \frac{\rho_2 V_{P2} - \rho_1 V_{P1}}{\rho_2 V_{P2} + \rho_1 V_{P1}} \\ T_{PP} &= \frac{2\rho_1 V_{P1}}{\rho_2 V_{P2} + \rho_1 V_{P1}} \end{aligned}\right\} \tag{3.14}$$

The analytical Eq. (3.14) has the following conclusion: When the longitudinal wave normal is incident, there is no converted wave in the newly generated fluctuation, and only the corresponding reflected longitudinal wave and transmitted longitudinal wave are generated. And as long as the following conditions are met:

$$\rho_2 V_{P2} \neq \rho_1 V_{P1} \tag{3.15}$$

The corresponding reflection coefficient R_{PP} must not be zero, that is to say the reflected wave must exist. Therefore, Eq. (3.15) becomes a sufficient condition for the existence of reflected waves. Also, because the product ρV_P of the defined density ρ and the velocity V_P is called the wave impedance, and the reflected wave is generated at the wave impedance interface, so the wave impedance surface is also called the reflection interface. In addition, by analyzing the formula:

$$R_{PP} = \frac{\rho_2 V_{P2} - \rho_1 V_{P1}}{\rho_2 V_{P2} + \rho_1 V_{P1}} \tag{3.16}$$

Correspondingly, when $\rho_2 V_{P2} > \rho_1 V_{P1}$, the reflection coefficient R_{PP} is a positive value, indicating that the phase of the reflected wave is consistent with the phase of the incident wave. When $\rho_2 V_{P2} < \rho_1 V_{P1}$, the reflection coefficient R_{PP} is a negative value, indicating that the phase of the reflected wave is exactly opposite to the phase of the incident wave; that is, there is a phase difference of π in the corresponding phase. This phenomenon is called a half-wave loss in the physical waveform. Similarly, from the analysis of the third equation of Eq. (3.14), the transmission coefficient T_{PP} is always positive; that is, the transmitted wave is always in phase with the incident wave. The numerical relationship between the reflection coefficient and the transmission coefficient can also be derived from the equations, as follows:

$$T_{PP} = 1 - R_{PP} \tag{3.17}$$

That is to say, the sum of the two coefficients is always equal to one.

When the incident wave is obliquely incident, that is, the corresponding incident angle α is not 0°, the corresponding energy distribution is much more complicated, so only two more common situations can be analyzed and discussed.

The first case is the case where a longitudinal wave is incident from a high-speed medium into a low-speed medium, as shown in Fig. 3.14. Then there is a wave speed ratio $V_{P2}/V_{P1} = 0.5$, a density ratio $\rho_1/\rho_2 = 0.8$, and Poisson's ratio $\sigma_1 = 0.30$ and $\sigma_2 = 0.25$.

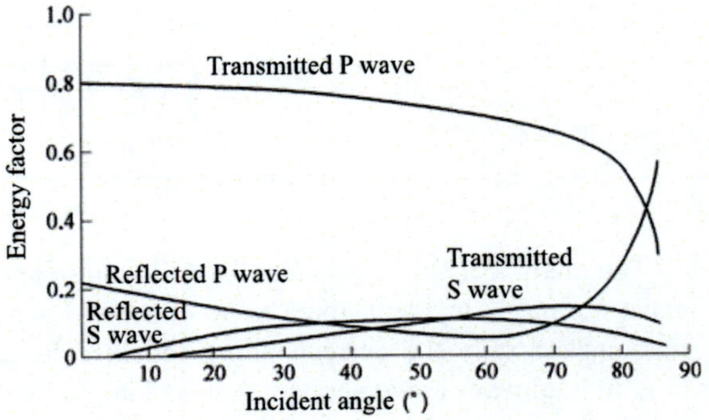

Fig. 3.14 Schematic diagram of the relationship between the energy distribution of the longitudinal wave from the high-speed medium into the low-speed medium

According to the figure, it can be seen that at the incident angle $\alpha < 20°$, even the entire process energy is mainly distributed to the transmitted longitudinal wave TP, only a small amount of reflected longitudinal wave RP energy, and the corresponding shear wave energy is small (especially when α tends to 0°, the corresponding RS and TS tend to be zero; that is, there is no transverse wave, which is exactly the same as the normal incidence analysis result). In the process of increasing the incident angle, part of the longitudinal wave energy will be converted to the transverse wave RS and TS, and the energy is still concentrated on RP and TP at the initial stage. But when α is between 40°and 60°, the intensity of shear wave RS and TS will exceed the intensity of longitudinal wave RP; that is, it is relatively easy to receive the shear wave generated by reflection when it is received far from the seismic source or is incident at a large inclination. When the incident angle α approaches 90°, the corresponding incident wave direction is almost parallel to the interface, and the transverse wave will disappear, then the energy will all focus on reflected longitudinal wave.

The second case is exactly the opposite of the former, which is the case when the longitudinal wave enters the high-speed medium from a low-speed medium, as shown in Fig. 3.15. Then there is a wave speed ratio $V_{P2}/V_{P1} = 2.0$, a density ratio $\rho_1/\rho_2 = 0.5$, and Poisson's ratio $\sigma_1 = 0.30$ and $\sigma_2 = 0.25$. It is easy to get $V_{P2}\rho_2 = V_{P1}\rho_1$, which means that the wave impedance of the upper and lower media are

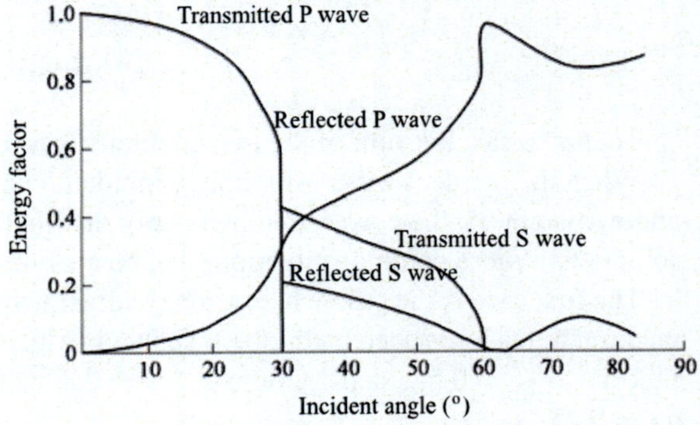

Fig. 3.15 Schematic diagram of the relationship between the energy distribution of the longitudinal wave from the low-speed medium into the high-speed medium

equal at this time. At this time, the critical angles of the longitudinal wave and the transverse wave are correspondingly: The first critical angle is expressed as $i_P = \arcsin V_{P1}/V_{P2}$, and the second critical angle is expressed as $i_S = \arcsin V_{S1}/V_{S2}$.

Analyzing Fig. 3.15, it can be seen that in the case of incident angle $\alpha < i_P$, which is equivalent to normal incidence, the corresponding longitudinal wave energy tends to zero, and the corresponding transverse wave and energy also tend to zero; that is, no reflected longitudinal wave or any transverse wave is generated at this time. When the incident angle α approaches i_P, the incident longitudinal wave will slide along the interface, and the corresponding TP will also approach zero. The reflected longitudinal wave RP, the reflected transverse wave RS and the scattered transverse wave TS will increase rapidly. The dramatic changes in the corresponding wave intensity also significantly reflect the conversion of the corresponding wave energy. Combined with the previous introduction to refracted waves, new fluctuations occur at this time—the refracted waves. In the case of the incident angle $\alpha > i_P$, the reflected transverse wave RS and the transmitted transverse wave TS are gradually weakened, and the reflected longitudinal wave RP is gradually increasing. In the extreme case, when the incident angle $\alpha = i_S$, the two transverse waves RS and TS energy will both tend to zero, and the reflected longitudinal wave RP will reach the maximum.

3.2.1.4 Factors Influencing the Propagation Velocity of Elastic Wave

In order to obtain better exploration results of the elastic seismic wave, it is crucial to accurately grasp the propagation velocity of elastic waves in the rock and soil mass. It is generally believed that the propagation velocity of elastic wave is closely related to the elastic properties of the medium. In the classical elastic wave theory, the following parameters can be used to describe the media-related elastic properties, including Young's modulus E, Poisson's ratio σ, shear modulus μ, volumetric compressive modulus K, and Lame coefficient λ, etc. The numerical relationship between them and elastic wave velocity is expressed as follows:

$$\sigma = \frac{V_P^2 - 2V_S^2}{2\left(V_P^2 - V_S^2\right)} \tag{3.18}$$

$$E = \frac{\rho V_S^2\left(3V_P^2 - 4V_S^2\right)}{V_P^2 - V_S^2} \tag{3.19}$$

$$K = \rho\left(V_P^2 - \frac{4}{3}V_S^2\right) \tag{3.20}$$

$$\lambda = \rho\left(V_P^2 - 2V_S^2\right) \tag{3.21}$$

$$\mu = \rho V_S^2 \tag{3.22}$$

where ρ, $V_{P,}$ and V_s—represent rock density, corresponding longitudinal wave velocity, and shear wave velocity, respectively.

It can be seen from the above formula that other rock elastic parameters can be calculated under the conditions of known rock density, longitudinal wave velocity, and transverse wave velocity. In combination with actual conditions, rock density often changes very little. Therefore, in the study of the numerical relationship between the elastic properties of related rock masses and its physical properties, only the compressional wave velocity and shear wave velocity and their wave velocity ratios can be considered, as well as the numerical relationship between the three and the corresponding physical properties of the rock mass.

(1) **Common wave velocity characteristics in media**

Due to the difference in elastic properties, the waves have different propagation wave velocities and wave impedances in different media. It is generally believed that the wave velocity in igneous and metamorphic rocks is greater than the wave velocity in sedimentary rocks, and the wave velocity in the corresponding sedimentary rocks is small and has a large variation range. The reason for this is that the sedimentary rock structure is generally complicated, and the compactness and water content of the rock have a great influence on the wave velocity. In particular, in very dry modern sediments and accumulations, the corresponding wave velocity is even lower than sound waves. Table 3.1 shows the corresponding longitudinal wave velocities and wave impedances in some common media.

The table shows that there are some differences in wave velocity and wave impedance between different strata and rock layers. These different lithological interfaces just become the corresponding reflection or refracting interface of seismic

Table 3.1 Longitudinal wave velocity and corresponding wave impedance of common media

rock	VelocityV_P (m/s)	Wave impedance (× 10^4 g/cm^{2}S)	rock	VelocityV_P (m/s)	Wave impedance (× 10^4 g/cm^{2}S)
Weathered zone	100–150	1.2–9	Pelitic schist	2700–4800	65–135
Dry sand gravel	100–600	2.8–14	Limestone dolomite	2000–6250	35–180
Mud	500–1900	3.8–30	Anhydrite rock salt	4500–6500	110–140
Wet sand	200–2000	3–40	Coal	1600–1900	20–36
Clay	1200–2800	15–65	Air	310–360	0.004
Loose sand	1500–2500	27–60	Oil	1300–1400	12–15
Marl	2000–4700	20–120	Water	1430–1590	14–16
Tight sandstone	1800–4300	40–116	Ice	3100–4200	30–45

waves, so it is also an important geological foundation for seismic exploration research.

(2) **Influence of rock physical properties on wave velocity**

The elastic wave velocity is determined by the elastic parameters and density of the rock, and the corresponding elastic parameters of the rock depend not only on the various mineral components that make up the rock, but also on the properties of the corresponding pore fluids in the rock. Of course, they are also related to various factors such as environmental pressure and temperature. In other words, environmental factors, porosity, and pore fluids all further affect speed by influencing rock mass density and related elastic parameters. It is the influence of the above factors that causes overlapping phenomena in the corresponding wave velocity range of the rock mass. Therefore, judging the lithology by the wave velocity data will produce multiple solutions. However, the speed ratio of the longitudinal wave to the transverse wave can be used, and the judgment is more accurate. Then there is the following formula, assuming that in the uniformly infinite elastic medium, the corresponding longitudinal and transverse wave speeds are expressed as follows:

$$\left.\begin{aligned} V_P &= \sqrt{\frac{(\lambda + 2\mu)}{\rho}} = \sqrt{\frac{E}{\rho}\frac{(1-\sigma)}{(1+\sigma)(1-2\sigma)}} \\ V_S &= \sqrt{\frac{\mu}{\rho}} = \sqrt{\frac{E}{2\rho}\frac{1}{(1+\sigma)}} \end{aligned}\right\} \tag{3.23}$$

The ratio of the wave velocity of the longitudinal wave to the transverse wave is expressed as follows:

$$\frac{V_P}{V_S} = \sqrt{\frac{1-\sigma}{0.5-\sigma}} \tag{3.24}$$

It can be seen from the above equation that the wave speed of longitudinal wave and shear wave are determined by three and two elastic parameters correspondingly, while the velocity ratio of the two waves is determined by one elastic parameter, which is more convenient for analysis and has a definite relationship. Therefore, the wave velocity ratio of the longitudinal wave and the transverse wave is used for comprehensive judgment, and the obtained lithology judgment is more in line with the actual situation. Specifically, the relationship between Poisson's ratio of the medium and the speed ratio is utilized to achieve mutual derivation, and the correspondence between the two is given in Table 2.3.

It can be seen from Table 3.2 that during the change of σ from 0 to 0.5, the corresponding wave velocity will change from $\sqrt{2}$ to ∞. Poisson's ratio of most consolidated rocks is roughly 0.25, and the wave speed ratio of the corresponding longitudinal and transverse waves is $\sqrt{3}$. Moreover, the table shows that the minimum wave velocity of the shear wave will drop to zero, and the maximum wave velocity

Table 3.2 Correspondence between medium wave velocity ratio V_P/V_S and Poisson's ratio

σ	0	0.1	0.2	0.25	0.3	0.4	0.5
V_p/V_s	1.41	1.50	1.63	1.73	1.7	2.45	∞

will be close to 70% of the longitudinal wave. The two extreme cases correspond to the two media conditions of liquid and extremely hard rock.

It is generally believed that the wave velocity ratio of the bedrock is $1.7 < V_P/V_S < 2.0$, and Poisson's ratio is $0.33 < \sigma < 0.5$. Relatively speaking, the loose layer with no consolidation has a wave speed ratio of $2.4 < V_P/V_S < 10$, and the corresponding Poisson's ratio is $0.4 < \sigma < 0.5$. There is obviously a big difference in the corresponding parameters of the bedrock. In the case of other rocks, the wave velocity ratio of limestone is the most stable, ranging from 1.8 to 2.0, and the average value is 1.9. Correspondingly, the wave velocity ratio of mudstone and shale has a relatively large range of variation, ranging from 1.8 to 3 0.0. Because of the obvious influence of the argillaceous composition, the wave velocity ratio of sandstone will vary greatly, and it has a closer relationship with its porosity. In particular, in the case of a liquid medium, since σ is zero, the corresponding V_S is also zero, which also proves from another angle that the transverse wave is not propagated in the liquid.

(3) **Influence of rock density on wave velocity**

In fact, Gardner et al. have conducted a series of experiments on the relationship between the longitudinal wave velocity and the rock density under different rock conditions, and the corresponding empirical formulas are expressed as follows:

$$\rho = 0.31V^{0.25} \tag{3.25}$$

where ρ represents the density of the rock and V represents the velocity of the longitudinal wave. It can be seen from the above equation that the increase of the density of the general medium will gradually accelerate the wave velocity of the elastic wave propagating in it.

(4) **Influence of rock porosity on wave velocity**

Rock porosity directly affects the density of the medium and indirectly affects other relevant elastic parameters, so porosity is also an important factor affecting the seismic wave velocity. For the convenience of research, the solid rock can be structurally decomposed into two parts, one of which is the mineral particle itself, which can be regarded as a porous rock skeleton, and the other part is a pore filled with a fluid or a gas. It can be seen from formula (3.25) that in the pores filled with liquid or gas, the seismic wave velocity is generally smaller than the seismic wave velocity in the rock skeleton. In general, in this two-phase medium, the elastic wave velocity is inversely proportional to the porosity. In 1956, Wyllie proposed the time-average equation in his paper, which simplified the numerical relationship between P-wave velocity and rock porosity. The corresponding expression is as follows:

$$\frac{1}{V}=\frac{\varphi}{V_1}+\frac{1-\varphi}{V_{\mathrm{m}}} \tag{3.26}$$

where φ—the porosity of the rock;

V—the wave velocity of the longitudinal wave in the rock;

V_1—the wave velocity of the longitudinal wave in the pore fluid;

V_{m}—the wave velocity of the corresponding longitudinal wave in the rock skeleton.

(5) **Influence of pore fluid saturation and temperature on wave velocity**

Not only the porosity and lithology of the rock will affect the amplitude and wave velocity of the longitudinal wave, but also the type, temperature, and saturation of the fluid in the rock pores will also have an effect on the propagation of the longitudinal wave. In 1976, Gregory proposed the conclusion of his own experiment: Under a certain pressure, the rock wave velocity will change correspondingly with the change of the fluid in the rock pores, and the maximum amplitude of the corresponding longitudinal wave amplitude may reach 25%. At the same time, while keeping the pressure constant and increasing the temperature from 24 to 750 °C, the change in longitudinal wave velocity in dry rocks is approximately 63%, while in saturated rocks it is only 33%.

3.2.1.5 Influence of the Del Defect of Structural Plane of Rock Mass on the Propagation of Elastic Wave

Rock mass is a combination of structural body (rock) and structural plane. In addition to the influence of the characteristics of the rock structure on the fluctuations, the geological discontinuities of various structural planes, voids, and fracture zones in the rock body are particularly obvious to change the elastic wave propagation path and energy dissipation, which is called structural plane effect.

(1) **Influence of structural plane**

In the rock mass, there are a variety of two-dimensional planar discontinuous geological interfaces with different sizes and relatively low mechanical strength, which are called structural planes, such as bedding (surface), fissures, cleavage, schistosity, and fracture. In the process of elastic wave propagation, when it encounters a structural surface, the phenomenon of wave reflection, refraction or diffraction will occur, which will affect the propagation speed and amplitude intensity of the wave.

Let's consider the plaster model with several cracks present (Fig. 3.16). If the crack is filled with air, at the solid–gas interface, the reflection reaches almost 100%, and the wave cannot transmit through the crack, but walks around the end of the crack, which is a round trip, showing a reduction in propagation speed.

When the crack is filled with liquid or other solid matter, the transmitted wave can pass, but its energy will be weakened to varying degrees, showing a significant decrease in amplitude intensity. Most of the fluctuating energy will be suddenly lost

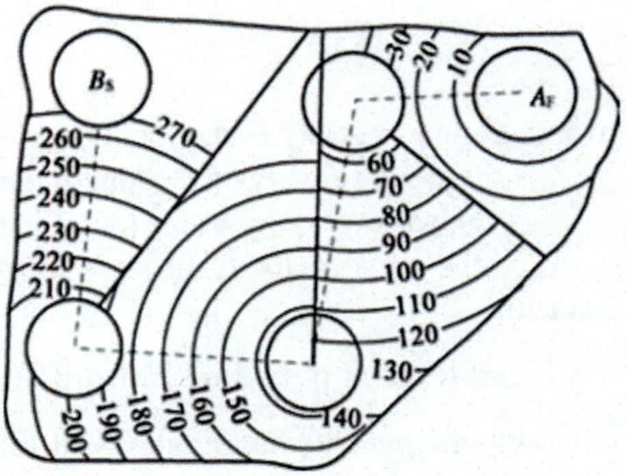

Fig. 3.16 Effect of cracks on the formation of fluctuations

in a certain crack width, and its width across the crack is related to the frequency. The lower the frequency, the greater the width across the crack (Fig. 3.17).

In rock masses with cracks, interlayers and broken zones, the apparent velocity of elastic wave propagation and the degree of energy absorption vary according to the nature, width and filling material of these structural planes. The greater the number of cracks, the greater the thickness of the interlayer, and the smaller the apparent wave velocity (Figs. 3.18 and 3.19).

Rock mass is anisotropic because of diagenetic condition, structural plane and in-situ stress, so elastic wave propagation in rock mass is anisotropic as well. Taking bedding type in structural plane as an example, the wave velocity (V) perpendicular to the direction of bedding is low, while the wave velocity (V_n/V) parallel to the direction of bedding is high. The ratio of the two (the anisotropic coefficient of wave velocity V_n/V) varies according to the combination degree of lithology and bedding (Table 3.3).

(2) **The influence of cavity**

When an elastic wave encounters a cavity in the propagation process, the apparent velocity of diffracting the same cavity varies with the length of the survey line. The shorter the survey line, the more intense the diffraction, the longer the survey line,

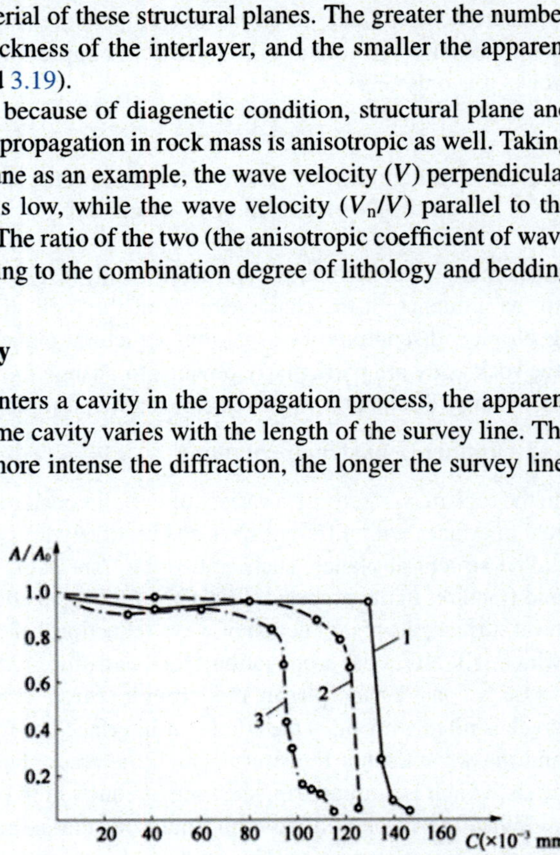

Fig. 3.17 Relationship between amplitude ratio (A/A_0) and spanning crack width (C)

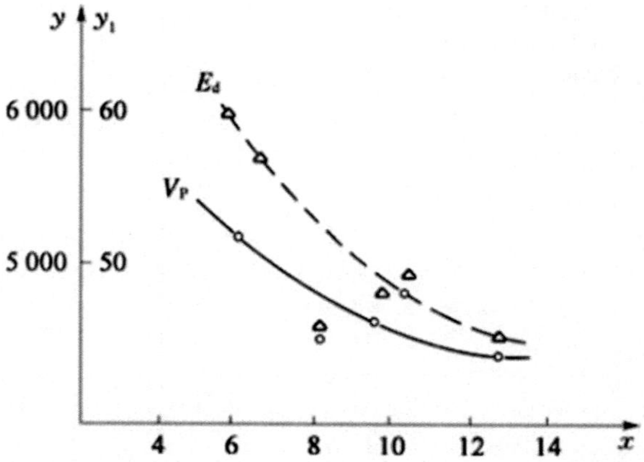

Fig. 3.18 Relationship between fracture density and V_P and E_d of granodiorite

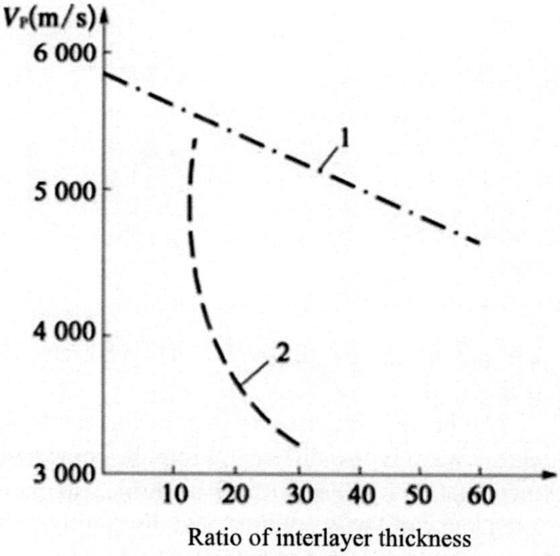

Fig. 3.19 Effect of the interlayer on V_P (1—the thickness of the interlayer occupies the length ratio of the measuring section; 2—the ratio of the carbonized shale to the thickness of the interlayer).

Table 3.3 Wave speeds of different lithologies and structures

Lithology	Wave velocity		Wave velocity anisotropy coefficient V_n/V
	Vertical bedding V (m/s)	Parallel bedding V_n(m/s)	
Limestone	3620	5540–6060	1.53–1.67
Medium grained sandstone	1550–1830	2400–2540	1.39–1.55
Quartz sandstone	3660	4420	1.21
Clay rock	3000–3400	3500–3800	1.21–1.3

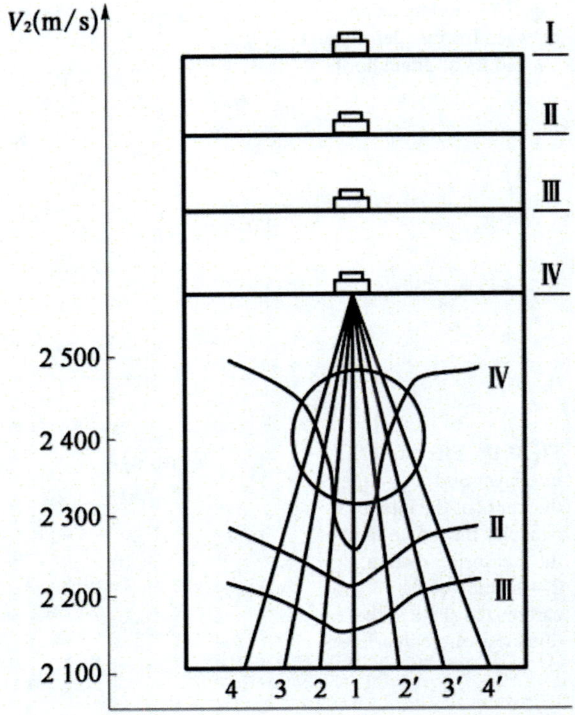

Fig. 3.20 Apparent velocity distribution of waves diffracting a cavity

and the weaker the diffraction. They all have different relatively low-speed areas (Fig. 3.20).

The energy dissipation effect of the cavity filled with different material on the elastic wave is also different. From the comparison of the indoor simulation experiment that cavity filled with air and brine: No matter what the situation, the cavity has a tendency of low amplitude, and the amplitude attenuation of the filled air is much more severe than that of the filled brine (Fig. 3.21).

3.2.2 Three-Dimensional Tunnel Geological Prediction Theory of Seismic Wave Method

The general idea of the three-dimensional propagation principle of seismic waves in tunnels is based on numerical simulation analysis technology and spatial three-point positioning principle. The finite element analysis software is used to establish a three-dimensional model based on the theoretical basis of seismic wave reflection method and the principle of dynamic finite element method. The specific method is to set up unfavorable geological bodies such as aquifer and fracture zone at a certain distance in front of the tunnel face, carry out research on the propagation law of elastic

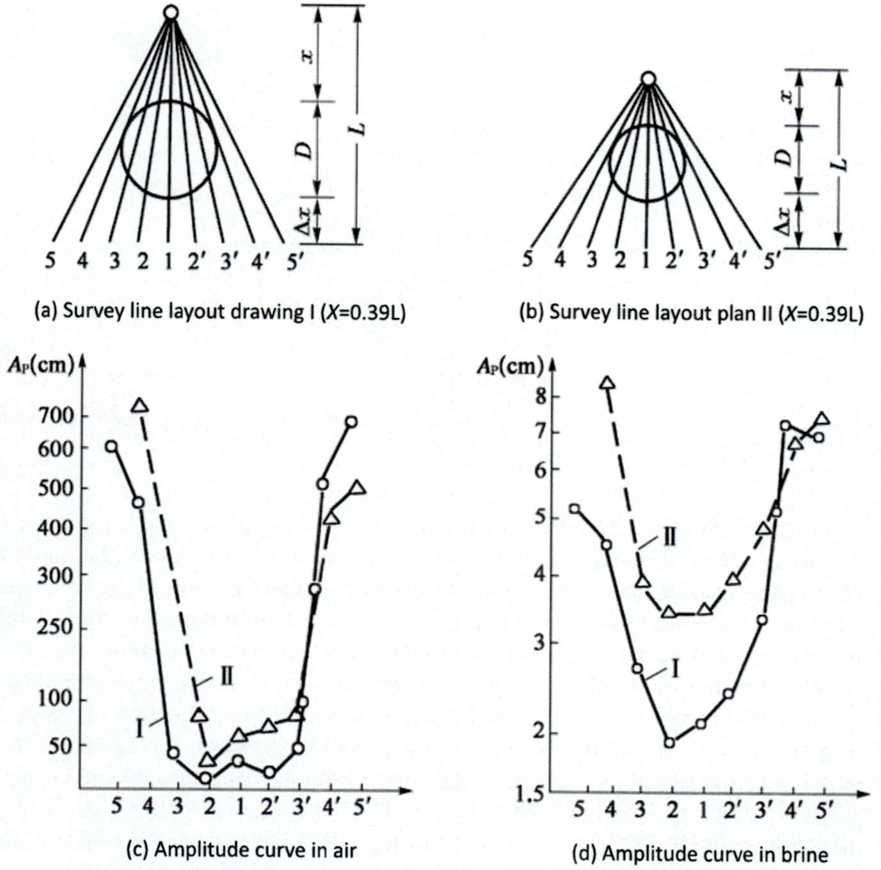

Fig. 3.21 Effect of cavity on amplitude (A_P)

waves inside the rock mass, and analyze the characteristics and identification methods of reflected signals under the influence of different types, locations and ranges of unfavorable geological bodies. The seismic source adopts the tunnel surrounding rock loading and hammering the seismic load to simulate the propagation process of the seismic wave in the tunnel surrounding rock. By selecting multiple receiving points at different locations, receiving seismic wave reflection records, analyzing and studying the propagation and reflection laws of seismic waves under different conditions, the characteristics and laws of seismic wave propagation changes in the tunnel are obtained. Based on the characteristics and laws of seismic waves and combined with the geological features of the tunnel, the three-dimensional tunnel advanced geological prediction analysis technology and three-dimensional imaging technology can be used to obtain the features such as location, shape and scale of the unfavorable geological bodies of the tunnel, as shown in Fig. 3.22.

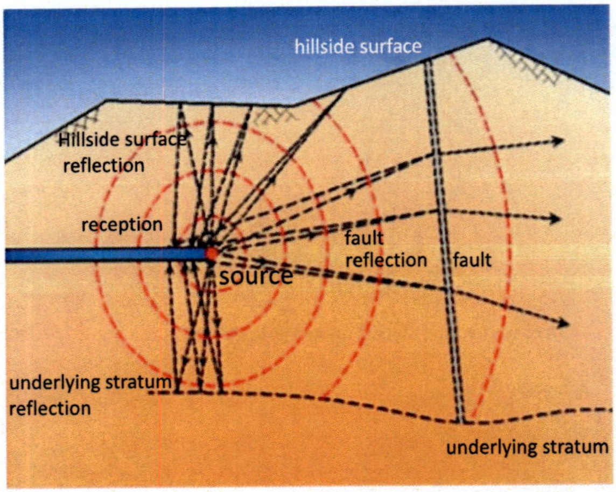

Fig. 3.22 Three-dimensional imaging tunnel advanced geological prediction schematic

In the actual operation process, due to the different geological conditions of different engineering projects, there are two choices of seismic source: hammering and explosives. If the geological conditions are poor, it is necessary to use the explosive source to excite seismic waves. If the geological conditions are good, the seismic wave should be excited by hammering. In the process of data acquisition, the three-point positioning principle is adopted, so the installation of the three-component geophone is required to meet a certain requirements, and the direction of the three detection components *X*, *Y*, *Z* of the three-component geophone should be installed in the direction set. In the process of data processing, because the collected signals come from all directions, *F-K* filtering is needed to extract the reflected waves from the front of the tunnel face and filter out the interference waves. In the process of data processing, the imaging speed and other parameters of three-dimensional imaging should be calculated accurately. The 3D imaging tunnel advanced geological prediction system analyzes the geological conditions in front of the tunnel face by comparing longitudinal waves and transverse waves and uses the difference in propagation characteristics of longitudinal waves and transverse waves under different geological conditions to improve the accuracy of tunnel geological forecasts.

3.2.2.1 3D Simulation Calculation Model

(1) **Unit selection**

The SOLID164 unit is a three-dimensional solid element with eight nodes. Each node has 9° of freedom, namely UX, UY, UZ, VX, VY, VZ, AX, AY, AZ, but only displacement is a physical degree of freedom in a practical sense. By default, the unit uses a single-point integration algorithm, or it can be set to a full integration algorithm in the unit properties. This can avoid the hourglass problem, but it will greatly increase the calculation time. Each node of the SOLID164 unit can have displacement, velocity, and acceleration. Some nodes can coincide. The unit can be a six-node wedge unit, a four-node tetrahedral unit, and a five-node pentahedral unit. Therefore, this type of unit is highly adaptable and can be well-adapted to numerical simulations of deformation problems such as impact. In the simulation analysis, the SOLID164 unit is used. The surrounding rock deformation is within the elastic deformation range. The surrounding rock is selected from the line elastic material. The sweep network division method is adopted, and the global size of the unit is set to 1 m. The parameters of the surrounding rock material are: density $\rho = 2500$ kg/m^3, elastic modulus $E = 2.25 \times 10^{10}$ N/m^2, Poisson's ratio $\sigma = 0.22$. According to the calculation formula of longitudinal wave velocity $V_P = \sqrt{\frac{E}{\rho}\frac{(1-\sigma)}{(1+\sigma)(1-2\sigma)}}$, the propagation velocity of longitudinal wave in the rock mass is 3200 m/s.

(2) **Blank material model**

The Gruneisen equation of state defines the equation of state of a material under impact compression as:

$$p = \frac{\rho_0 C^2 \mu\left[1 + \left(1 - \frac{\gamma_0}{2}\right)\mu - \frac{a}{2}\mu^2\right]}{\left[1 - (S_1 - 1)\mu - S_2\frac{\mu^2}{\mu+1} - S_3\frac{\mu^3}{(\mu+1)^2}\right]} + (\gamma_0 + \alpha\mu)E \tag{3.27}$$

The equation of state when the material expands is

$$p = \rho_0 C^{2\mu} + (\gamma_0 + \alpha^{\mu})E \tag{3.28}$$

where C—the intercept of the curve V_S-V_P;
S_1, S_2, and S_3—the slope coefficient of the curve V_S-V_P;
γ_0—Gruneisen constant;
a—the first-order volume correction of γ_0 and $\mu = \rho/\rho_0 - 1$.

The parameters that the equation of state model needs to input include C, S_1, S_2, S_3, γ_0, E_0, and V_0. The null blank material model can be defined in conjunction with this equation of state.

In the simulation analysis, the blank material model and the Gruneisen equation of state are selected to simulate liquid water, where C is the speed of sound in the water. The blank material model used and the state equation parameter settings are shown in Table 3.4.

(3) **Condition of Non-reflective boundary**

When simulating a geodynamic system, it is often necessary to use a finite field to represent underground space or large rock mass. For such problems, in order to avoid

Table 3.4 Blank material model and state equation parameters

Material model	Density (kg/m^3)	Sound velocity in water (m/s)	S_1	S_2	S_3	γ_0	α	E_0(J/m^3)
Water	1000	1480	2.86	1.886	0.247	0.5	0	0

the influence of the reflection of the wave at the boundary on the solution domain, a condition of non-reflective boundary can be applied to the finite field surface to simulate an infinite space.

The LS DYNA program is used to set a finite-domain artificial damping boundary to process the non-reflective boundary, so that the isolated area can still ensure the radiation characteristics of the wave energy propagating from the near area to the far area, and the energy will not be artificially limited in a limited area, causing the reflection of the stress wave at the boundary, which will interfere with the actual stress distribution of the structure.

(4) **Loading of hammering seismic wave**

In the advance prediction of tunnel, hammering is performed at a plurality of hammer points to excite seismic waves. During the process of hammering, the surrounding rock in the unexcavated area will not be broken, and the surrounding rock of the tunnel is elastically deformed. The hammering seismic wave propagates outward in the form of a sphere with the hammer point as the center. The form of spherical diffusion brings difficulties to the loading of the hammering seismic wave during the simulation calculation. With the propagation of the spherical wave, the seismic waves in the medium can be approximately regarded as plane waves in the distance from the hammer point. Therefore, in this book, the Ricker wavelets commonly used in seismic exploration are used for loading to evenly act on the tunnel section (Fig. 3.23).

The Ricker wavelet can be uniquely determined by only one signal parameter f, and its mathematical expression is as follows:

$$\text{Ricker}(t) = \left[1 - 2(\pi f t)^2\right]\exp\left[-(\pi f t)^2\right] \tag{3.29}$$

where t—time;

f—the peak frequency.

(5) **Establishment of 3D model of tunnel**

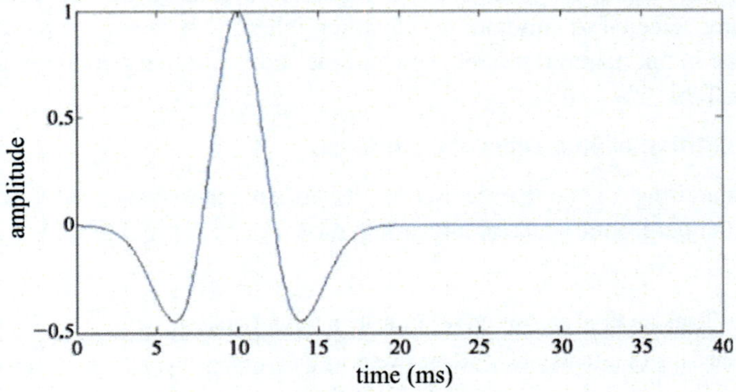

Fig. 3.23 Waveform of Ricker wavelet

In the three-dimensional simulation finite element calculation model of tunnel advance prediction, the coordinate system *Z*-direction is the tunnel direction, the *Y*-direction is the vertical direction, and the *X*-direction is the horizontal direction orthogonal to the tunnel. In actual operation, the measured coordinate system directions *X*, *Y*, *Z* correspond to the simulation analysis coordinate system directions *X*, *Y*, *Z* one-to-one. The coordinate system directions described below are all measured coordinate system directions.

In the calculation and analysis model, the section length of the formed and supported tunnel is 40 m, and the length of the unexcavated rock mass in front of the tunnel is 50 m; that is, the length of the tunnel strike is 90 m. The overall length of the tunnel in the vertical direction is 68.83 m, of which the height of the tunnel is 8.83 m, and the upper and lower boundaries are 30 m from the vault and the bottom of the tunnel, respectively. The overall length in the horizontal direction orthogonal to the tunnel is 60 m, of which the tunnel width is 11.6 m; that is, the boundary between the calculation model and the tunnel in the horizontal direction is 24.2 m away from the tunnel sidewall.

In the process of finite element analysis such as tunnel excavation and advanced support, the overall modeling should be established strictly in accordance with the construction drawings in order to ensure the accuracy of calculation. The tunnel's transverse size should be greater than 5 times the diameter of the tunnel, taken vertically upward to the surface, and taken vertically downward at least 4 times the diameter of the tunnel. In the analysis of this book, it is not necessary to consider the stress change and gravity influence of the whole mountain. Only the influence of the unfavorable geological bodies such as the aquifer existing in front of the tunnel face on the seismic wave is considered, that is, only the reflection of the seismic wave when it encounters the elastic interface on the front of the tunnel is considered. The horizontal and vertical directions orthogonal to the tunnel direction are considered to be uniform and infinite, and non-reflective boundary conditions are applied to the four side boundaries. The model size of the non-reflective boundary model can be appropriately reduced, generally 2.3 times the hole diameter.

The calculation analysis model has a symmetrical structure and symmetric stress conditions, which is a symmetry problem. In the actual process of analysis and calculation, the actual model established is to take 1/2 model along the symmetry plane in order to facilitate calculation and save machine time. The smaller size of the established 3D model will not have a significant impact on the calculation results, but it can save a lot of calculation time and storage space, which is convenient for calculation and analysis.

The three-dimensional simulation analysis model of the tunnel advance prediction is shown in Fig. 3.24, in which the purple part is the aquifer with a thickness of 5 m and a distance of 45 m from the tunnel face.

The cloud image of the seismic wave propagation at four different times is shown in Figs. 3.25, 3.26, 3.27, and 3.28.

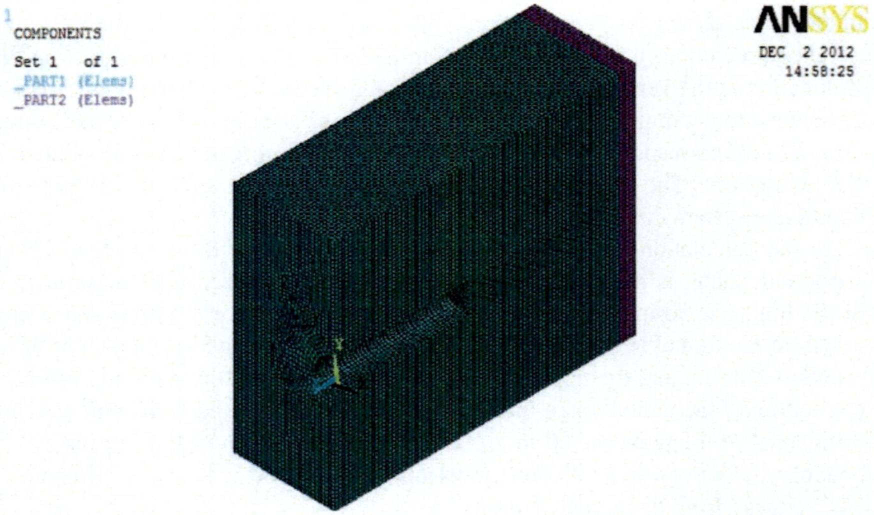

Fig. 3.24 Three-dimensional simulation calculation model of tunnel

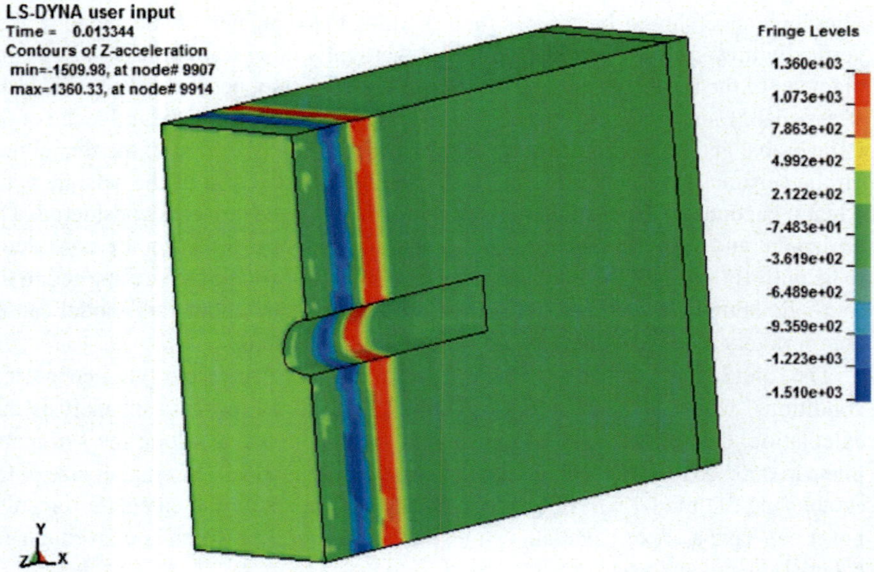

Fig. 3.25 Cloud image of seismic wave propagation at $t = 0.013344$ s

3.2.2.2 Analysis of 3D Simulation Results

(1) **The influence of the change of the receiving point position on the waveform**

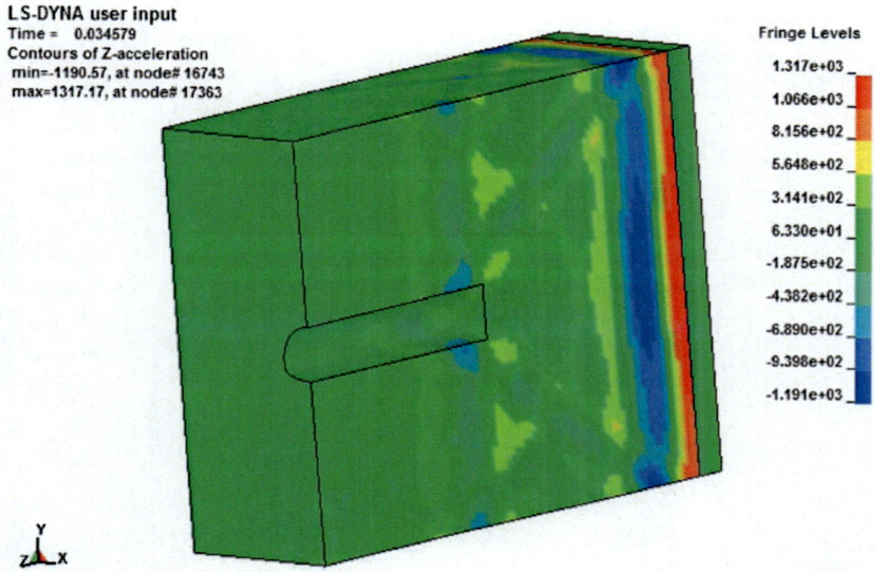

Fig. 3.26 Cloud image of seismic wave propagation at $t = 0.034579$ s

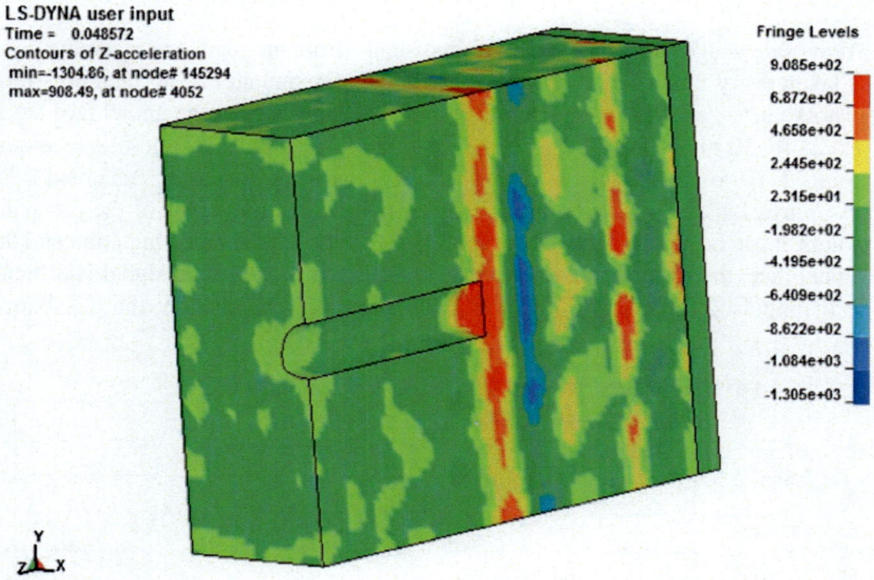

Fig. 3.27 Cloud image of seismic wave propagation at t =0.048572 s

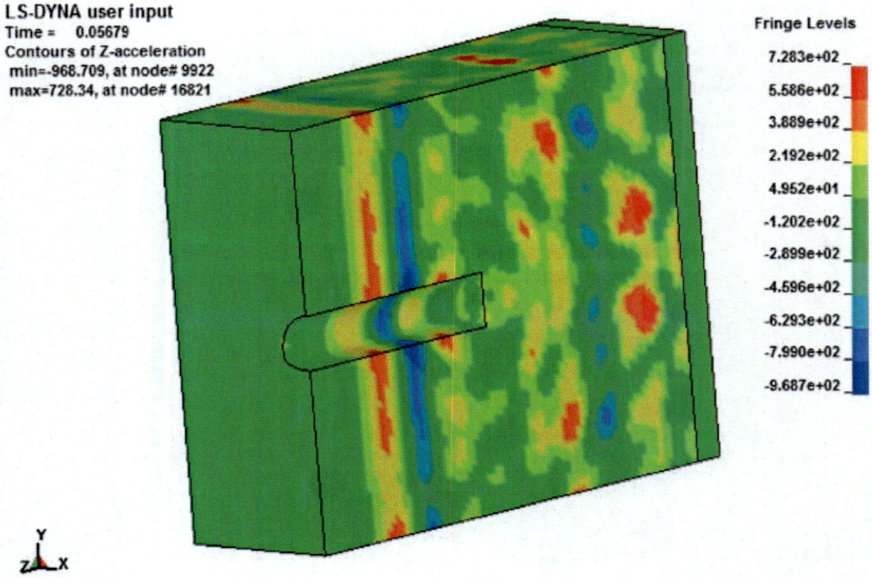

Fig. 3.28 Cloud image of seismic wave propagation at $t = 0.05679$ s

At the side wall of the tunnel at a certain distance from the road, one receiving point is taken every 2 and 30 m is taken continuously to simulate the acceleration–time history curves of 15 receiving points in total. The distances to the tunnel face are 2, 4…28 m, 30 m in turn. The acceleration–time history curves of the five receiving points 2, 10, 16, 24 m, and 30 m are shown in Figs. 3.29, 3.30, 3.31, 3.32, and 3.33.

Comparing and analyzing the acceleration–time history curves of the receiving points, it can be seen that the closer the distance between the receiving point and the tunnel face, the stronger the interference signal, and the reflected signal is difficult to distinguish. Therefore, in the actual tunnel advance prediction, the installation

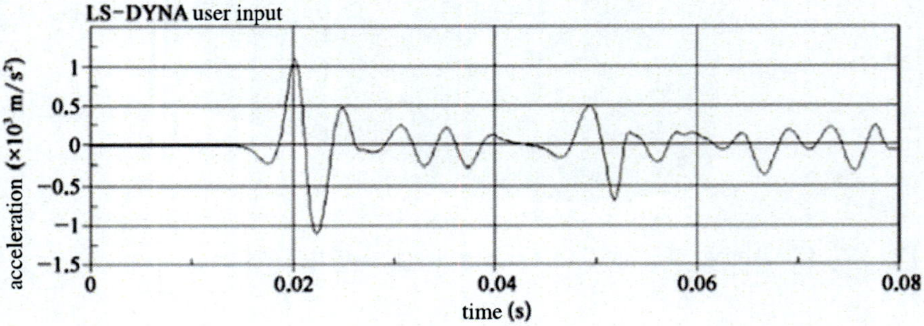

Fig. 3.29 Acceleration–time history curve of the receiving point 2 m away from the tunnel face

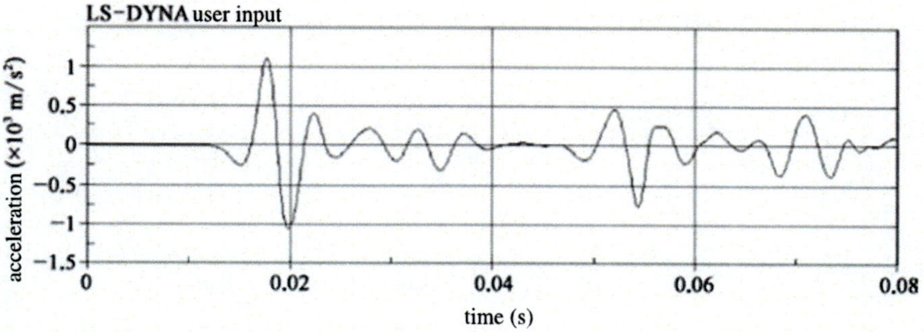

Fig. 3.30 Acceleration–time history curve of the receiving point 10 m away from the tunnel face

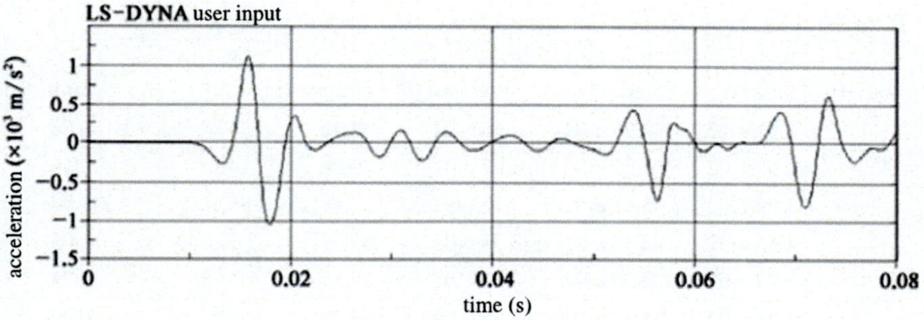

Fig. 3.31 Acceleration–time history curve of the receiving point 16 m away from the tunnel face

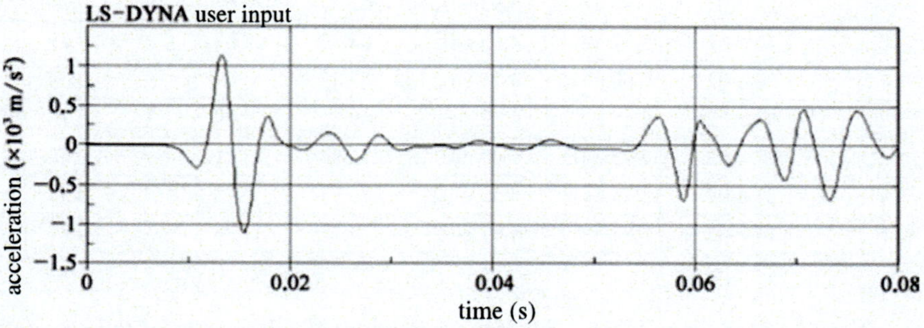

Fig. 3.32 Acceleration–time history curve of the receiving point 24 m away from the tunnel face

position of the geophone should be kept a certain distance from the excitation point to ensure the signal-to-noise ratio of the reflected wave signal.

According to acceleration–time history curves of the above-mentioned receiving points, the distance between each node from the aquifer and the error with the set distance are calculated, as shown in Table 3.5.

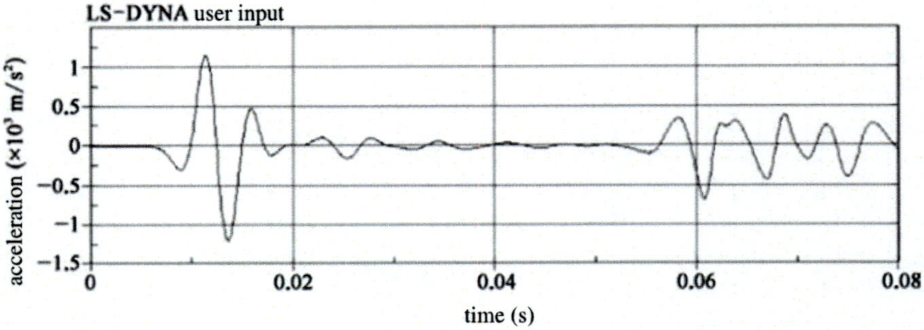

Fig. 3.33 Acceleration–time history curve of the receiving point 30 m away from the tunnel face

Table 3.5 Calculated distance, set distance, and error of each receiving point node

Serial number	First arrival time (s)	Reflection time (s)	Course time difference (s)	Calculated distance (m)	Set distance (m)	Relative error (%)
1	2.02 e-02	4.98 e-02	2.96 e-02	47.40	47	0.84
2	1.95 e-02	5.02 e-02	3.07 e-02	49.08	49	0.16
3	1.89 e-02	5.11 e-02	3.22 e-02	51.48	51	0.94
4	1.84 e-02	5.15 e-02	3.31 e-02	52.98	53	0.03
5	1.77 e-02	5.23 e-02	3.46 e-02	55.41	55	0.75
6	1.71 e-02	5.26 e-02	3.56 e-02	56.89	57	0.20
7	1.65 e-02	5.33 e-02	3.68 e-02	58.84	59	0.27
8	1.58 e-02	5.41 e-02	3.83 e-02	61.31	61	0.51
9	1.53 e-02	5.46 e-02	3.93 e-02	62.87	63	0.20
10	1.46 e-02	5.52 e-02	4.05 e-02	64.83	65	0.27
11	1.38 e-02	5.60 e-02	4.21 e-02	67.43	67	0.64
12	1.33 e-02	5.65 e-02	4.32 e-02	69.12	69	0.18
13	1.27 e-02	5.73 e-02	4.46 e-02	71.34	71	0.47
14	1.21 e-02	5.76 e-02	4.55 e-02	72.77	73	0.32
15	1.15 e-02	5.82 e-02	4.67 e-02	74.72	75	0.37

It can be seen from Table 3.5 that the maximum absolute error of the calculated distance and the set distance at the interface between the receiving point and the aquifer is 0.48 m, and the minimum absolute error is 0.08 m. The maximum relative error is 0.94%, and the minimum relative error is 0.03%. It can be seen that when the tunnel geological conditions are good, it is more accurate to estimate the position of the aquifer in front of the tunnel face by the reflected wave travel time.

(2) Waveform comparison of the receiving point of the side wall and the vault

At the distance of 2, 4, …, 28, 30 m from the tunnel face, 15 different receiving points are set on the side wall and the vault, among which the acceleration–time history

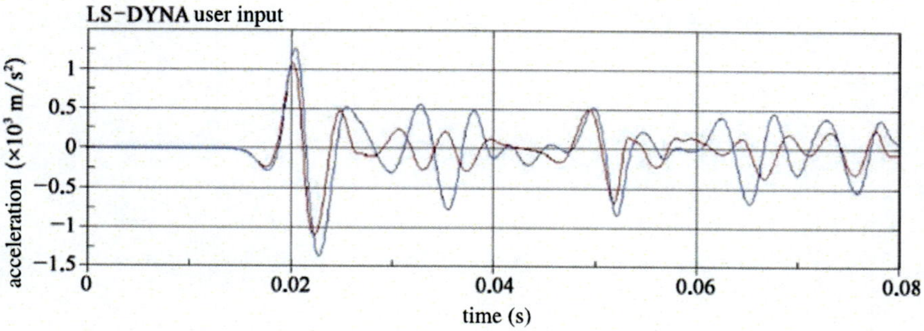

Fig. 3.34 Acceleration–time history curve of the receiving point of the side wall and vault at a distance of 2 m from the tunnel face

curves of the receiving points at 2, 10, 16, 24, and 30 m on the side wall and the vault are shown in Figs. 3.34, 3.35, 3.36, 3.37, and 3.38. The red line represents the side wall and the blue line represents the vault.

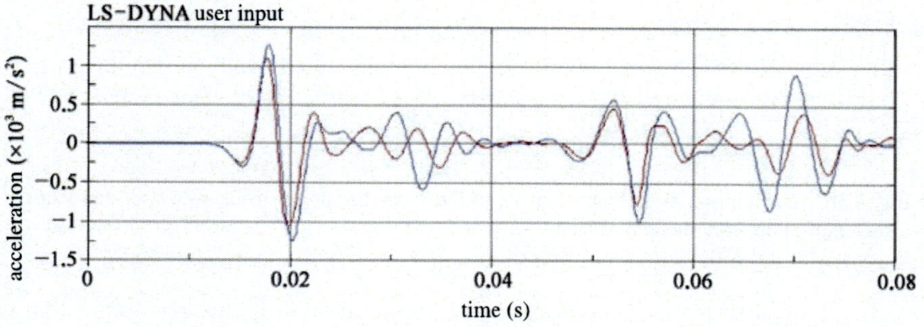

Fig. 3.35 Acceleration–time history curve of the receiving point of the side wall and vault at a distance of 10 m from the tunnel face

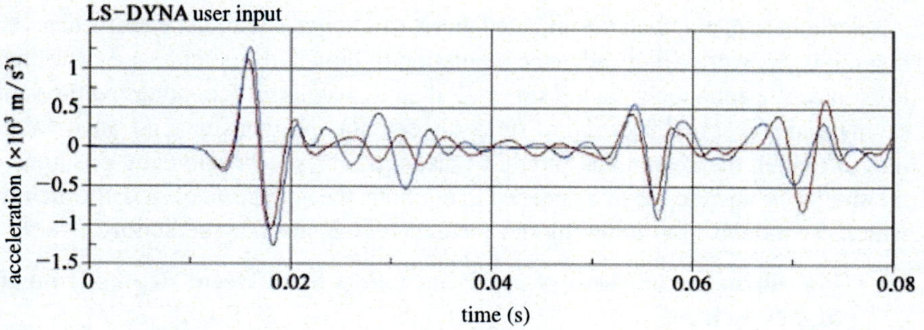

Fig. 3.36 Acceleration–time history curve of the receiving point of the side wall and vault at a distance of 16 m from the tunnel face

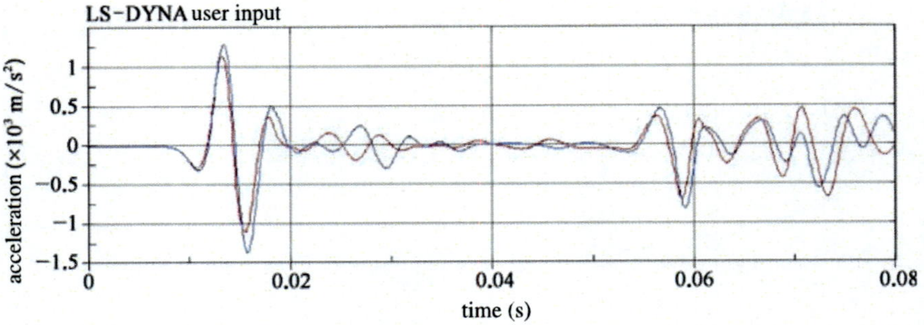

Fig. 3.37 Acceleration–time history curve of the receiving point of the side wall and vault at a distance of 24 m from the tunnel face

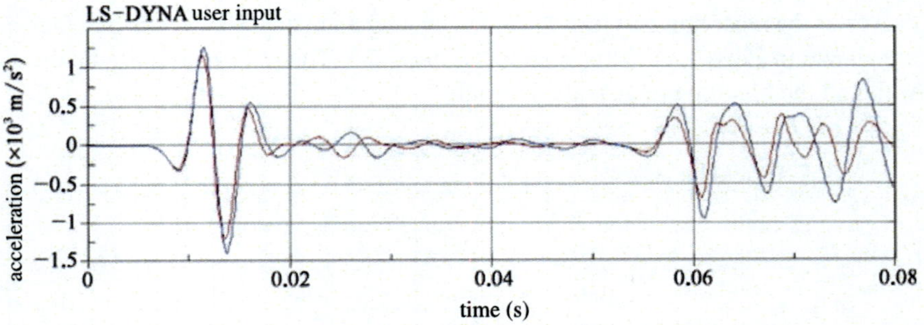

Fig. 3.38 Acceleration–time history curve of the receiving point of the side wall and vault at a distance of 30 m from the tunnel face

It can be seen from the comparison of acceleration–time history curves that the peak acceleration of the reflected wave at the receiving point of the vault is greater than the peak acceleration of the receiving point at the side wall; that is, the reflected wave has stronger energy and is more conducive to receiving. It can also be seen from the figure that the interference wave at the receiving point of the vault is strong, and it is difficult to distinguish the reflected wave. The actual situation needs further verification in the work. In the advance prediction of tunnel, it is simple and convenient to arrange the geophones on the side wall than to install the geophones on the vault, saving manpower and material resources. In the three-dimensional advance prediction of tunnel, the geophones should be arranged at the corresponding positions of the side walls and the vaults as needed to facilitate the collection of rich information, which is conducive to improving the accuracy of the advance prediction.

(3) **Waveform comparison of receiving points at different depths from the tunnel surface**

Different receiving points are selected at the distance of 0, 1, 2, 3, 4, and 5 m from the tunnel surface to obtain the acceleration–time history curves of the receiving points, as shown in Figs. 3.39, 3.40, 3.41, 3.42, 3.43, and 3.44.

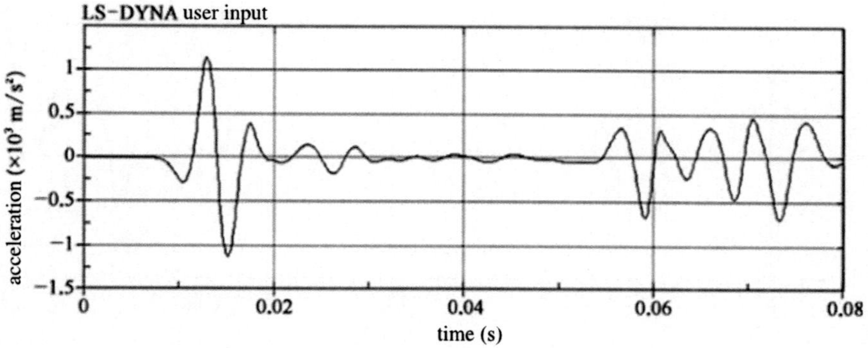

Fig. 3.39 Acceleration–time history curve of the receiving point 0 m away from the tunnel surface

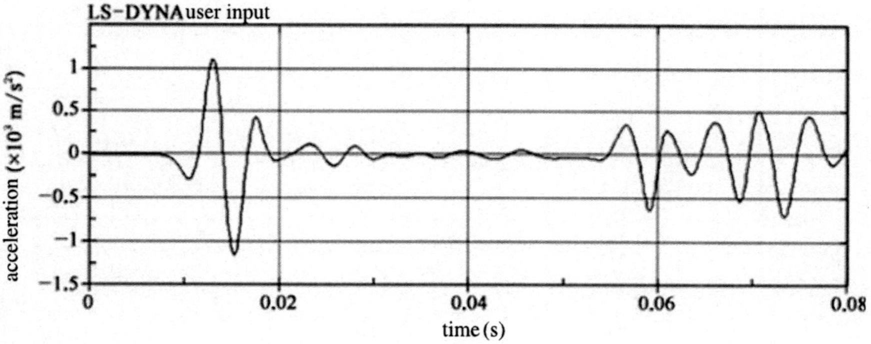

Fig. 3.40 Acceleration–time history curve of the receiving point 1 m away from the tunnel surface

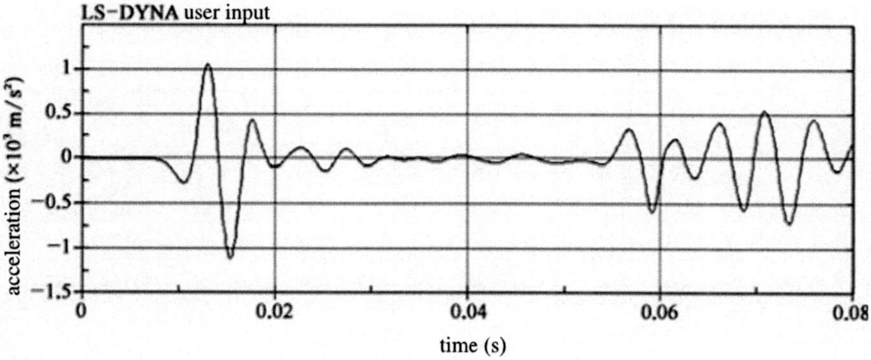

Fig. 3.41 Acceleration–time history curve of the receiving point 2 m away from the tunnel surface

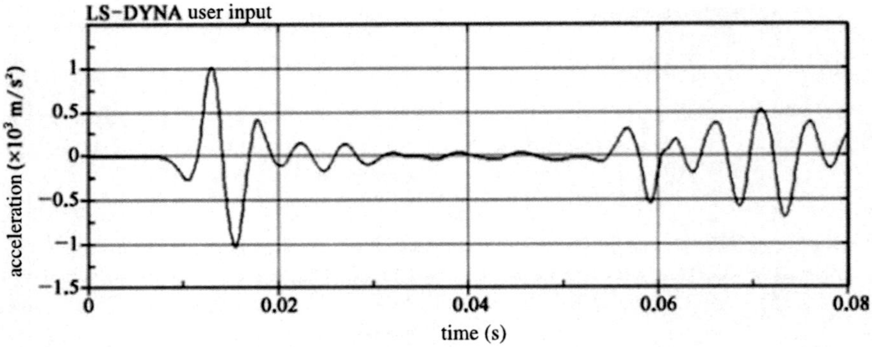

Fig. 3.42 Acceleration–time history curve of the receiving point 3 m away from the tunnel surface

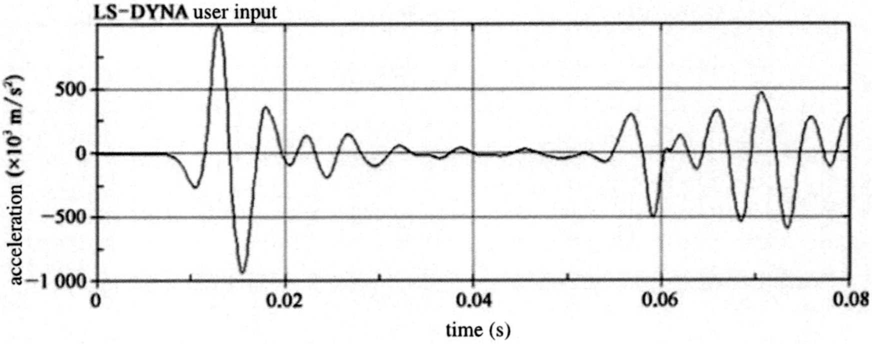

Fig. 3.43 Acceleration–time history curve of the receiving point 4 m away from the tunnel surface

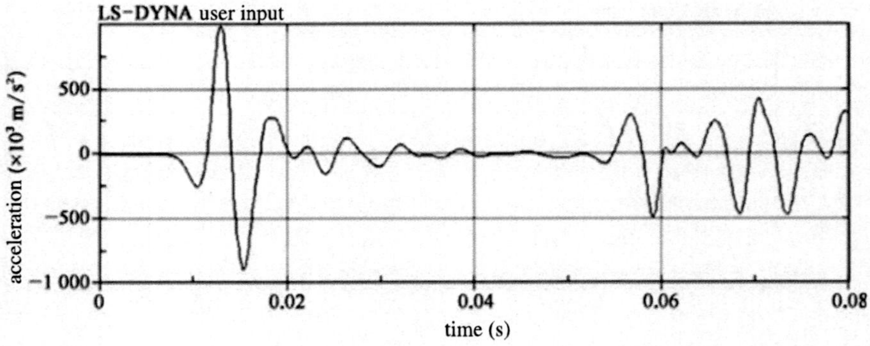

Fig. 3.44 Acceleration–time history curve of the receiving point 5 m away from the tunnel surface

As can be seen from Figs. 3.39, 3.40, 3.41, 3.42, 3.43, and 3.44, as the distance between the receiving point and the tunnel surface increases, the signal strength decreases. Therefore, the geophone should not be installed too deep to facilitate better signal reception and improve the accuracy and reliability of the advance prediction.

(4) **The influence of the change of the distance of the unfavorable geological body on the waveform**

The section size and excavation length of the model remain unchanged. The longitudinal length of the model is, respectively, taken as 80, 90, 100, 110, 120 m, and the distance of the aquifer is 35, 45, 65, 75, 85 m from the tunnel face, respectively. The acceleration–time history curve of the receiving point 25 m away from the tunnel face in each model is shown in Figs. 3.45, 3.46, 3.47, 3.48, and 3.49.

From Figs. 3.45, 3.46, 3.47, 3.48, and 3.49, it can be seen that when the distance between the aquifer and the tunnel surface is small, the energy of the reflected wave is larger, and the waveform of the reflected wave is obvious, which is easy to judge.

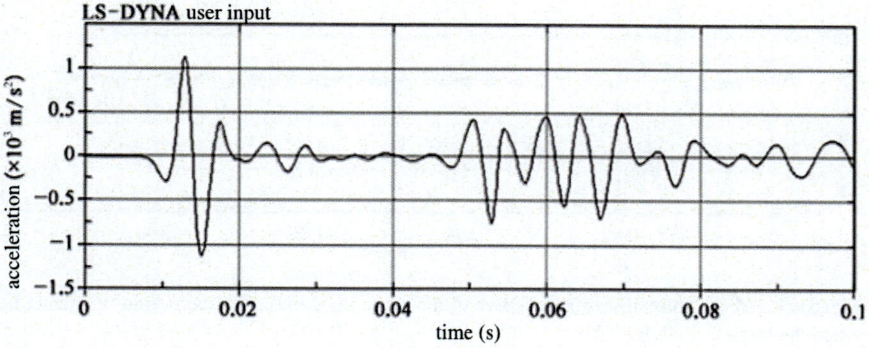

Fig. 3.45 Acceleration–time history curve of the receiving point 25 m away from the tunnel face when the model length is 80 m

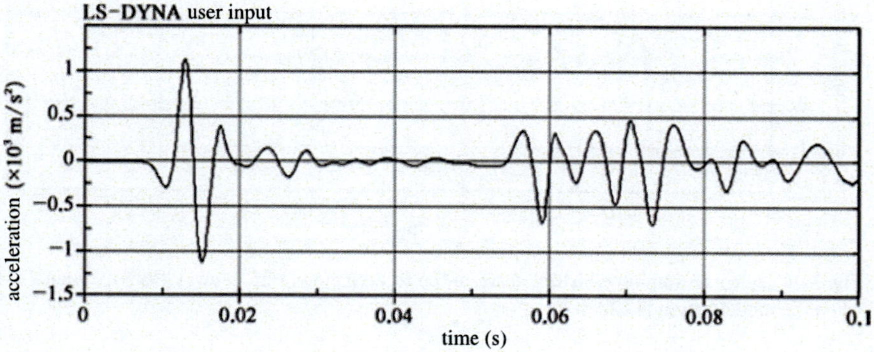

Fig. 3.46 Acceleration–time history curve of the receiving point 25 m away from the tunnel face when the model length is 90 m

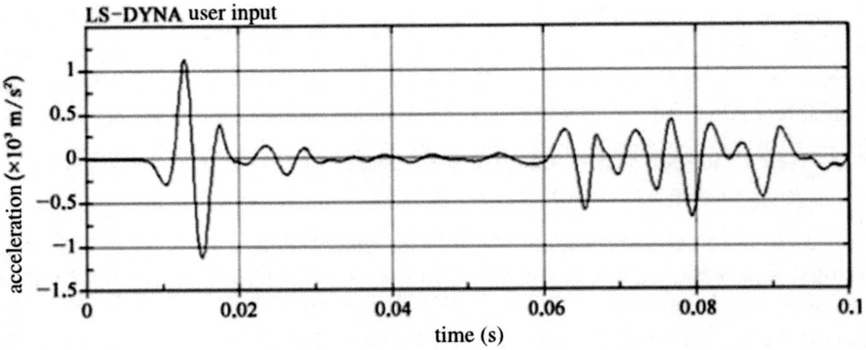

Fig. 3.47 Acceleration–time history curve of the receiving point 25 m away from the tunnel face when the model length is 100 m

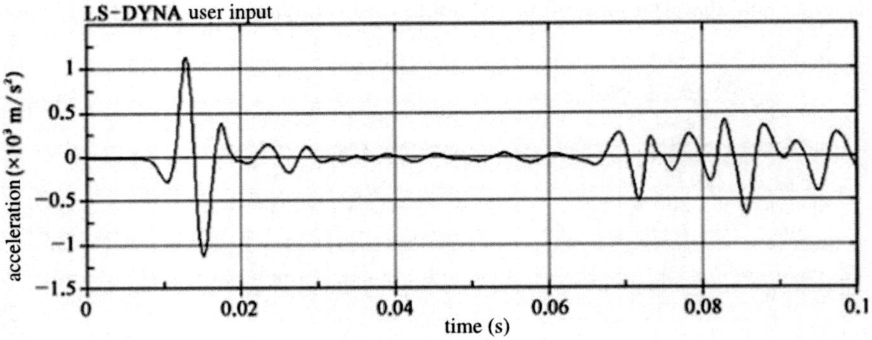

Fig. 3.48 Acceleration–time history curve of the receiving point 25 m away from the tunnel face when the model length is 110 m

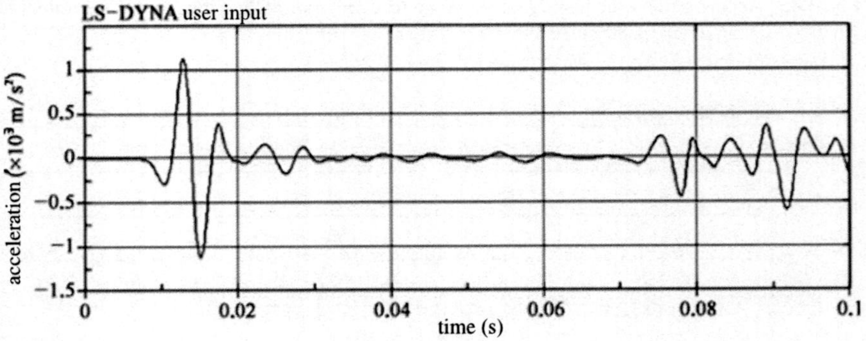

Fig. 3.49 Acceleration–time history curve of the receiving point 25 m away from the tunnel face when the model length is 120 m

As the distance between the aquifer and the tunnel face increases, the travel distance of the seismic wave increases, the travel time increases, the energy loss increases, and the energy of the reflected wave decreases. Therefore, when the seismic wave method is used to make advance predictions of unfavorable geological bodies ahead, there are certain limitations on the prediction distance.

(5) The influence of material parameter changes on the waveform

The values of the elastic modulus E and Poisson's ratio σ of the rock mass are changed, respectively, to discuss the influence of the change of the lithological material parameters on the waveform.

Under the condition that the values of other parameters of the rock mass remain unchanged, different elastic modulus of the rock mass is taken. The elastic modulus of the rock mass is, respectively, taken as 1.5×10^{10} N/m^2, 2.25×10^{10} N/m^2, and 3.0×10^{10} N/m^2, and the acceleration–time history curves of the receiving point 25 m away from the tunnel face are shown in Figs. 3.50, 3.51, and 3.52.

Under the condition that the values of other parameters of the rock mass remain unchanged, different Poisson's ratio σ of the rock mass is taken. Poisson's ratio σ of the rock mass is, respectively, taken as 0.12, 0.22, and 0.32, and the acceleration–time history curves of the receiving point 25 m away from the tunnel face are shown in Figs. 3.53, 3.54, and 3.55.

From the comparison of Figs. 3.50, 3.51, and 3.52, it can be found that as the elastic modulus of the rock mass increases, the signal of the reflected wave gradually weakens, and the travel time of the reflected wave is gradually shortened. From the comparison of Figs. 3.53, 3.54, and 3.55, it can be found that as Poisson's ratio increases, the travel time of the reflected wave will gradually decrease, and the signal intensity will gradually decrease. Through comprehensive analysis, it can be found that the change of elastic modulus and Poisson's ratio of the rock mass has a great influence on the signal reception. The better the surrounding rock, the longer the predictable distance. According to the travel time and signal strength of the reflected wave, the quality of the surrounding rock in front can be judged.

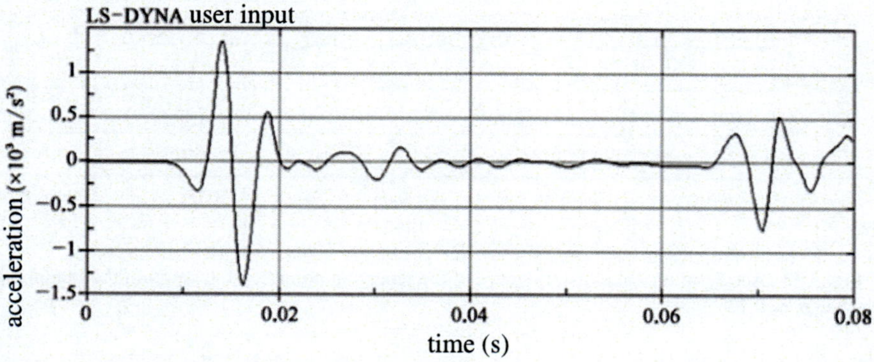

Fig. 3.50 Acceleration–time history curve of the receiving point 25 m away from the tunnel face when $E = 1.5 \times 1010$ N/m^2

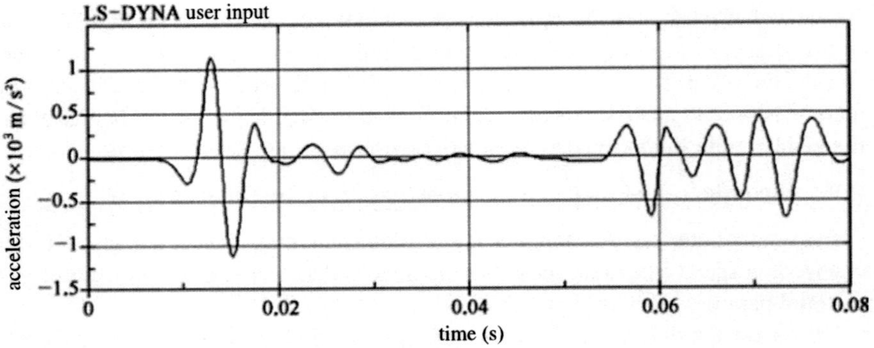

Fig. 3.51 Acceleration–time history curve of the receiving point 25 m away from the tunnel face when $E = 2.25 \times 1010$ N/m^2

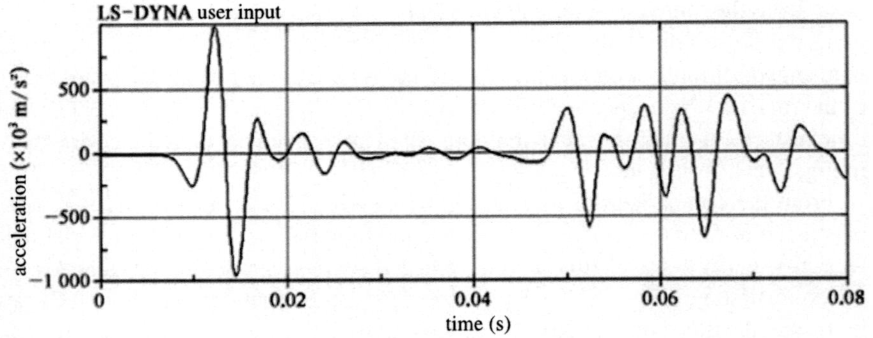

Fig. 3.52 Acceleration–time history curve of the receiving point 25 m away from the tunnel face when $E = 3.0 \times 1010$ N/m^2

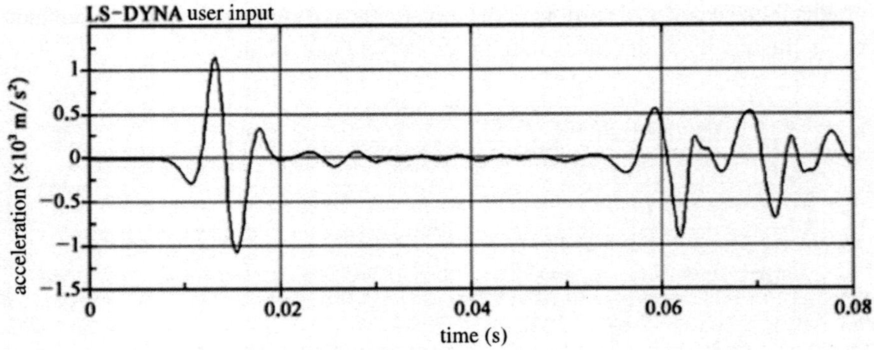

Fig. 3.53 Acceleration–time history curve of the receiving point 25 m away from the tunnel face when $\sigma = 0.12$

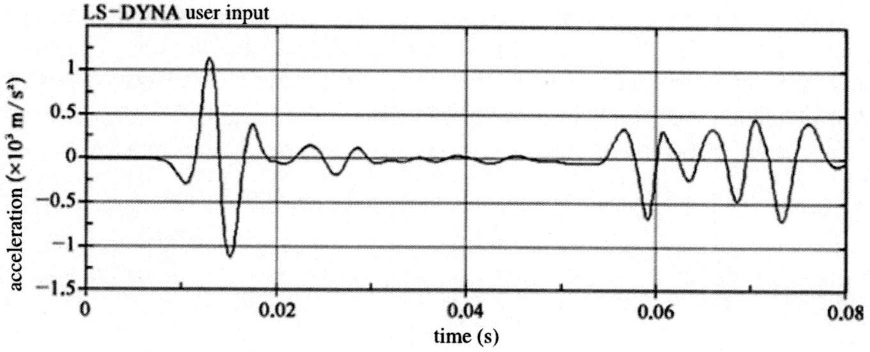

Fig. 3.54 Acceleration–time history curve of the receiving point 25 m away from the tunnel face when $\sigma = 0.22$

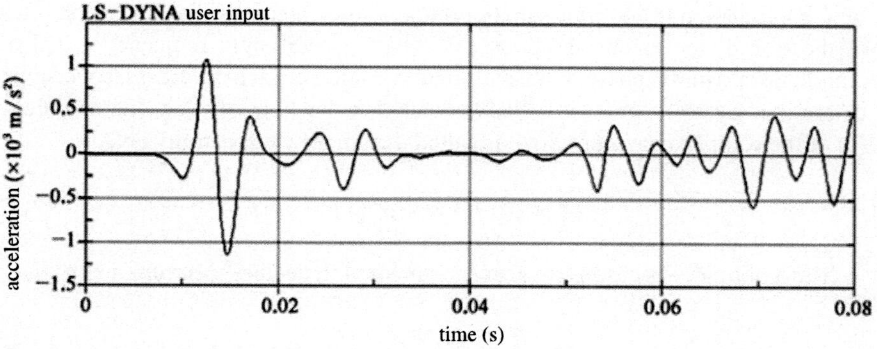

Fig. 3.55 Acceleration–time history curve of the receiving point 25 m away from the tunnel face when $\sigma = 0.32$

3.3 Processing and Imaging of Signal of Three-Dimensional Tunnel Geological Advance Prediction

3.3.1 Signal Processing of 3D Tunnel Geological Advance Prediction

3.3.1.1 Theoretical Models of One-Dimensional Frequency Filtering

(1) **The principle of one-dimensional frequency filtering**

One-dimensional filtering is to filter out certain frequency components in the seismic record and can only filter out interference waves that are different in frequency from the effective wave, so as to suppress the interference wave, strengthen the primary wave, and improve the signal-to-noise ratio.

Seismic waves are formed by the superposition of harmonics of different frequencies. The principle of superposition is based on the fact that the wave equation is linear. Through Fourier transform, the seismic signal can be transformed from the time domain to the frequency domain for analysis.

$$X(f) = \int_{-\infty}^{\infty} x(t)e^{-\mathrm{i}2\pi ft}dt \tag{3.30}$$

$$x(t) = \int_{-\infty}^{\infty} X(f)e^{\mathrm{i}2\pi ft}df \tag{3.31}$$

Equation (3.30) and (3.31) are one-dimensional Fourier transform pairs, $X(f)$ is a positive transform of the time domain signal $x(t)$, and $x(t)$ is the inverse transform of $X(f)$. Equation (3.31) shows that a non-periodic vibration pulse can be composed of the sum of the tuned vibrations $e^{i2\pi ft}$ of many different frequencies, and the amplitude and initial phase of the harmonic vibration of each different frequency are determined by $X(f)$. $X(f)$ is called the complex spectrum of $x(t)$. The modulus $A(f)$ of the complex spectrum $X(f)$ is called the amplitude spectrum, i.e.,

$$A(f) = |X(f)| = \sqrt{a^2(f) + b^2(f)} \tag{3.32}$$

The amplitude spectrum can also be expressed in decibel spectrum, i.e.,

$$B(f) = 20\lg A(f) \tag{3.33}$$

where $X(f) = a(f) + ib(f)$, $a(f)$, and $b(f)$ represent the real and imaginary parts of $X(f)$, respectively.

The angle $\varphi(f)$ of the complex spectrum $X(f)$ is called the phase spectrum, i.e.,

$$\varphi(f) = \arctan\frac{b(f)}{a(f)} \tag{3.34}$$

The purpose of band-pass filtering is to obtain the amplitude spectrum $A(f)$ of the seismic record by spectral analysis of the seismic record $x(t)$. According to the spectral difference between the signal and the noise in the seismic record, appropriate parameters are selected to carry out band-pass filtering, thereby eliminating noise, highlighting effective signals, and improving the signal-to-noise ratio of seismic record. Figure 3.56 is a schematic diagram of the principle of band-pass filtering. The equation for filtering realization is shown in the following equation:

$$X_1(f) = X_0(f)H(f) \tag{3.35}$$

where $X_0(f)$, $X_1(f)$—the spectrum of the input signal and the output signal;

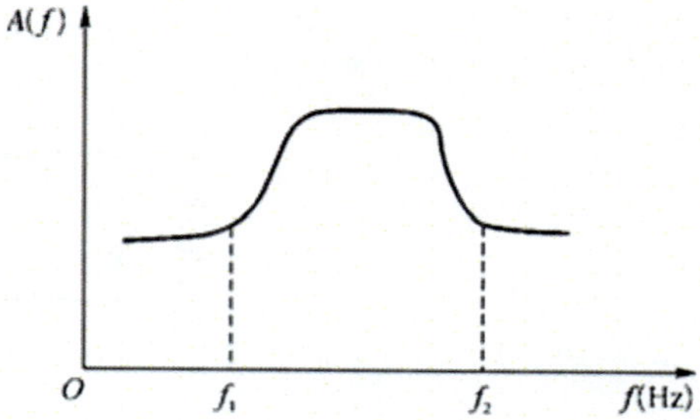

Fig. 3.56 Principle of band-pass filtering

$H(f)$—the filter factor.

The inverse Fourier transform is then performed, which is the filtered signal.

In order to reduce the influence of the truncation factor at the end of the selected time window (Gibbs phenomenon), in the spectrum analysis, the chrome processing should be performed on both sides of the time window. Note: The cutoff frequency cannot be too large or too small. If it is too small, the edging loses its meaning and the Gibbs effect is obvious. If it is too large, the filtering effect is not ideal. It is suggested that the cutoff frequency can be determined by spectrum analysis and multiple trial calculation results. The implementation process of spectrum analysis and band-pass filtering is shown in Fig. 3.57.

(2) *Theoretical Model I*

As shown in Fig. 3.58, Model I is a geological model constructed by two inclined planes, and survey lines are arranged on the same observation plane. Using the one-shot multiple-collection system, the composite record is shown in Fig. 3.59.

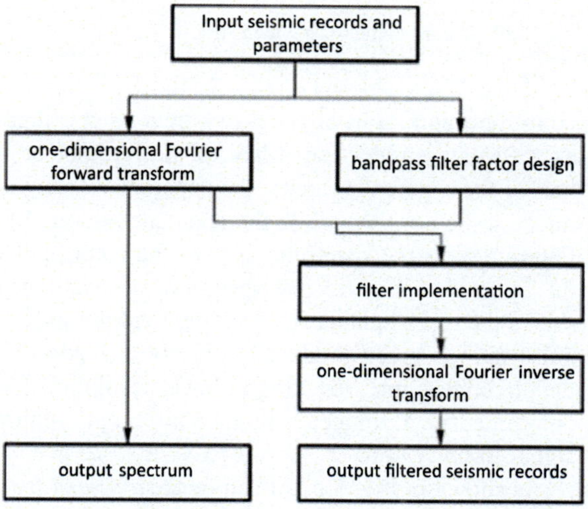

Fig. 3.57 Flowchart of spectrum analysis and band-pass filtering

Fig. 3.58 Geological model of inclined tectonic surface

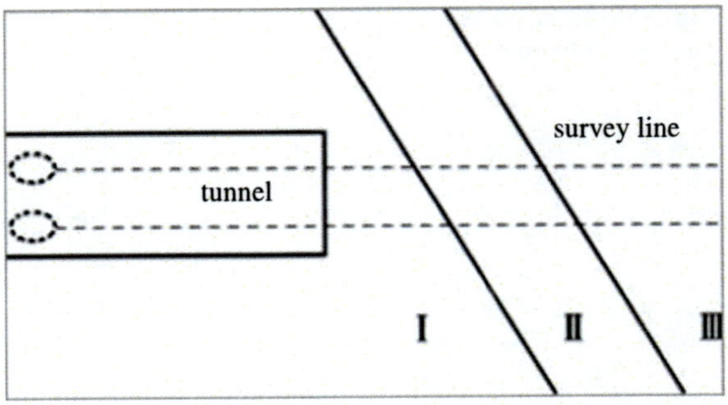

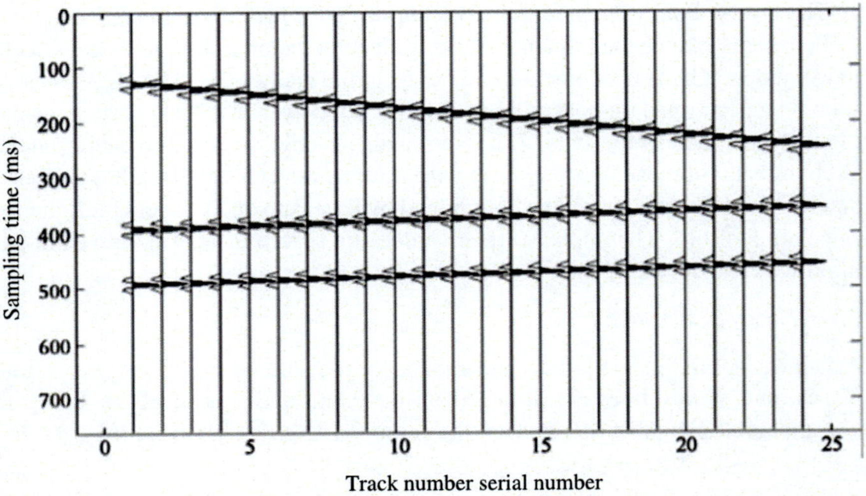

Fig. 3.59 Record gather of inclined structural plane model

In the figure, a in-phase axis (direct wave) tilted to the right and two in-phase axes (negative reflection) tilted to the left can be seen. According to the definition of the apparent velocity, the apparent velocity of the direct wave can be determined to be positive, while the apparent velocity of the reflected wave is negative. Figure 3.60 is the spectrum chart of the record gather. It can be clearly seen from the figure that the effective spectrum range is between 400 and 600. Figure 3.61 shows the record gather after adding random noise. It is obvious that the signal-to-noise ratio is greatly reduced after noise is added, and the noise almost obscures the effective signal, resulting in lower resolution. The following will filter through one-dimensional frequency domain filtering to achieve the purpose of improving the signal-to-noise ratio.

As shown in Fig. 3.62, it can be clearly seen from the spectrogram after adding random noise that the spectrum of random noise is distributed over the entire frequency domain and overlaps with the spectrum of the effective wave. When

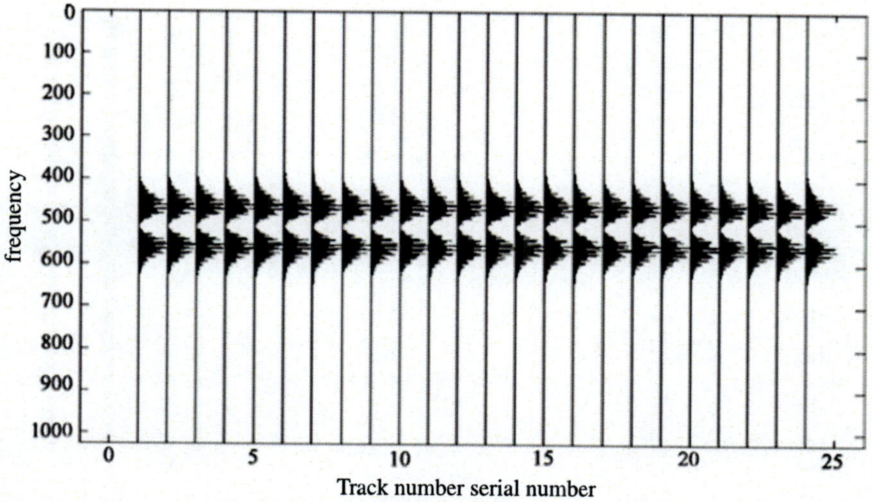

Fig. 3.60 Spectrogram of the record gather

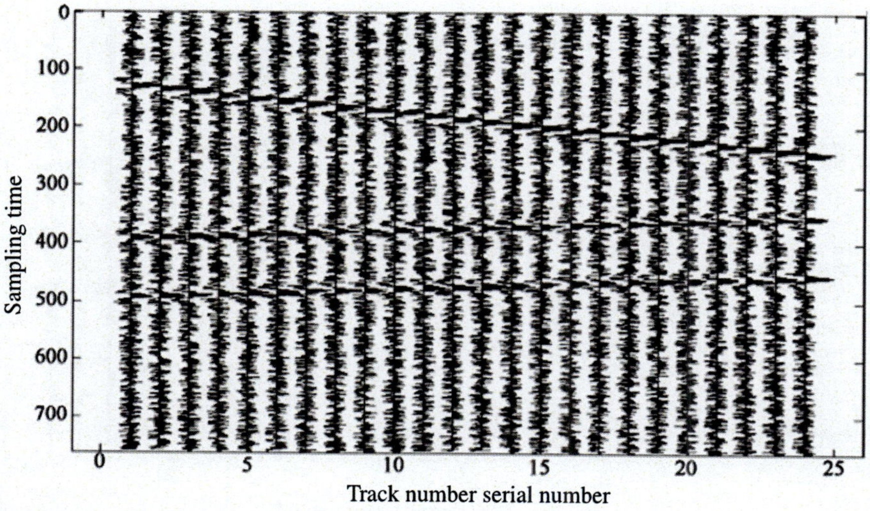

Fig. 3.61 Record gather after adding random noise

performing frequency domain filtering, it is only necessary to set the spectral range of the effective wave and then perform inverse Fourier transform.

Figure 3.63 shows the record gather after one-dimensional frequency filtering. Compared with the noise-added record gather shown in Fig. 3.61, the random noise is obviously suppressed very well, and the signal-to-noise ratio is greatly improved. Figure 3.64 is the spectrum after filtering. Compared with the noise-added spectrum in Fig. 3.62, it can be clearly seen that the spectrum of random noise is well-suppressed,

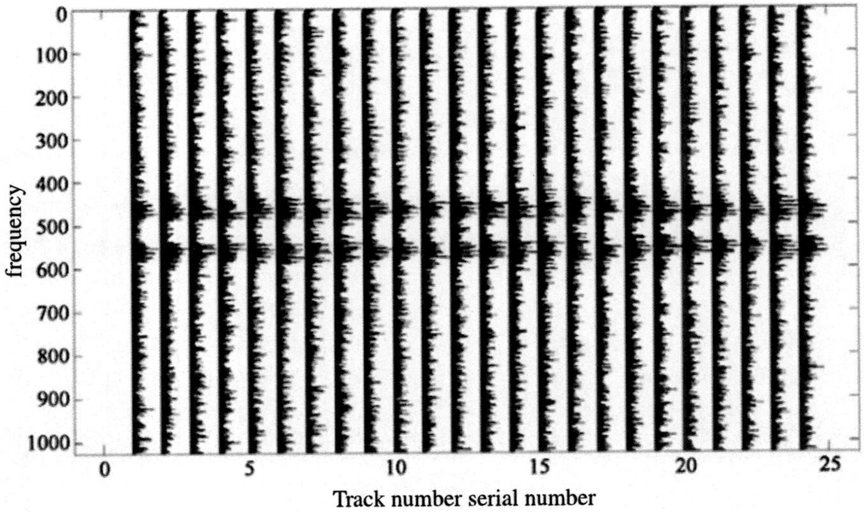

Fig. 3.62 Spectrogram after adding noise

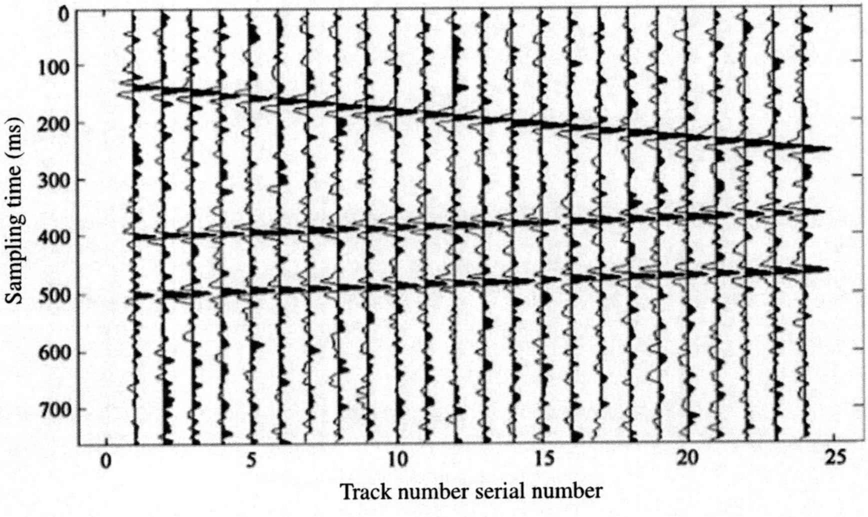

Fig. 3.63 Record gather after one-dimensional frequency filtering

and there is still a small amount of noise. But in general, the filtering effect is achieved, which also shows that one-dimensional frequency filtering can achieve a certain filtering effect and improve the signal-to-noise ratio of seismic record.

The following continues to verify the one-dimensional filtering theoretical model and establish a three-dimensional theoretical model for one-dimensional filtering.

(3) **Theoretical model II**

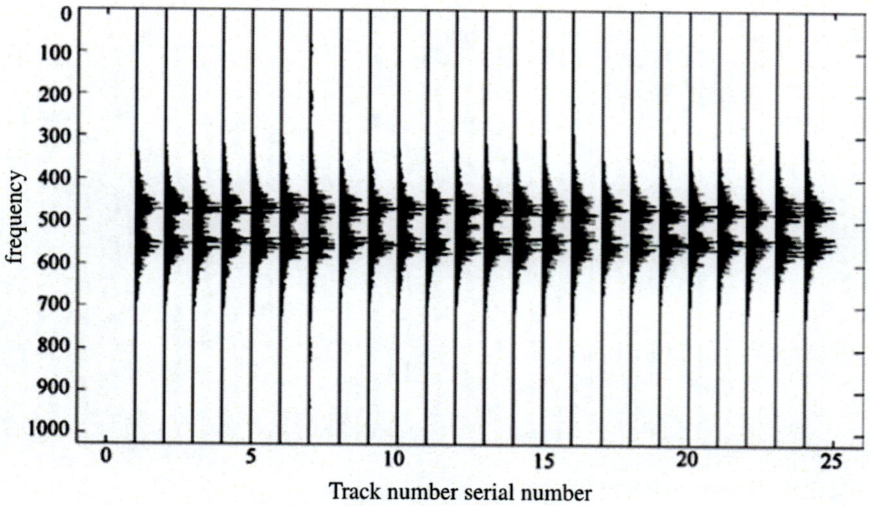

Fig. 3.64 Spectrogram after filtering

Model II mainly simulates the actual tunnel model, and the design is a three-dimensional model. The model mainly sets three abnormal spatial positions. The abnormal points are located at the upper point P_1 (−50, 10, 6) behind the space, the upper left point P_2 (50, 20, −5) in front of the space, and the lower right point P_3 (100, 0, 12) in front of the space. The schematic diagram of the model is shown in Fig. 3.65.

Figure 3.66 shows the original profile with noise. As can be seen from the profile, the noise almost obscures the effective signal, and the signal-to-noise ratio is very low. If no filtering is performed, the effective signal cannot be analyzed at all.

Figure 3.67 shows the longitudinal wave profile with noise obtained after the separation of the longitudinal and transverse waves. It can be seen from the profile that the noise exists in the entire time domain and the signal-to-noise ratio is very low, which requires filtering.

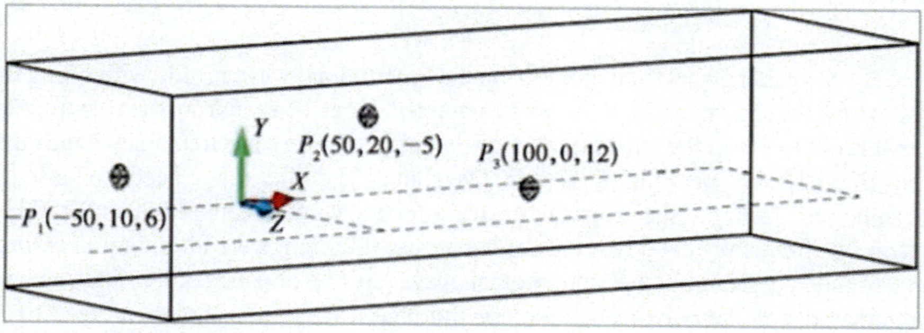

Fig. 3.65 Schematic diagram of Model II

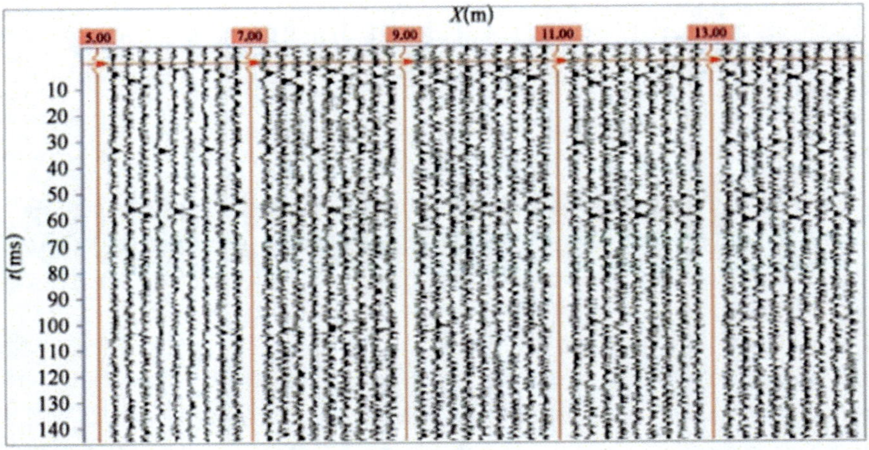

Fig. 3.66 Original profile with noise

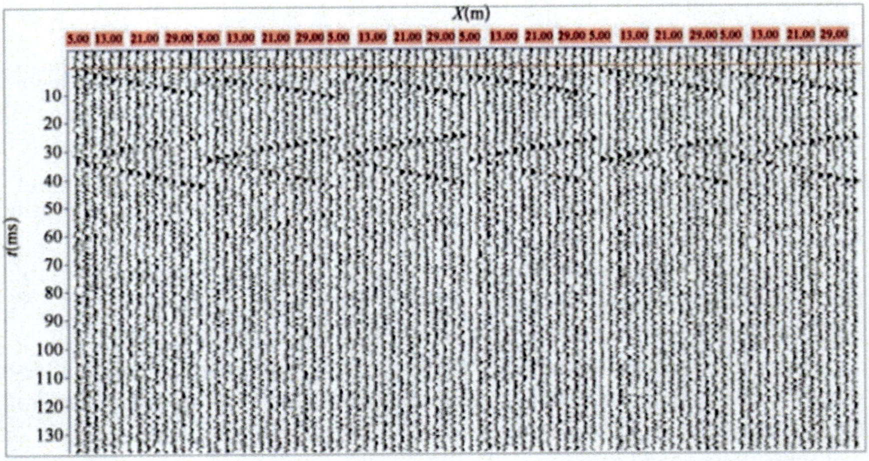

Fig. 3.67 Longitudinal wave profile with noise after separation of longitudinal and transverse waves

By performing spectrum analysis on the longitudinal wave profile with noise, the spectrogram shown in Fig. 3.68 can be obtained. It can be seen from the spectrogram that noise exists in the entire frequency domain, and there are frequency bands that overlap with the spectrum of the effective signal. The filter range needs to be set. It can be seen from the spectrogram that the effective wave is between 80 and 280 Hz, then this frequency band is set as the band-pass filter range for filtering processing.

Figure 3.69 shows the longitudinal wave profile obtained after filtering and compared with the noise-containing longitudinal wave profile shown in Fig. 3.67, it can be clearly seen that the noise has been greatly suppressed and the signal-to-noise

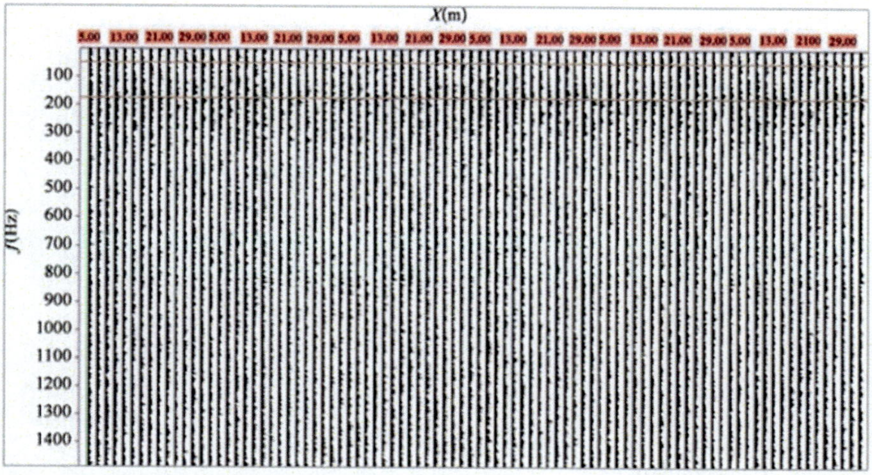

Fig. 3.68 Spectrogram of a longitudinal wave with noise

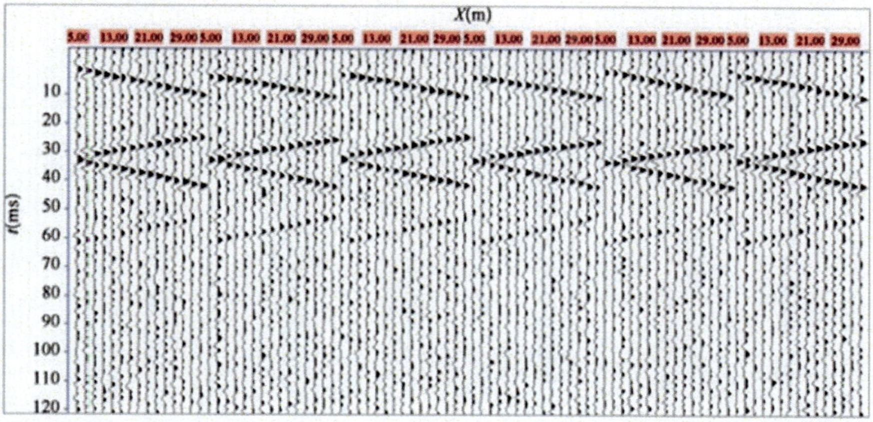

Fig. 3.69 Longitudinal wave profile obtained after filtering

ratio has been greatly improved. It can be seen from the filtered spectrogram that the spectrum of the effective signal is also preserved.

Figure 3.70 shows the transverse wave profile with noise obtained after the separation of the longitudinal and transverse waves. It can be seen from the profile that the noise exists in the entire time domain and the signal-to-noise ratio is very low, which requires filtering to improve the signal-to-noise ratio.

By performing spectrum analysis on the transverse wave profile with noise, the spectrogram shown in Fig. 3.71 can be obtained. It can be seen from the spectrogram that noise exists in the entire frequency domain, and there are frequency bands that overlap with the spectrum of the effective signal. The filter range needs to be set.

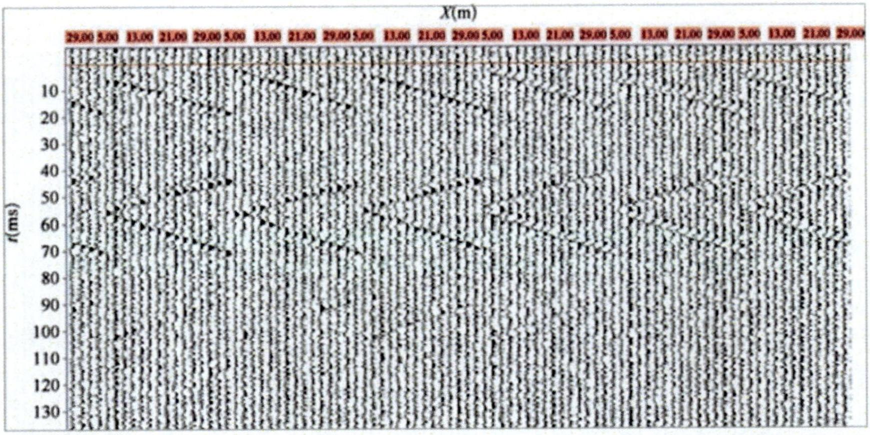

Fig. 3.70 Transverse wave profile obtained after separation of longitudinal and transverse waves

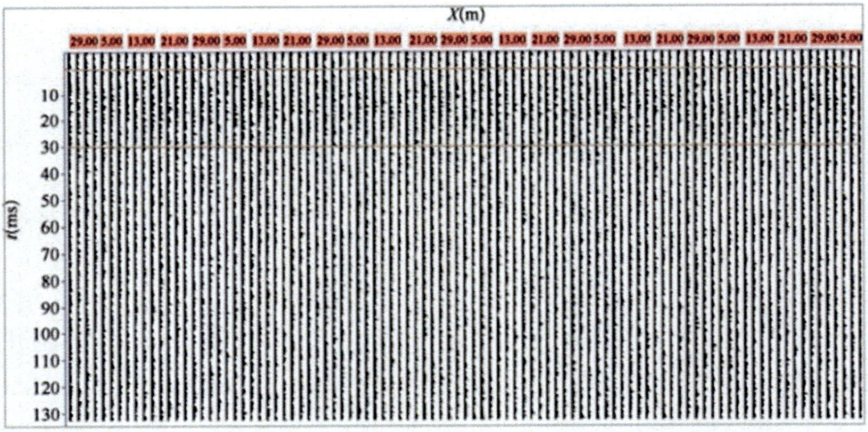

Fig. 3.71 Spectrogram of a transverse wave with noise

Figure 3.72 shows the transverse wave profile obtained after filtering and compared with the noise-containing transverse wave profile, it can be clearly seen that the noise is greatly suppressed and the signal-to-noise ratio is improved.

The above two theoretical models verify that the one-dimensional frequency filtering can improve the signal-to-noise ratio of the signal to a certain extent, but it cannot effectively separate the direct wave, the reflected wave, and other effective waves. In the three-dimensional advance prediction, the information of reflected wave often contains information about the unfavorable geological bodies, so it is particularly important to extract the information of reflected wave effectively. By extracting the information of the direct wave, the velocity of the longitudinal wave

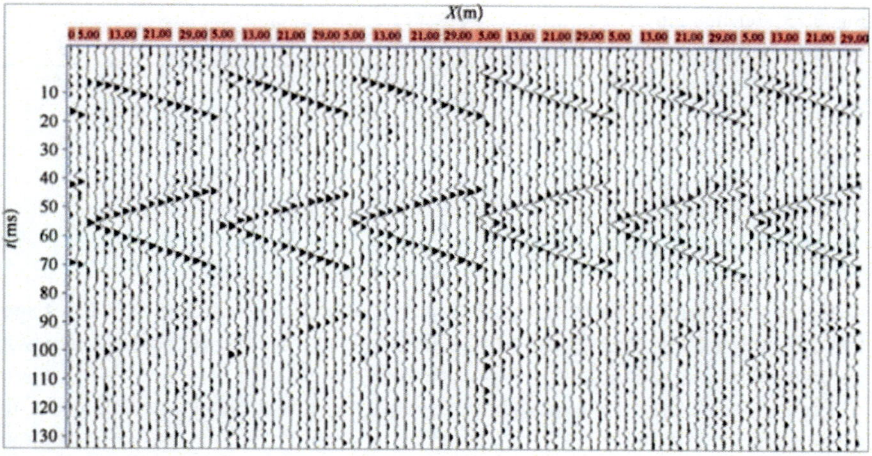

Fig. 3.72 Transverse wave profile obtained after filtering

and the transverse wave can be obtained more accurately, which is also very important for accurately locating the position of the unfavorable geological bodies ahead. A better way to separate different types of waves is to use the two-dimensional *F-K* filter mentioned below.

3.3.1.2 Principle and Theoretical Models of Two-Dimensional *F-K* Filtering

(1) Fundamental

In the case of linear array reception, the shot point and the receiving geophone are on the same straight line. When excited at the shot point, each geophone receiving unit not only receives the reflected wave signal from the underground object, but also receives the direct wave interference signal propagating along the ground surface. Due to the small distance between the shot point and the receiver unit of the geophone (usually 2–4 times the geophone distance), the reflected wave and the direct wave arrive at the geophone almost simultaneously in time. The direct wave forms a strong interference to the reflected wave, and this interference is difficult to separate it from the reflected wave signal in the time domain. In addition, since the spectral characteristics of the direct wave signal and the reflected wave signal are basically the same, the use of traditional time domain filtering or frequency domain filtering cannot obtain satisfactory results. However, considering the relationship between the propagation direction and the receiving array of the reflected wave and the direct wave, there is a big difference in the velocity domain.

Since the seismic wave field is very complex, and the geological structure information in front of the tunnel face is completely reflected in the reflected wave, the effective extraction of the reflected wave is one of the key technologies, which is

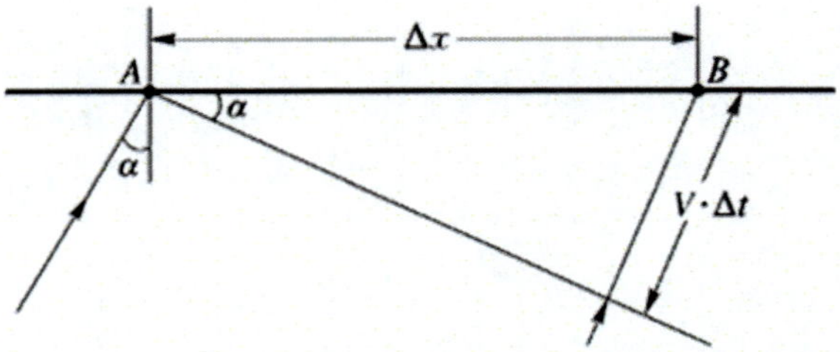

Fig. 3.73 Relationship between reflected wave and apparent speed

different from the key technology of conventional reflection seismic. In this paper, the data processing mainly uses FK filtering to extract the reflected signal with negative apparent velocity.

Figure 3.73 shows the relationship between the direction of propagation of the reflected wave and the survey line. In order to simplify the analysis, it is assumed that the reflected wave is a plane wave, which has a propagation velocity V in the uniform layer and an angle α with the normal line of the survey line AB, which is called the incident angle.

When the reflected wave propagates in the direction shown in the figure at the incident angle α, the time difference between the plane wave front reaching A and B is Δt, and the velocity observed in the direction of the survey line AB is

$$V^* = \frac{\Delta x}{\Delta t} = \frac{V}{\sin\alpha} \tag{3.36}$$

where V^*—the apparent velocity of the reflected wave propagating in the direction of the tunnel line AB;

Δx—the distance between the two geophones on the survey line.

The direct wave propagates in the direction of the survey line AB, and its propagation velocity is assumed to be V_D. Usually, the velocity of the reflected wave is greater than the velocity of the direct wave. It can be seen from Eq. (3.36) that since there is an incident angle α between the direction of propagation of the reflected wave and the normal of the survey line AB, the apparent velocity V^* of the reflected wave propagating in the direction of AB is greater than its true velocity V, i.e., $V^* > V$. Obviously $V^* > V_\text{D}$, that is, the apparent velocity of the reflected wave propagating in the direction of the survey line AB is greater than the velocity of the direct wave. If the wave velocity V_D of the direct wave is selected as the filter threshold of the velocity domain, the velocity component smaller than V_D is filtered out, and the velocity component larger than V_D is retained, so that the direct wave interference can be eliminated. Two-dimensional filtering is carried out on this basis.

In seismic exploration, the signal recorded by one survey line can be expressed as a binary function of the survey line distance x and time t:

$$g = g(x, t) \tag{3.37}$$

To perform a two-dimensional Fourier transform on Eq. (3.37), its two-dimensional spectrum G (k, f) can be obtained, which can be expressed as the following two-dimensional Fourier transform pair:

$$\left.\begin{aligned} g(x,t) &= \int_{-\infty}^{\infty}\int_{-\infty}^{\infty} G(k,f)e^{i(2\pi kx+2\pi ft)}\mathrm{d}k\mathrm{d}f \\ G(k,f) &= \int_{-\infty}^{\infty}\int_{-\infty}^{\infty} g(t,x)e^{-i(2\pi kx+2\pi ft)}\mathrm{d}x\mathrm{d}t \end{aligned}\right\} \tag{3.38}$$

where k—corresponds to the number of waves of x;

f—corresponds to the frequency of time t.

In the actual application process, a discrete form of FFT is used. The discretized expression of the function g (x, t) is:

$$\begin{aligned} g_{\mathrm{mn}} &= g(m\nabla, n\Delta) \\ m &= 0, 1, \ldots, M-1 \\ n &= 0, 1, \ldots, N-1 \end{aligned} \tag{3.39}$$

where m—the number of geophones on a survey line;

∇—the distance between the geophones, called the spatial sampling interval;

$m\nabla$—the length of the survey line;

n—the number of sampling points in the time domain;

Δ—the sampling interval in the time domain;

$n\Delta$—the recording length of the signal.

Equation (3.39) has the following Fourier transform:

$$\begin{cases} G_{\mathrm{mn}} = \sum\limits_{p=0}^{M-1}\sum\limits_{q=0}^{N-1} g_{\mathrm{mn}} e^{-i2\pi\left(\frac{m}{M}p+\frac{n}{N}q\right)} \\ g_{\mathrm{mn}} = \dfrac{1}{\mathrm{MN}}\sum\limits_{p=0}^{M-1}\sum\limits_{q=0}^{N-1} G_{\mathrm{mn}} e^{i2\pi\left(\frac{m}{M}p+\frac{n}{N}q\right)} \end{cases} \quad (m = 0, 1, \ldots, M-1; n = 0, 1, \ldots, N-1) \tag{3.40}$$

$\{G_{\mathrm{mn}}\}$ is called the discrete Fourier transform of $\{g_{\mathrm{mn}}\}$, also called the wave spectrum of $\{g_{\mathrm{mn}}\}$.

The digital record $\{g_{\mathrm{mn}}\}$ is a two-period sequence with $X = M\nabla$ and $T = N\Delta$ as a period. Its time domain fundamental frequency is $d = \frac{1}{T}$, and the spatial domain fundamental frequency is $e = \frac{1}{X}$. The two independent variables of the two-dimensional wave spectrum $\{G_{\mathrm{mn}}\}$ are wave number $k = \mathrm{me} = \frac{m}{M\nabla}$ and frequency $f = nd = \frac{n}{N\Delta}$. The ratio of frequency to wave number is called the apparent velocity, and its size can be determined by:

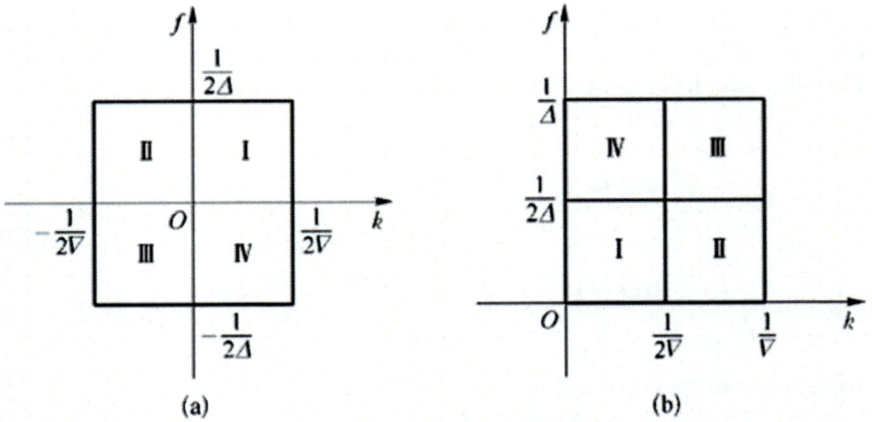

Fig. 3.74 Periodic representation of the two-dimensional spectrum

$$V^* = \frac{f}{k} = \frac{nX}{mT} \tag{3.41}$$

As can be seen from the foregoing discussion, the apparent speeds of direct and reflected wave are quite different. Reflected on the *F-K* plane by two-dimensional FFT, the direct wave and the reflected wave are located in different areas. Given the filtering threshold of the velocity domain, the range of filtering can be calculated by Eq. (3.41).

Since the two-dimensional spectrum $G\,(k, f)$ has periodicity, its wave spectrum is usually calculated only in the region $|k| \leq \frac{1}{2\nabla}, |f| \leq \frac{1}{2\nabla}$, which is shown in Fig. 3.74a. Using the periodicity of the two-dimensional spectrum, the wave spectrum shown in Fig. 3.74a can be transformed into the form of Fig. 3.74b. The wave spectrum values of the corresponding four blocks I, II, III, and IV in the two figures are equal. Thus, the wave spectrum $G\,(k, f)$ can be calculated in the region $0 \leq k \leq \frac{1}{\nabla}, 0 \leq f \leq \frac{1}{\Delta}$ using Fig. 3.74b.

For the convenience of calculation, $\{G_{mn}\}$ in Eq. (3.40) is written as follows:

$$G_{\mathrm{mn}} = \sum_{p=0}^{M-1} \left[\sum_{q=0}^{N-1} g_{\mathrm{mn}} e^{-i2\pi \cdot \frac{n}{N} q} \right] e^{-i2\pi \cdot \frac{m}{M} p} \tag{3.42}$$

let

$$a(p, n) = \sum_{q=0}^{N-1} g_{\mathrm{mn}} e^{-i2\pi \cdot \frac{n}{N} q} \tag{3.43}$$

and

$$G_{mn} = \sum_{p=0}^{M-1} a(p, n) e^{-i2\pi \cdot \frac{m}{M} p} \tag{3.44}$$

Formula (3.43) is a one-dimensional Fourier transform in the time domain. In the calculation, one-dimensional FFT is performed for the time series recorded by each array element with Formula (3.44), and then the one-dimensional spectrum with the same number n is arranged according to the number *p*. One-dimensional FFT is performed with Formula (3.44) again to obtain the two-dimensional spectrum $\{G_{mn}\}$ of the time series $\{g_{mn}\}$.

With a two-dimensional spectrum, the next step is to choose the appropriate filter gate parameters. The selection of the filter gate parameters is mainly based on the acoustic parameters of the measurement site. As can be seen from the foregoing discussion, the wave number k, the frequency f, and the apparent velocity V^* of the two-dimensional spectrum $\{G_{mn}\}$ have the following relationship with its sequence numbers *m* and *n*:

$$\left.\begin{aligned} f &= \frac{n}{N\Delta} \\ k &= \frac{m}{M\nabla} \\ V^* &= \frac{nM\nabla}{mN\Delta} \end{aligned}\right\} \tag{3.45}$$

The corresponding sequence numbers $n_1 = N\Delta f_1$ and $n_2 = N\Delta f_2$ can be calculated according to the filter gate parameters f_1 and f_2 of the frequency domain. Similarly, two lines on the *f*-*k* plane can be calculated according to the filter gate parameters V_1^* and V_2^* in the velocity domain:

$$n = \frac{N\Delta}{M\nabla} V_1^* m \text{ and } n = \frac{N\Delta}{M\nabla} V_2^* m \tag{3.46}$$

The filtering region of the velocity domain is determined by them. The *f*-*k* plane filter gate derived from the given parameters f_1, f_2, $V^*{}_1$, $V^*{}_2$ is shown in Fig. 3.75. The oblique line in the figure represents the two-dimensional spectrum to be retained, and the remaining part represents the part to be filtered out.

Normally, two-dimensional filtering is performed in the frequency domain. The spectral values within the filtering gate are retained, and the spectral values outside the filtering gate are made to be zero. The two-dimensional FFT inverse transform is performed on the filtered wave spectrum, so that the filtered seismic record can be obtained.

In the selection of the filter, the fan filter is a commonly used two-dimensional filter, which can suppress the interference whose visual speed is lower than the effective wave and suppress the interference whose frequency is higher than the effective wave.

The frequency domain factor of the fan filter is

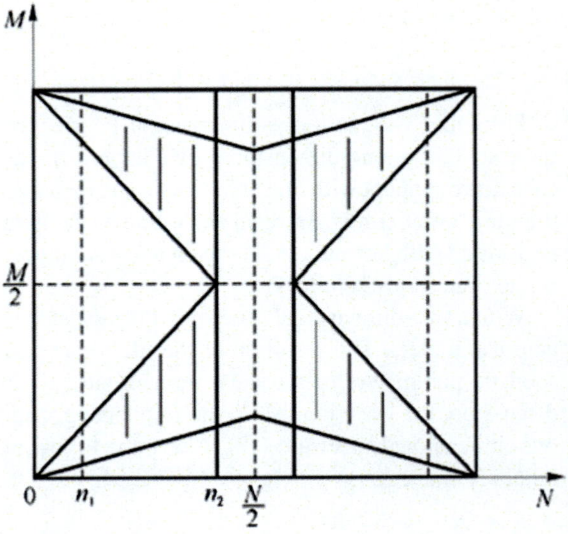

Fig. 3.75 Filter gate with given parameters

$$H(f, k_x) = \begin{cases} 1, \text{ when}\left|\frac{f}{k_x}\right| > V^*, |f| < f_c \\ 0. \text{ others} \end{cases} \tag{3.47}$$

That is, in Fig. 3.76, the red area $H\,(f, k_x) = 1$.

The space–time domain factor of the fan filter is

$$h(t, x) = \frac{1}{2\pi^2 x}\left[\frac{1 - \cos 2\pi\left(\frac{x}{V^*} + t\right) f_c}{\frac{x}{V^*} + t} + \frac{1 - \cos 2\pi\left(\frac{x}{V^*} - t\right) f_c}{\frac{x}{V^*} - t}\right] \tag{3.48}$$

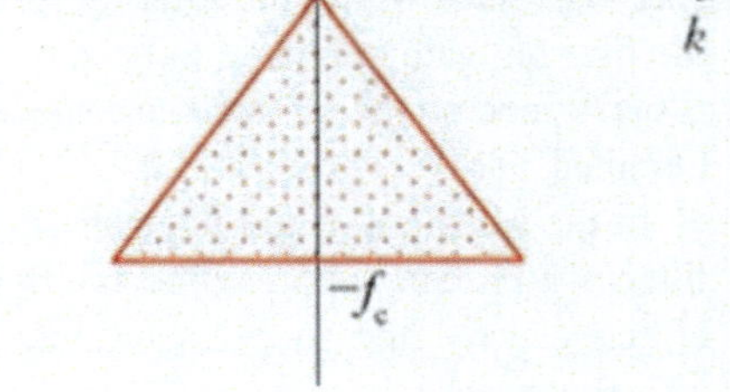

Fig. 3.76 Frequency wave diagram of fan filter

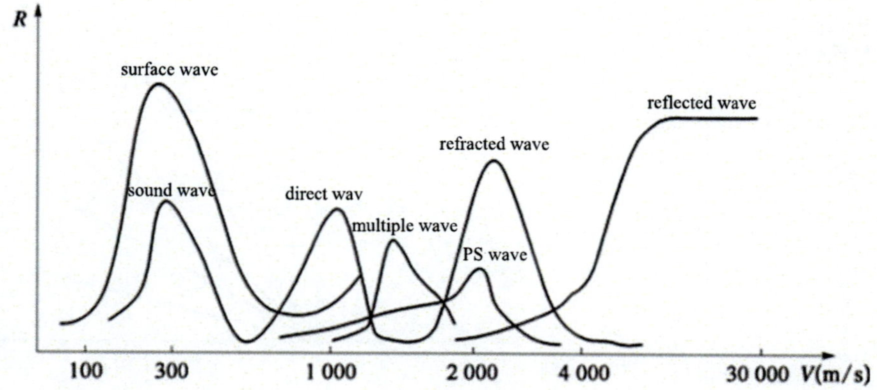

Fig. 3.77 Distribution diagram of seismic wave signal in apparent velocity domain

According to the characteristics of different types of signals of seismic waves, the general distribution law of the apparent velocity domain of all types of signals is shown in Fig. 3.77. The general distribution law of the frequency domain is shown in Fig. 3.78, and the general distribution law of frequency and apparent velocity in *F*-*K* domain is shown in Fig. 3.79. It is difficult to separate signals with overlapping frequencies or wave numbers by one-dimensional frequency or wave number filtering, but *F*-*K* domain filtering can make full use of the advantages of both, effectively separating different types of signals from each other, and achieving the purpose of wave field separation.

(1) The law of apparent velocity distribution.

① Apparent velocity of surface wave (m/s): $100 < V < 800$.
② Sound waves (m/s): $200 < V < 400$.
③ Direct wave (m/s): $400 < V < 1500$.
④ Multiple wave (m/s): $800 < V < 2000$.
⑤ PS-wave (m/s): $800 < V < 4000$.
⑥ Refracted wave (m/s): $1500 < V < 5000$.
⑦ Reflected wave (m/s): $V > 3600$.

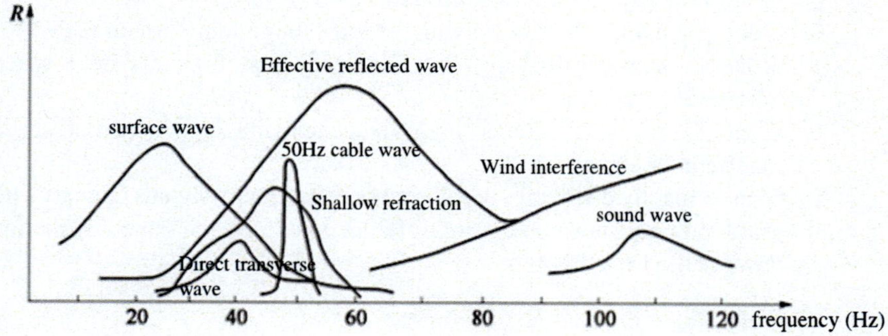

Fig. 3.78 Distribution diagram of seismic wave signal in frequency domain

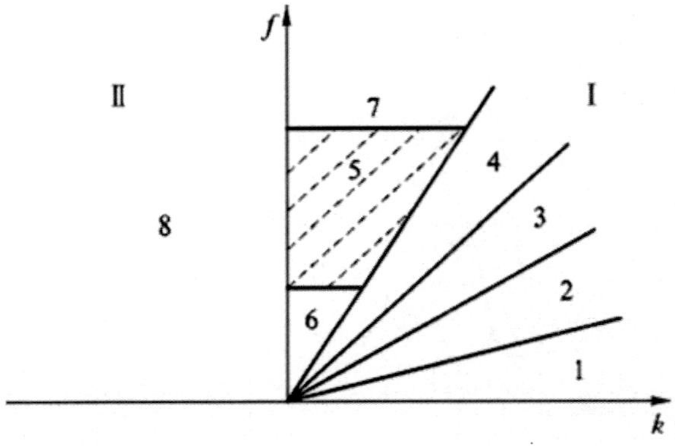

Fig. 3.79 *F-K* signal distribution overview

In addition, according to the characteristics of apparent velocity, the apparent velocity of reflected wave received by the record track near the shot point is relatively high, while the apparent velocity received by the record track far from the shot point is relatively low. Different observation systems, different working areas, and different acquisition parameters may have slight differences in the distribution of apparent velocity.

(2) Frequency distribution law

① Surface wave (Hz): $5 < f < 60$.
② Direct transverse wave (Hz): $15 < f < 70$.
③ Shallow refracted wave (Hz): $20 < f < 60$.
④ 50 Hz cable wave (Hz): $45 < f < 55$.
⑤ Wind disturbance (Hz): $f > 60$.
⑥ Sound wave (Hz): $f > 100$.
⑦ Effective reflected wave (Hz): $f > 15$.

(3) Distribution law of signal in *F-K* domain

① Zone 1 mainly distributes low frequency and low visual velocity interference signals.
② Zone 2 mainly distributes surface waves.
③ Zone 3 mainly distributes reflected transverse waves and direct waves.
④ Zone 4 mainly distributes converted waves (PS).
⑤ Zone 5 is mainly distributed with reflected longitudinal waves (PP).
⑥ Zone 6 mainly distributes low frequency and high apparent speed interference signals.
⑦ Zone 7 mainly distributes high frequency and high apparent speed interference signals.
⑧ Zone 8 mainly distributes negative apparent velocity signals (signals with different frequency domains of different types have the same distribution laws as the I-quadrant).

When the spatial sampling is appropriate, the in-phase signals with different inclination angles in the *t-x* plane can be automatically separated in the *F-K* plane due to their different inclination angles. For this reason, according to the frequency of the signal and the law of the apparent speed, different filters can be designed according to the nature of the signal to be extracted, and the signals of different apparent velocity and different frequencies are filtered or retained to realize the two-dimensional signal filtering. The selection of cutoff frequency and cutoff apparent velocity is very important. Do not choose them too big or too small. If it is too big, it will not provide good filtering effect or even achieve the filtering effect; if it is too small, the Gibbs effect will appear, and the expected filtering effect will not be achieved. Of course, the cutoff frequency and the cutoff apparent velocity depend on the specific situation, and the interactive trial effects can be used to determine the appropriate parameters. *F-K* filtering flowchart is shown in Fig. 3.80.

It can be known from the theory that *F-K* filtering is mainly based on the difference in apparent velocity of various waves, so it is theoretically possible to extract different types of waves.

(2) Theoretical model I

Model I is a geological model constructed with two inclined planes, as shown in Fig. 3.81, and survey lines are arranged on the same observation plane. Using the one-shot multiple-collection system, the composite record is shown in Fig. 3.82.

In Fig. 3.59, an in-phase axis (direct wave) tilted to the right and two in-phase axes (negative reflection) tilted to the left can be seen. According to the definition of the apparent velocity, the apparent velocity of the direct wave can be determined to be positive, while the apparent velocity of the reflected wave is negative. It can be seen from the model's *F-K* spectrogram (Fig. 3.83) that the spectrum of the direct

Fig. 3.80 *F-K* filtering flowchart

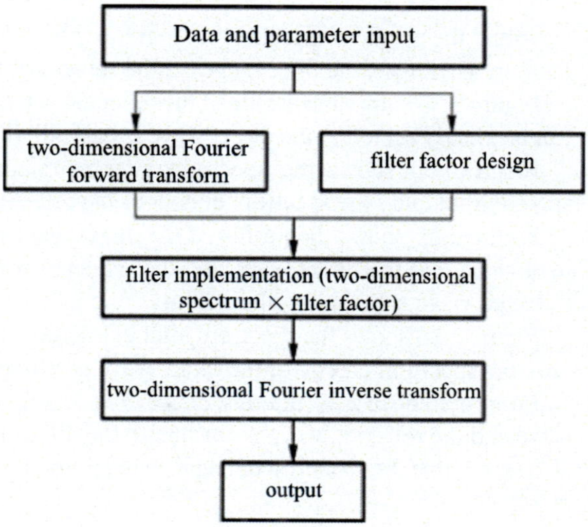

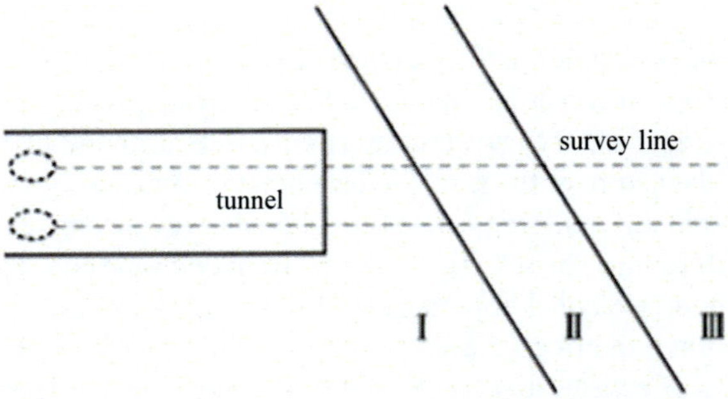

Fig. 3.81 Geological model of inclined tectonic plane

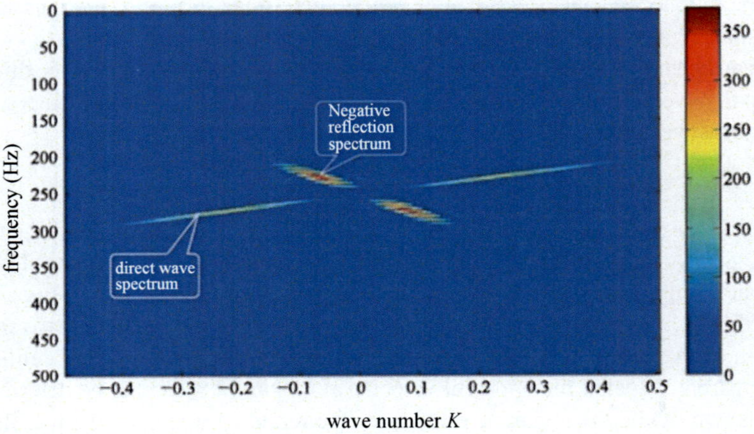

Fig. 3.82 *F-K* spectrum of the model (theoretical model I)

wave and the negative reflection in the frequency domain are distributed in different regions. Therefore, the filter range can be set so that the desired wave is retained.

Figure 3.84 shows the profile of the reflected wave extracted after *F-K* filtering. It can be clearly seen that the direct wave is basically filtered out and is well-pressed.

Figure 3.84 shows the spectrum after extracting the reflected wave. In the spectrogram, only the spectrum energy of the reflected wave is strong.

Figure 3.85 shows the profile of the direct wave extracted after *F-K* filtering. It is obvious that the direct wave and the reflected wave are well-separated, and the reflected wave is basically filtered out.

Figure 3.86 shows the spectrum after extracting the direct wave. In the spectrum, only the spectrum energy of the direct wave is strong.

From the above *F-K* filtering effect, *F-K* filtering can better separate the direct wave and the reflected wave according to the difference in apparent velocity, which also shows that the spectrum of negative reflection is located in the second and fourth quadrants.

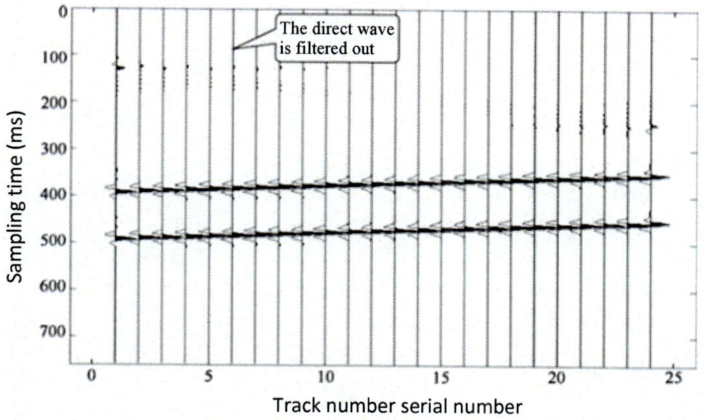

Fig. 3.83 Negative reflection extracted by *F-K* filtering (theoretical model I)

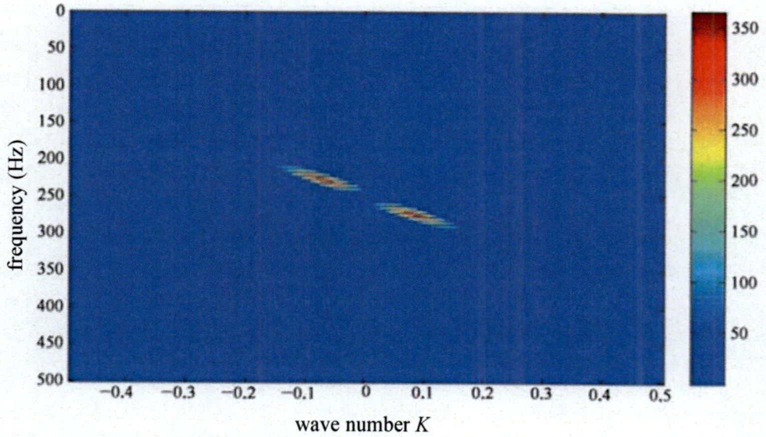

Fig. 3.84 *F-K* spectrogram after extracting reflected wave (theoretical model I)

(3) Theoretical Model II

Model II is a composite record of reflected waves at common shot points simulating horizontal layered media, including direct waves, refracted waves, and reflected waves. The composite record is shown in Fig. 3.87.

As can be seen from the composite record in Fig. 3.87, the direct and refracted waves are in a straight-line shape, and the reflected wave is in a hyperbolic form. The direct, refracted, and reflected waves cannot be separated in the time domain. However, due to the difference of the slope in the time profile, that is, the difference in apparent velocity, the apparent velocities of direct wave and reflected wave are fixed values, while the apparent velocity of reflected wave is variable, and they can be seen in different areas in the frequency domain.

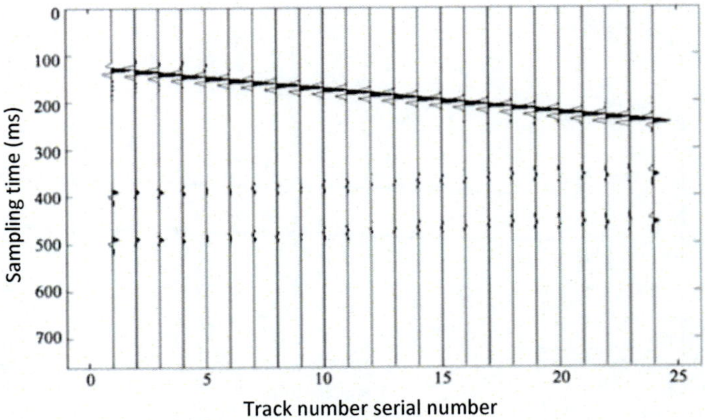

Fig. 3.85 Direct wave extracted by *F-K* filtering (theoretical model I)

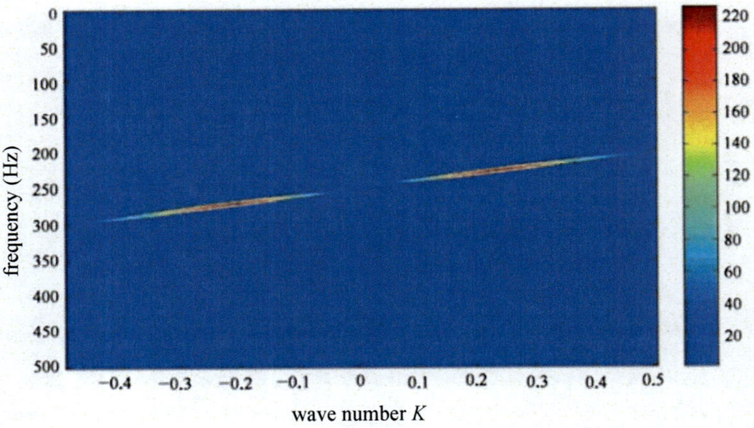

Fig. 3.86 *F-K* spectrogram after extracting direct wave (theoretical model I)

As shown in Fig. 3.88, this allows the filter range to be set so that the desired wave remains.

Figure 3.89 shows the profile of direct wave extracted by *F-K* filtering. It is obvious that the refracted and reflected waves are basically filtered out and are well pressed.

Figure 3.90 shows the spectrum after extracting the direct wave. In the spectrum, only the spectrum energy of the direct wave is strong.

Figure 3.91 is the profile of reflected wave extracted after *F-K* filtering. It is obvious that the refracted wave and direct wave are basically filtered out and well suppressed.

Figure 3.92 is the spectrum diagram after the reflection wave is extracted. In the spectrum diagram, only the spectrum energy of the reflection wave is strong.

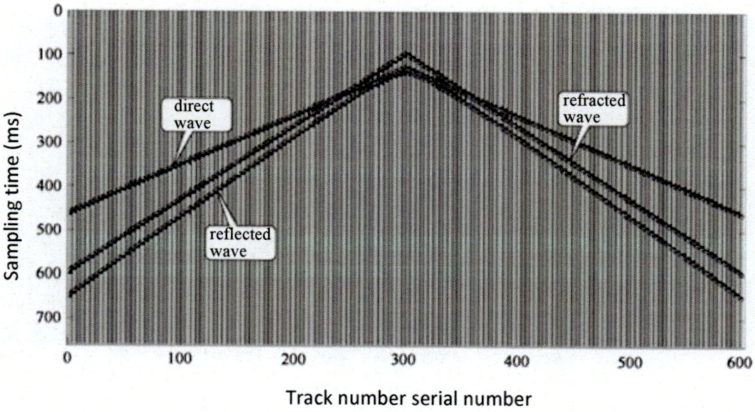

Fig. 3.87 Composite record of common shot points reflection wave (theoretical model II)

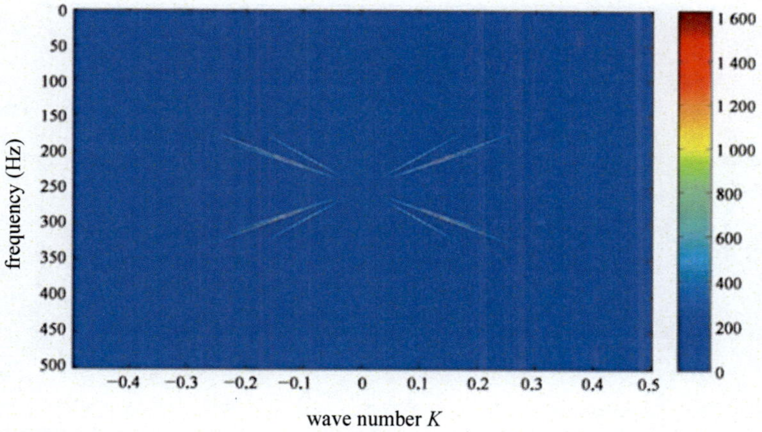

Fig. 3.88 *F-K* spectrum of the composite record (theoretical model II)

The model further verifies that *F-K* filtering has a better function of separating different types of waves.

The following two-dimensional *F-K* filtering theoretical model verification is carried out to establish a three-dimensional theoretical model for two-dimensional filtering.

(4) Theoretical Model III

Model III mainly simulates the actual tunnel model, and the design is a three-dimensional model. The model mainly sets three abnormal spatial positions. The abnormal points are located at the upper point P_1 (−50, 10, 6) behind the space, the upper left point P_2 (50, 20, −5) in front of the space, and the lower right point P_3 (100, 0, 12) in front of the space. The schematic diagram of the model is shown in Fig. 3.93.

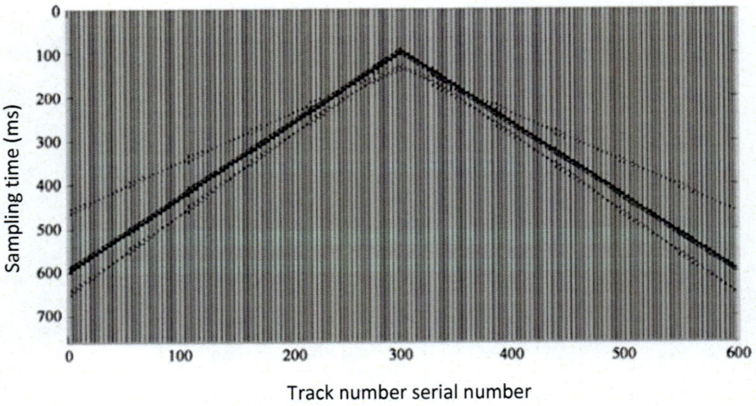

Fig. 3.89 Direct wave extracted by *F-K* filtering (theoretical model II)

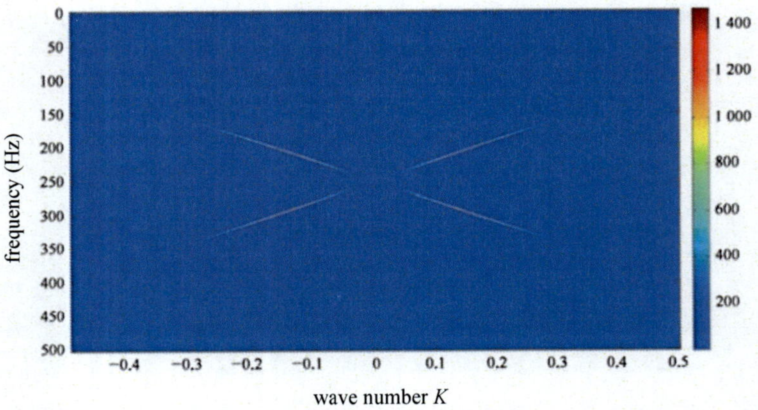

Fig. 3.90 *F-K* spectrogram after extracting direct wave (theoretical model II)

Figure 3.94 is a theoretical model without noise. Figure 3.95 shows the longitudinal wave profile after the separation of the longitudinal and transverse waves. It can be seen from the figure that the direct wave first arrives at each geophone, and then the reflected wave of the abnormal body at the upper left point P_2 in front of the space arrives, followed by the reflected wave of the abnormal body at the upper point P_1 behind the space. Finally, the reflected wave of the abnormal body at the lower right point P_3 in front of the space arrives.

Figure 3.96 shows the *F-K* spectrum diagram of the longitudinal wave profile, which can be obtained by setting different filtering ranges.

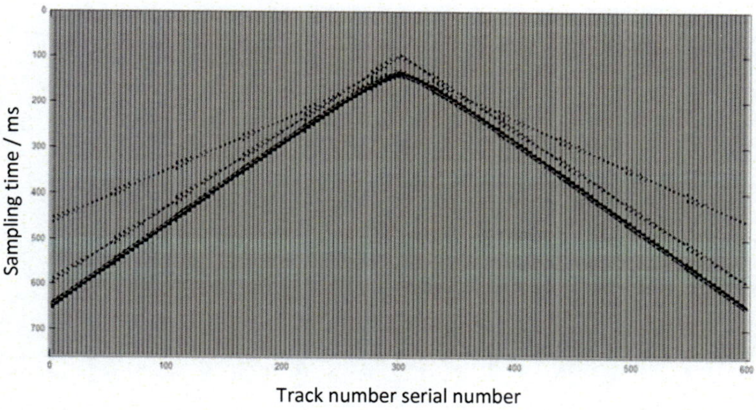

Fig. 3.91 Reflected wave extracted by *F-K* filtering (theoretical model II)

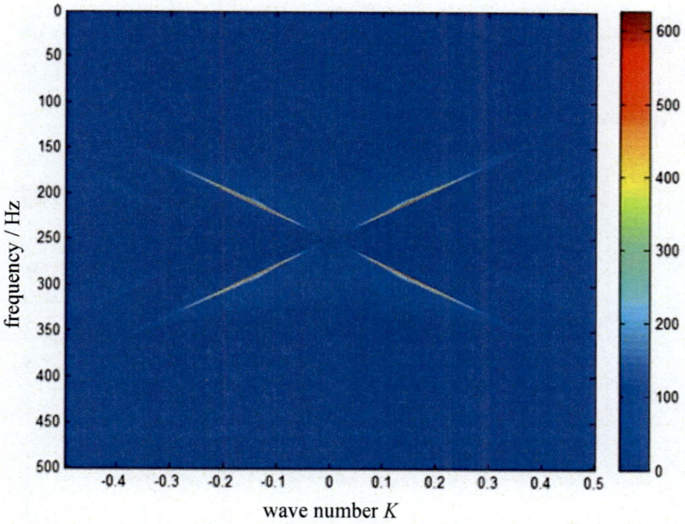

Fig. 3.92 *F-K* spectrum after extracting the reflected wave (theoretical model II)

Extracting the direct wave shown in Fig. 3.97, it can be seen from the figure that the extracted waves are not entirely direct wave, but also the reflected wave of the abnormal body at P_1. This is because the apparent velocities of the reflected wave and the direct wave are basically the same, and the *F-K* filtering is based on the difference of apparent velocity to achieve the purpose of separating different types of waves, so the two waves cannot be completely separated. From the extracted direct wave, it is easy to calculate the wave velocity of the longitudinal wave at about 3771 m/s, and in the three-dimensional imaging, the position of the abnormal body can be estimated based on the velocity of the longitudinal wave.

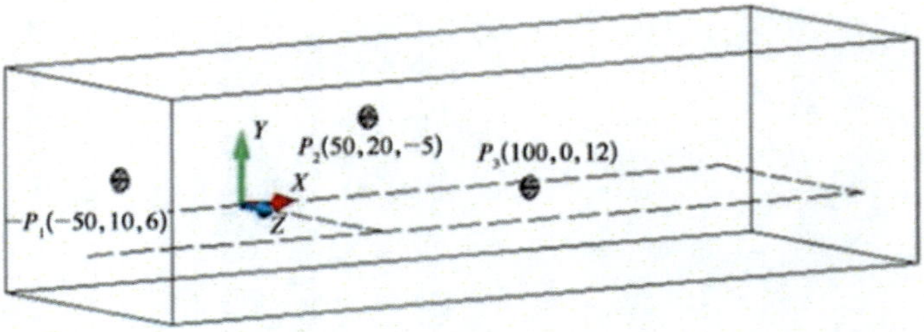

Fig. 3.93 Schematic diagram of Model III

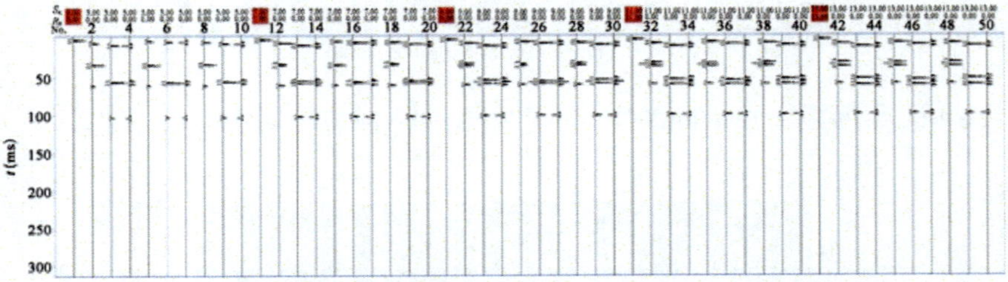

Fig. 3.94 Theoretical model without noise

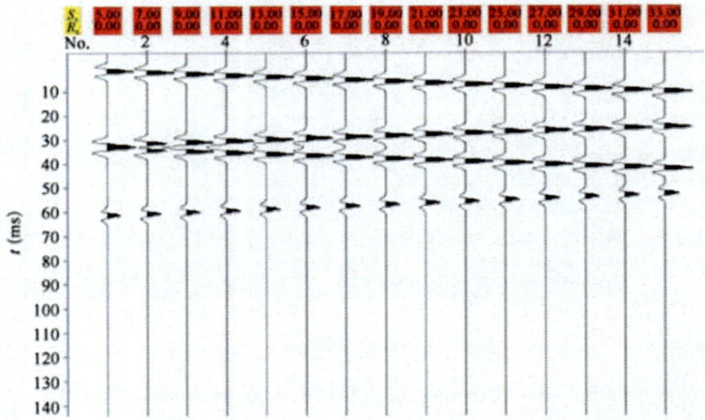

Fig. 3.95 Longitudinal wave profile after separation of longitudinal and transverse waves

Figure 3.98 shows the reflected longitudinal waves at P_2 and P3 extracted. The transverse wave analysis after separation is performed below.

Figure 3.99 shows the transverse wave profile obtained after the separation of the longitudinal and transverse waves. The velocity of the transverse wave is lower than that of the longitudinal wave. It can be seen from the figure that the time for the transverse wave to reach the geophone is slightly longer than that of the longitudinal wave.

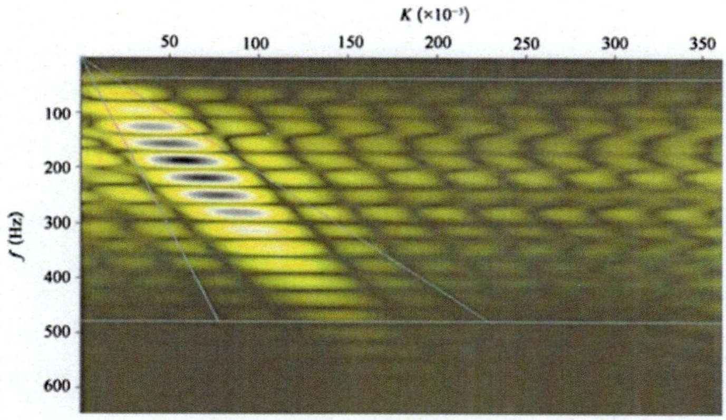

Fig. 3.96 *F-K* spectrogram of longitudinal wave profile

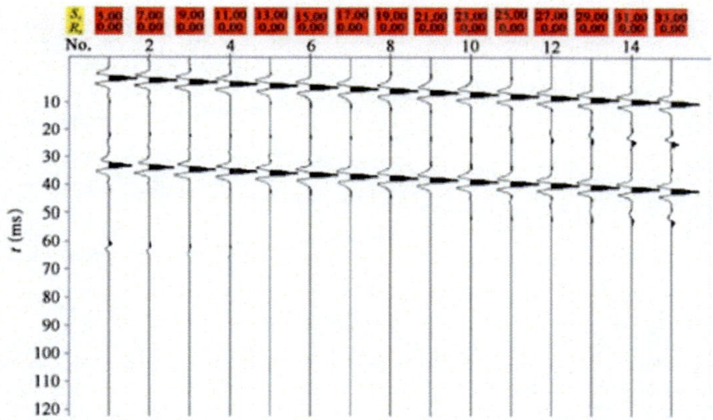

Fig. 3.97 Direct wave extracted by *F-K* filtering (from the longitudinal wave profile of theoretical model III)

Figure 3.100 shows the *F-K* spectrum of the transverse wave profile. By setting different filtering ranges, different types of waves can be obtained separately.

By setting different filtering ranges, the direct wave of the transverse wave as shown in Fig. 3.101 can be extracted. It can be seen from the figure that the extracted waves are not entirely direct wave, but also the reflected wave of the abnormal body at P_1. This is because the apparent velocities of the reflected wave and the direct wave are basically the same, and the *F-K* filtering is based on the difference of apparent velocity to achieve the purpose of separating different types of waves, so the two waves cannot be completely separated. From the extracted direct wave, it is easy to calculate the wave velocity of the transverse wave at about 2460 m/s.

Figure 3.102 shows the reflected transverse waves at P_2 and P3 extracted.

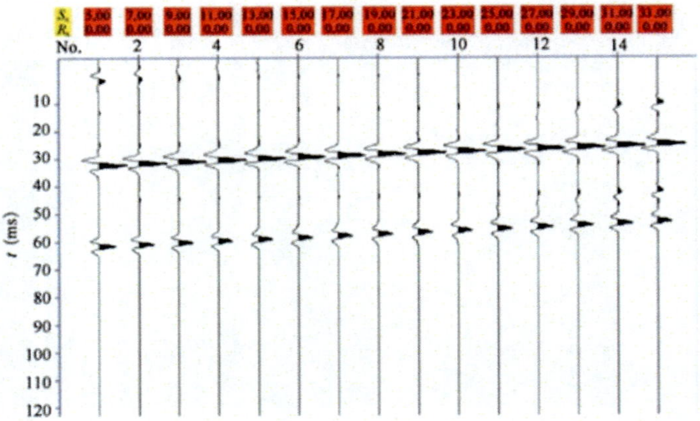

Fig. 3.98 Reflected wave extracted by *F-K* filtering (from the longitudinal wave profile of theoretical model III)

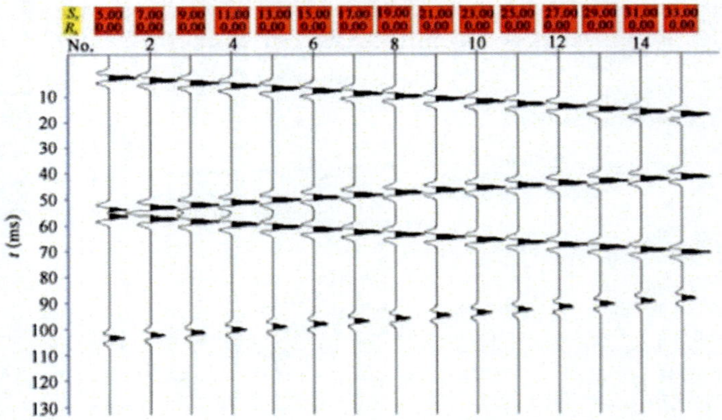

Fig. 3.99 Transverse wave profile after separation of longitudinal and transverse waves

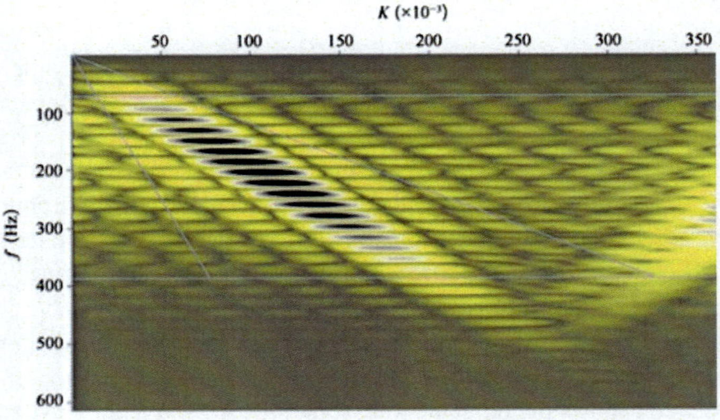

Fig. 3.100 *F-K* spectrum of transverse wave profile

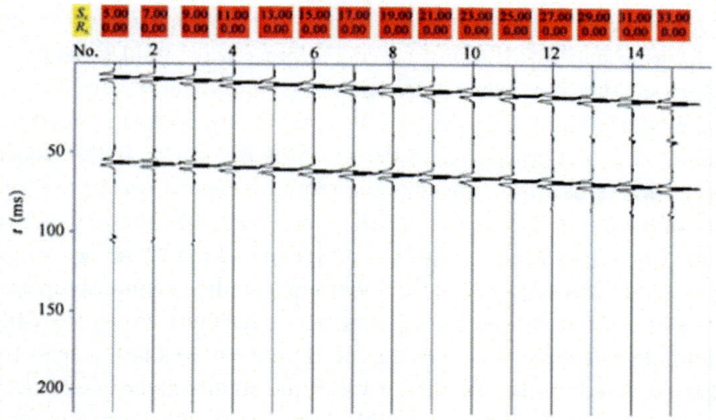

Fig. 3.101 Direct wave extracted by *F-K* filtering (from the transverse wave profile of theoretical model III)

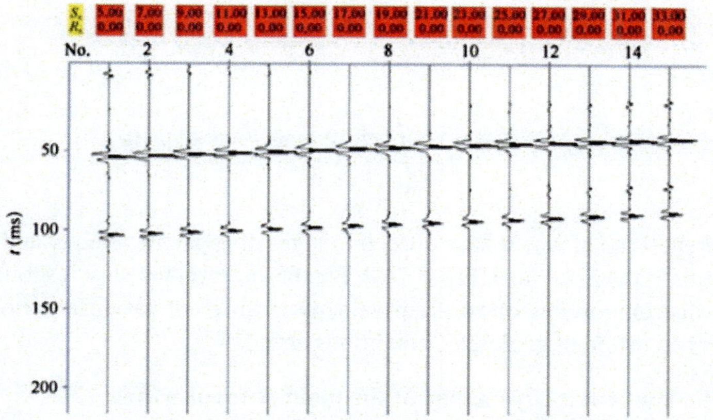

Fig. 3.102 Reflected wave extracted by *F-K* filtering (from the transverse wave profile of theoretical model III)

3.3.1.3 Three-Dimensional Wave Field Separation Technology in Tunnel Surrounding Rock

For tunnel seismic advance detection, except for the reflected waves from the anomalous body in front of the working face, other signals are interference signals. These interference signals include direct waves, surface waves, scattered waves, and reflected waves generated by the tunnel face, and reflected waves generated by abnormal bodies behind the tunnel face. In the data processing of advanced seismic detection, the wave field must be separated first to filter out interference signals such as surface waves and reflected waves in other directions. Based on the separation of the wave field, wave field separation, velocity analysis and migration imaging of the forward reflected waves can be performed. In the tunnel detection environment, the *F-K* filtering and *T-P* filtering based on linear wave field separation are more suitable for tunnel advance detection, and the effect of identification and separation of the wave field is better. In particular, the spatial observation method of the three-dimensional seismic wave advance detection of the tunnel with three-way offset proposed in this book fully considers the requirements of wave field separation in the design of the observation system. *F-K* filtering and *T-P* filtering are two effective filtering methods. However, due to the complexity of the tunnel wave field and the limitations of the filtering method itself, the single method cannot completely achieve the wave field separation. Therefore, in the actual data processing, *F-K* filtering and *T-P* filtering are often used to filter the interference waves in the seismic record to realize the extraction of the reflected wave of the complex wave field in the tunnel.

The principle of *F-K* filtering is based on the two-dimensional Fourier transform. For the two-dimensional seismic profile *g* (*t*, *x*), the corresponding frequency wave number spectrum *G*(ω, *kx*) can be obtained by two-dimensional Fourier transform.

$$G(\omega, k_x) = \frac{1}{2\pi} \int_{-\infty}^{\infty} g(t, x) e^{-j(\omega t - k_x x)} \mathrm{d}t \mathrm{d}x \tag{3.49}$$

$$g(t, x) = \frac{1}{2\pi} \int_{-\infty}^{\infty} G(\omega, k_x) e^{j(\omega t - k_x x)} \mathrm{d}\omega \mathrm{d}k_x \tag{3.50}$$

Different types of waves have different ranges of apparent velocity and frequency distribution in the *F-K* plane (Fig. 3.103). For the apparent velocity, it should be noted that the distribution law of apparent velocity is different for different observation methods and different geological conditions ahead.

(1) **The law of velocity change of the main types of waves**

① The velocity of surface wave is generally between 100–800 m/s.

Fig. 3.103 Schematic diagram of the distribution of the apparent velocity of the signal in the *F-K* domain

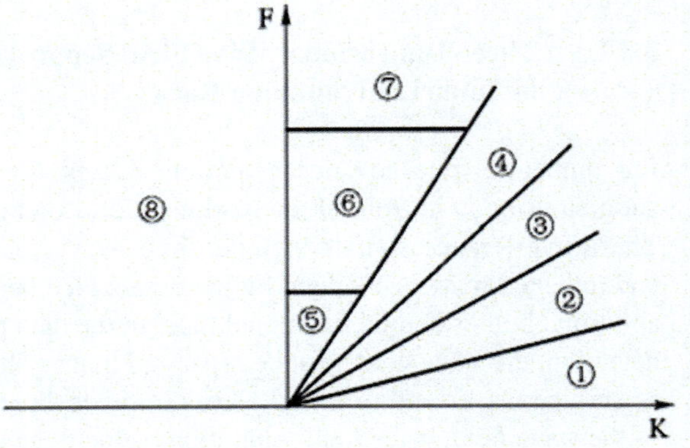

② The velocity of sound waves is about 300 m/s.
③ The velocity of direct wave is between 400–1500 m/s.
④ The velocity of multiple waves is between 800–2000 m/s.
⑤ The velocity of the converted wave is between 800–4000 m/s.
⑥ The velocity of the refracted wave is between 1500–5000 m/s.
⑦ The velocity of the reflected wave is generally greater than 2000 m/s.

The apparent velocity of each type of wave should be determined according to the actual situation (observation method, etc.).

(2) **Frequency distribution law of various seismic waves**

① The frequency of surface wave is generally 5–60 Hz.
② The frequency of sound wave is generally greater than 20 Hz.
③ The interference frequency of wires and cables is around 50 Hz.
④ The frequency of direct transverse wave is generally 15–70 Hz.
⑤ The frequency of refracted wave is generally 20–60 Hz.
⑥ The apparent velocity of the refracted wave is 1500–5000 m/s.
⑦ The frequency of reflected wave is generally greater than 15 Hz.

3.3.2 Imaging Theory of Three-Dimensional Tunnel Advance Geological Prediction

In order to obtain the most intuitive survey results, seismic reflection data needs to be processed by seismic imaging. The relevant imaging technology is relatively mature in petroleum exploration applications. The whole process of petroleum seismic signal processing includes energy gain, deconvolution, static correction, velocity analysis, and co-center point superposition. Finally, structural imaging is performed by offset and other methods. However, for the object of advance prediction in the tunnel, especially in most cases, it is necessary to detect the irregular abnormal body. From the theoretical analysis, it is obvious that the direct application of petroleum seismic imaging technology often has some errors.

According to the principle of seismic wave propagation, in the signal processing process, if only the reflected signal is considered, and the generated scattered wave is suppressed or eliminated as an interference signal, it will inevitably lead to the loss of many useful information. Moreover, the seismic resolution requirements of various engineering problems are much higher than those of petroleum exploration. If the shallow seismic reflection wave method directly draws on the petroleum seismic imaging method for imaging processing, the ideal effect will not be obtained. Therefore, it is necessary to comprehensively use seismic exploration, waveform signal analysis, probability statistics, and computer programming techniques and simultaneously perform combined imaging processing on shallow seismic reflection signals and scattered signals to achieve efficient and high-precision imaging, and actively avoid the error caused by dynamic and static correction. Further research on signal acquisition and data processing is required to design a practical shallow seismic imaging system. This research obviously has great academic and practical significance. The data processing system for 3D imaging geological advance prediction is based on seismic diffraction scanning superimposed migration imaging. Based on the common reflection surface element (CRS) superposition technique, the diffraction wave in seismic record is processed. This section introduces the principle of diffraction scanning superimposed migration imaging and the common reflection surface element superposition technique.

3.3.2.1 Diffraction Scanning Superposition Migration Imaging Principle

The ray theory is the theoretical basis of the diffraction scan offset superposition, and its essence is to automatically shift the reflected wave to its true position in space.

The method first divides the area of exploration into a grid and scans them one by one along the grid points. Assuming that a certain scanning point *P* is a reflection point, it should be the first to observe which geophones on the ground can record the reflected wave of the reflection point at what time. It is apparent that the position and recording timing of the geophone that receives the reflected wave of the reflection point on the ground is related to the inclination angle φ of the reflection interface of the point *P*, as shown in Fig. 3.104.

If $\varphi = 0$, then the detection point R_j that received the reflected wave is symmetric with respect to the shot point S_i relative to PP' (*P'* is the vertical projection point of

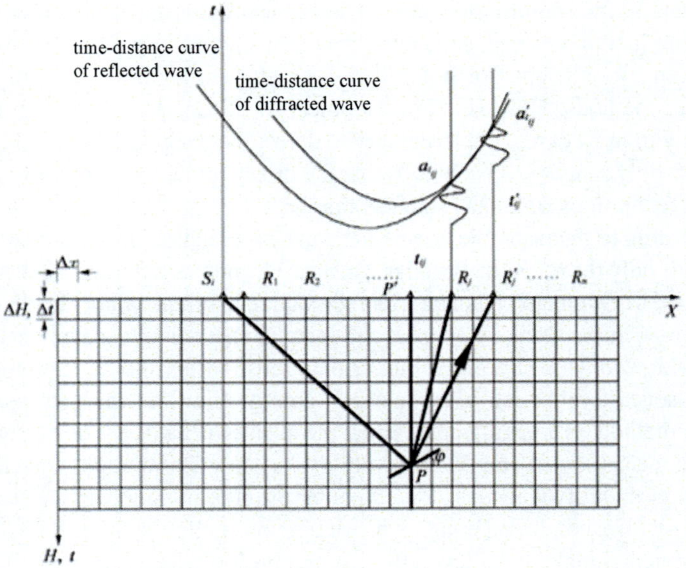

Fig. 3.104 Scanning offset superposition principle

the P point on the ground), and when the reflected wave travels,

$$t_{ij} = \frac{S_i P + PR_j}{V} \tag{3.51}$$

If the reflection interface is tilted to the left ($\varphi > 0$), the geophone R_j receiving the reflected wave of the point will be shifted to the left. Otherwise, if the reflection interface is tilted to the right ($\varphi < 0$), the geophone R_j receiving the reflected wave of the point will be shifted to the right. However, regardless of the inclination angle φ, the travel time of the reflected wave can be expressed by Eq. 3.52.

$$t_{ij} = \frac{\sqrt{H^2 + (X_P - X_{S_i})^2} + \sqrt{H^2 + (X_{R_j} - X_P)^2}}{V} \tag{3.52}$$

where H—the vertical depth of the scanning point P;

X_p, X_{Si}, X_{Rj}—the abscissa of the scanning point P, the shot point $S_{i,}$ and the geophone $R_{j,}$ respectively.

t—the vertical round trip time of scanning point P.

Equation (3.51) shows that the relationship between the position of the geophone that receives the reflected wave of the point P on the ground and the recording time is a hyperbolic curve of the diffraction wave with the point P as the diffraction point.

For each point on the grid, the travel time t_{ij} of diffracted wave of each geophone R_j at a shot point S_i is calculated by using Formula (3.51). Then, the amplitude value $a_{t_{ij}}$ is extracted at the corresponding t_{ij} on the record trace R_j, and put it on point P. A memory unit is allocated to each grid point on the grid, and the total amplitude value is obtained by superposition of all amplitude values f on the same scanning point P.

$$a_\Sigma = \sum_{j=1}^{n} (a_{t_{ij}})_P \tag{3.53}$$

where n—the number of tracks per shot.

If the scanning point P is a reflection point on the reflection interface, no matter how large φ is, at the geophone R_j that receives the reflected wave of the point P, the time interval curve of the diffracted wave of the point P and the time interval curve of the reflected wave of the point P are always tangent. At the tangent point R_j, the travel time of the diffracted wave of the point P is the same as the travel time t_{ij} of the reflected wave. Near the tangent point R_j, the travel time of the diffracted wave of the point p is similar to the travel time t_{ij} of the reflected wave. Therefore, the time–distance curve [Eq. (3.51)] of the diffracted wave is used to calculate the t_{ij} of the diffracted wave. The amplitude value $a_{t_{ij}}$ at the time of the t_{ij} is extracted from the record track R_j, and all the amplitude values at the same scanning point are superimposed so that the total amplitude value a_Σ is obtained, and the offset superposition is completed. Since the reflected waves of the record tracks near the geophone R_j are substantially in phase, the superimposed total amplitude value a_Σ

will increase. On the other hand, if the scanning point P is not a reflection point, the reflected waves of the record tracks in the vicinity of the geophone R_j are different phases, so that the amplitudes after the superposition have inevitably canceled each other, and the total amplitude value $a_{\sum}$ is relatively reduced.

If the data is processed multiple times, the corresponding amplitudes of all the record tracks of different shots are superimposed according to the above method to obtain the total amplitude value.

$$a_{\Sigma} = \sum_{i=1}^{m} \sum_{j=1}^{n} \left(a_{t_{ij}}\right)_P \tag{3.54}$$

where m—the total number of shots.

If the point P is a diffraction point, the travel time the diffracted wave is calculated by Eq. (3.54). The amplitude value $a_{t_{ij}}$ at the time t_{ij} is extracted on the record track R_j. The amplitude values of the same point P are superimposed to get the total amplitude value.

$$a_{\Sigma} = \sum_{j=1}^{n} \left(a_{t_{ij}}\right)_P \tag{3.55}$$

Since the same diffracted wave received by each track is superimposed in phase at point P, the total amplitude value a_{Σ} will increase. On the contrary, if the point P is not a diffraction point, since the amplitudes of the record tracks are not in phase, they cancel each other out in the process of random amplitude superimposition, and the total amplitude value after superimposition is relatively reduced.

At the same time, the data is covered multiple times, and all the record tracks of different shots are superimposed to obtain the total amplitude value.

$$a_{\Sigma} = \sum_{i=1}^{m} \sum_{j=1}^{n} \left(a_{t_{ij}}\right)_P \tag{3.56}$$

Therefore, in the depth and time profile superimposed by the scanning method, the total amplitude value increases where there are reflection points and diffraction points, and the total amplitude value decreases relatively where there are no reflection points or diffraction points, and the diffracted wave automatically converges to the diffraction point.

After processing by the above method, the total amplitude increases naturally where there is a diffraction point and decreases relatively naturally where there is no diffraction point. Therefore, the diffracted wave automatically converges to the diffraction point, showing the position of the real reflection interface.

3.3.2.2 Common Reflection Surface Element Superposition Technique

The idea of co-reflecting surface element superposition is firstly derived from the concept of radius of curvature R_{NIP} of the normal incident point wave proposed by Professor Hubral. Afterward, the common reflection element (CRE) superposition method based on two parameters and the CRS superposition method based on three parameters optimization were put forward successively in the two-dimensional case.

Generally speaking, the seismic wave received by the surface does not come from a certain reflection point in the underground, but from a Fresnel zone that can produce constructive interference reflection waves. All common reflection point gathers from the same Fresnel zone are called common reflection surface element gathers. Common reflection surface element superposition is to superimpose the reflection point signals in the common reflection surface element gather from the same Fresnel zone to obtain the zero-offset profile from the multiple coverage data, expand the number of superpositions, enhance the amplitude value of the effective signal, and improve the signal-to-noise ratio of the signal.

As shown in Fig. 3.105, the lower part of the figure is the curved reflection interface. Above the interface is the constant velocity medium. If multiple coverage data is collected according to the direction of the seismic line above the model, different common inspections can be obtained. From the time–distance curve, all the common offset time–distance curves form a common offset time–distance surface.

The theoretical basis of the common reflection surface element superposition is the ray theory. In Fig. 3.106, for the reflection point R, the common reflection surface element superposition is to superimpose the multiple coverage observation data from a tangent arc of point R (the second-order approximation of the reflection interface of point R) according to the travel time relationship described by the common reflection surface element superposition surface (the green part in Fig. 3.106).

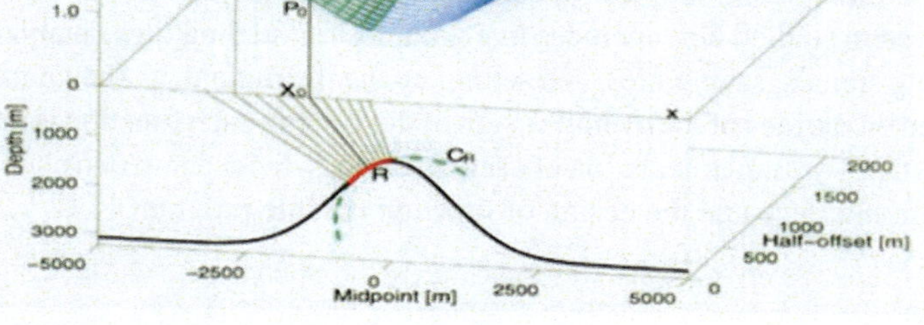

Fig. 3.105 Diagram of CRS superposition principle

The zero-offset travel time equation based on wave front curvature mainly includes parabolic travel time equations and hyperbolic travel time equations. In the semioffset coordinate system, the parabolic travel time equation based on the wave front curvatures $K_{\mathrm{N}}, K_{\mathrm{NIP}}$ and the incident angle α is a parabolic travel time approximation:

$$t(x_m, h) = t_0 + \frac{2\sin\alpha}{v_0}(x_m - x_0) + \frac{\cos^2\alpha}{v_0}\left[\frac{(x_m - x_0)^2}{R_{\mathrm{N}}} + \frac{h^2}{R_{\mathrm{NIP}}}\right] \tag{3.57}$$

Hyperbolic travel time approximation:

$$t^2(x_m, h) = \left[t_0 + \frac{2\sin\alpha}{v_0}(x_m - x_0)\right]^2 + \frac{2t_0\cos^2\alpha}{v_0}\left[\frac{(x_m - x_0)^2}{R_{\mathrm{N}}} + \frac{h^2}{R_{\mathrm{NIP}}}\right] \tag{3.58}$$

where x_{m}—midpoint coordinate;

h—half offset coordinate;

t—travel time;

(x_0, t_0)—any point in the profile with zero offset.

It is known from the above equation that CRS superposition can be realized by obtaining the values of the three parameters (α, R_{N}, R_{NIP}). Therefore, the key to CRS superposition is to optimize the three parameters (α, R_{N}, R_{NIP}).

3.4 Three-Dimensional Tunnel Geological Advance Prediction System

3.4.1 Signal Acquisition System of Three-Dimensional Tunnel Geological Advance Prediction

The design and development of the instrument system follows a design idea of "from the whole to the part, then from the part to the whole." Firstly, based on the requirements of an overall high-level analysis system, the design system needs to implement submodules, and then program and develop each function separately. After the functions of each submodule of the system are realized, they are connected by programming control to form a complete data acquisition system. The data acquisition system studied here includes five modules for reading files, analyzing signals, collecting signals, saving files, and setting up the instrument, as shown in Fig. 3.106.

Based on the idea of the overall design of the system, each function is implemented step by step. The implementation of each function is divided into two parts: the design of system interface and the design of function module program.

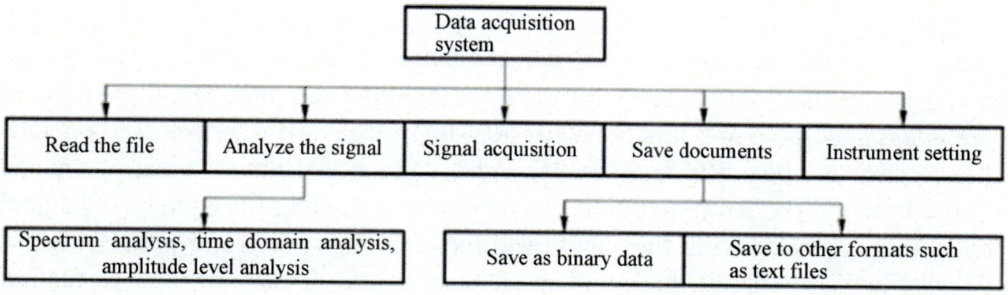

Fig. 3.106 Component module of data acquisition system

3.4.1.1 System Interface Design

The system interface is the window that the user ultimately faces directly, and it is the platform for the user to interact with the computer. There are two kinds of users to consider when designing the interface. One is the end user. The end user only faces the various functions provided by the front panel. The other is the program developer. The program developer not only faces the functions of the system interface, but also the modular subroutines. For the interface that the end user faces, there are two aspects to consider when designing. On the one hand, whether the interface is simple and easy to use and whether the end user can quickly locate the functions they need. On the other hand, aesthetics should also be considered. A good interface can give people a pleasant feeling, so that users can face the program interface designed by programmers for a long time without feeling bored or tired. For the modular subroutine panel, because the developer pays attention to the subroutine interface and the implemented functions, the front panel controls are required to be classified reasonably and neatly arranged. Figure 3.107 shows the main interface of the system.

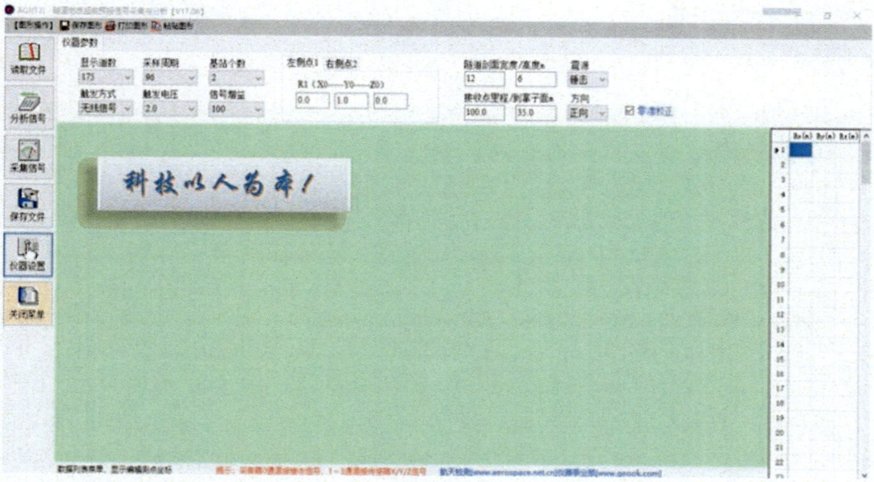

Fig. 3.107 Main interface of the system

The main interface of the system is a very important part of the system design. It integrates all the functions of the system. Conciseness and convenience are the characteristics of the front panel of this interface. In this main interface, three functions are integrated, including function selection, login welcome interface, and operation flow, which are beautiful and simple, and the functional flow is clear at a glance. In the function concentration area, the functions of reading files, analyzing signals, collecting signals, saving files, and instrument settings required by the system are listed. The right side of the interface is set as a data display window to facilitate timely viewing of data. When the waveform is displayed in the window, in order to be simple and concise, the full-screen display of the waveform can be achieved by clicking "Close Menu."

3.4.1.2 Function Module Program

The design of the function module program is the core part of the system development, including five modules for reading files, analyzing signals, collecting signals, saving files, and instrument settings. The following is a brief introduction to the design ideas, corresponding requirements, and functional implementation of the five module programs.

(1) **Module for reading files**

Execute the function of reading files and enter the signal selection window, as shown in Fig. 3.108. First select the folder, determine the signal file, and then read the signal record for which analysis is to be performed. Pressing the Ctrl key + clicking the mouse, or using the Shift + ↑/↓ key combination can select multiple records at the same time for processing. If a record is needed to be added in the processing process,

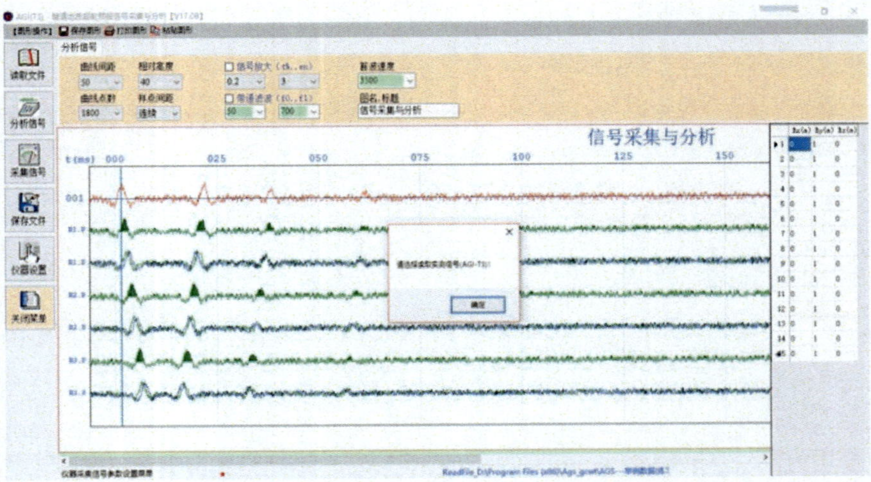

Fig. 3.108 Interface for reading binary files

add record can be clicked. After the record is read and returned, various methods can be used to analyze and process the signal.

(2) **Module for signal analysis**

The measured signal data is very complex, and only the effective signals that are beneficial to the research are cared about. In order to make the result closer to the real effect, the signal should be analyzed before data processing to filter out the noise unrelated to the effective signal. The function of the module for signal analysis is to retain the effective signal and eliminate the noise. When the specific function is implemented, the analysis methods such as spectrum measurement analysis, distortion measurement analysis, and amplitude level measurement of the signal are used. It is required to list the image name, curve spacing, relative height, number of curve points, parameter setting of sample point distance, display window, signal amplification, and band-pass filtering options in the analysis signal window.

Each subfunction module is relatively independent. A system must be an organic whole, requiring each system to be organically linked through a relationship. Structures and global variables in the C# language are needed here. Structures must be used in every slightly complex complete program, and this system is no exception. It is these structures that link the scattered subfunction programs together. In this system, the loop structure, the selection structure, the condition structure, and the sequence structure are mainly used. In addition, the flexible use of global variables can make system development save a lot of troublesome steps and can make the system run smoothly and not easy to make mistakes. These programming techniques are used to integrate the various functional modules of the system to form a comprehensive program flow diagram, as shown in Fig. 3.109.

After the data acquisition is completed, clicking the "Analyze Signal" button can get the interface for analyzing signal, as shown in Fig. 3.110. In the module of data analysis in this window, the collected signals can be analyzed and processed with curve spacing, waveform height, curve number, sample point spacing, etc., which is convenient to view the quality of the collected signal.

(3) **Module for signal acquisition**

The module for signal acquisition is the most important part in the design of signal acquisition software. The design idea can be roughly divided into three parts.

(1) Parameter setting function. In this section, the acquisition method of the signal acquisition system is given, that is, the layout of the measurement points in the work area, as shown in Fig. 3.111. According to the arrangement of the schematic diagram, the parameters such as the recording start point, the detection interval, the moving step, the surface wave offset, the reflection offset, and the base station arrangement are, respectively, set. The schematic diagram of the detection arrangement makes the arrangement of the measuring points more vivid. "Collection," "retest," "continue," "save," "clear," and "stop" buttons are also required to facilitate timely preservation in the process of data collection and retest of bad data, to achieve the effect of obtaining high-quality data through manual control.

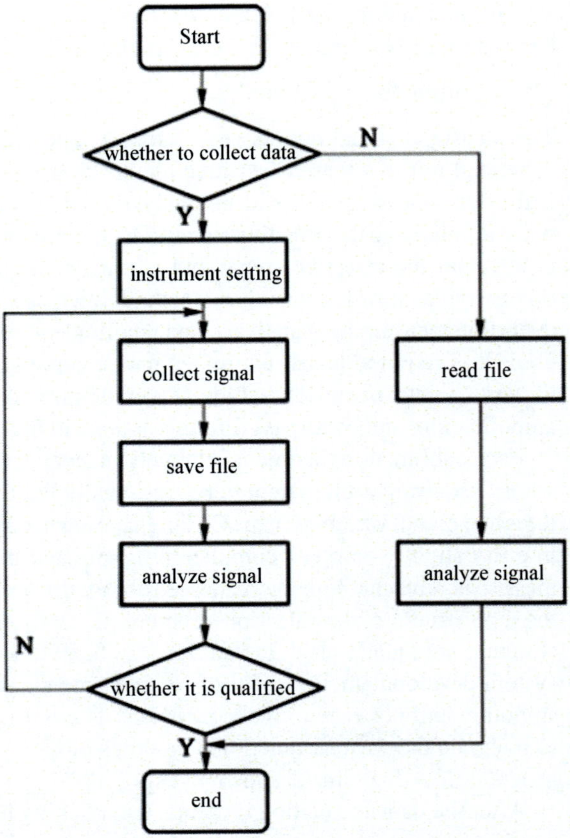

Fig. 3.109 Integrated block diagram

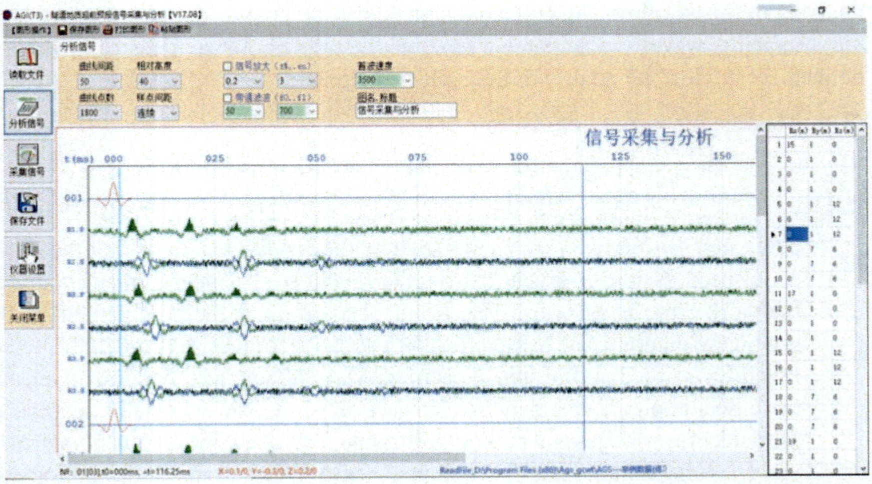

Fig. 3.110 Interface of module for signal analysis

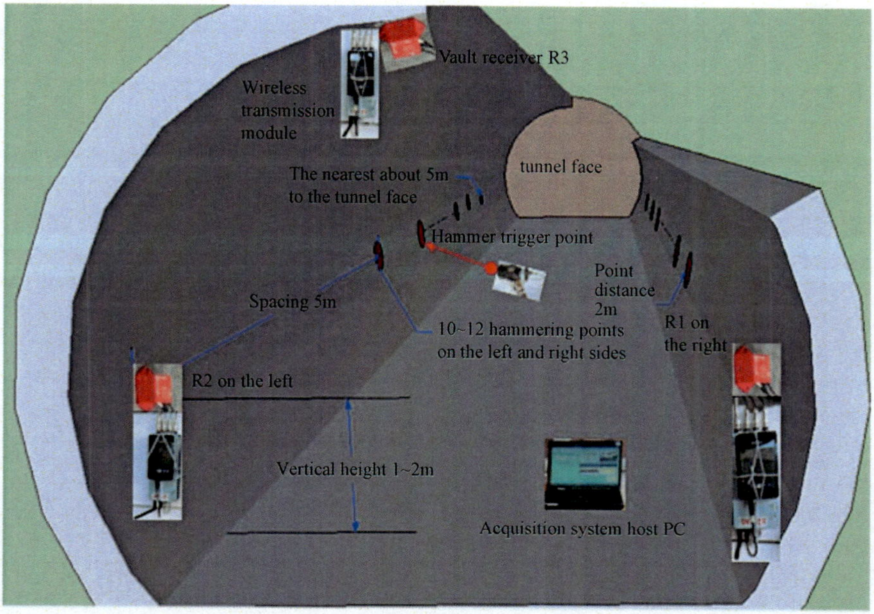

Fig. 3.111 Schematic diagram of the layout of the observation system in the tunnel

(2) The design for graphic display window. It is used to visually display the received waveform data. Due to the different speeds of *P*-wave and *S*-wave, the different propagation modes of SH-wave and SV-wave, *P*-wave, and *S*-wave are usually separated in data processing. The acquisition system uses a three-component geophone to receive signals. The signal components include *P*-wave, SH-wave, and SV-wave. The system displays the received coordinates of the three types of waves in the three geophones. In addition, according to the design features of the detection system, it is also required to display the waveform of the acceleration sensor to provide a trigger time standard for data processing.

(3) When starting to collect data, clicking the "Acquisition" button can switch to the acquisition interface for signal acquisition. When the acquisition stops, the waveform graph can be displayed on the window, as shown in Fig. 3.112. In the waveform display diagram, the P-wave is displayed separately, and the SH-wave and SV-wave are displayed in different colors on the same coordinate axis.

(4) **Module for file saving**

Data storage is also a very important part of the data acquisition system, especially in real-time monitoring systems, where many continuous data is saved to the hard disk before the user has time to observe or apply it, in preparation for future use.

In the design of data saving function of this system, only the measured signal data is saved as binary data. The interface of data saving is given, and the user can choose

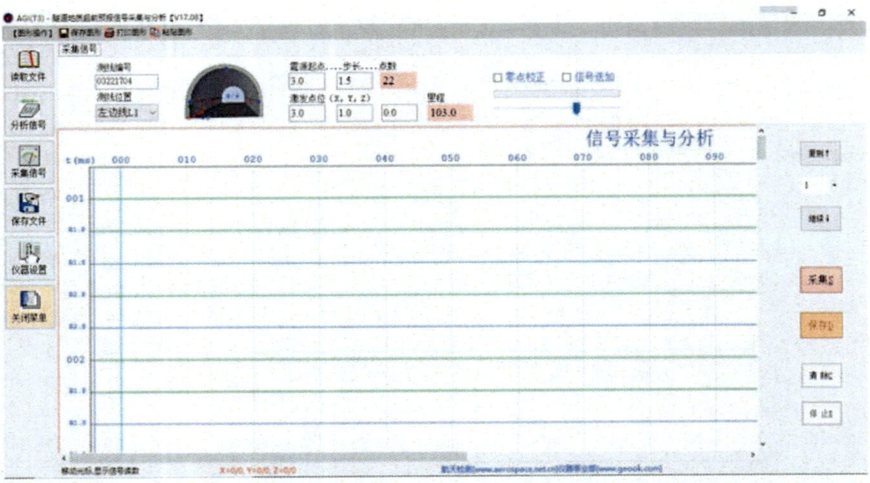

Fig. 3.112 Interface of module for signal acquisition

the path to be saved. After inputting the file name, the data will be saved as binary files on the hard disk.

(5) **Module for instrument setting**

The module for instrument setting is used to collect signals in this system. Different work areas have their own characteristics and different requirements for measurement. Therefore, before collecting signals, the parameters of different work areas should be set accordingly. These settings include the number of display channels, the sampling period, the number of base stations, the trigger mode, the trigger voltage, the detection method, and the gain segmentation, as shown in Fig. 3.113.

3.4.1.3 Wireless Data Transmission Technology and Wireless Synchronization Technology

The wireless data transmission adopts virtual instrument technology and Wi-Fi technology and adopts the C# language with simple syntax, high flexibility, and compatibility to research and develop the instrument. On the basis of not reducing performance, it reduces wiring costs and improves flexibility to meet the needs of practical engineering applications such as wireless structure diagnosis, environment and machine status detection, and high-efficiency data collection.

The Wi-Fi signal has a radius of up to 300 ft, about 100 m, and the transmission speed can reach 54 Mbps. The tunnel geological prediction system uses Wi-Fi transmission signals, which ensures the quality of data transmission, reduces the time of survey line laying, improves the efficiency of data acquisition, and saves a lot of costs for the tunnel construction.

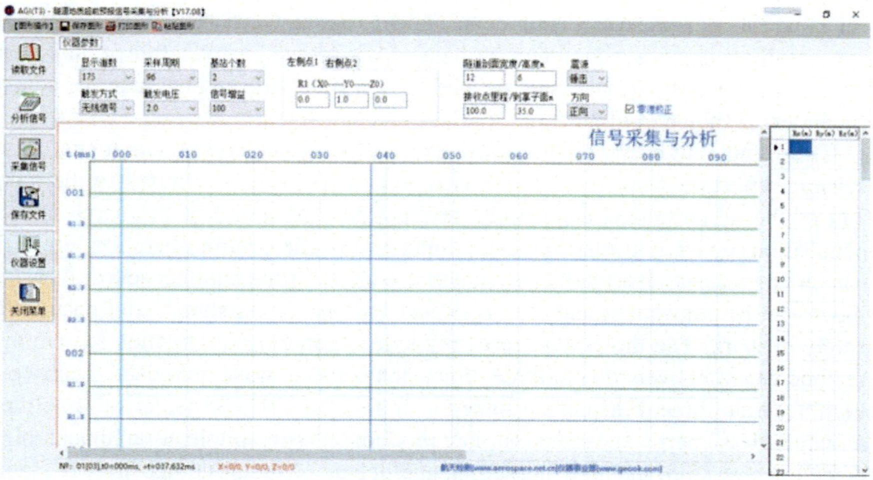

Fig. 3.113 Interface of module for instrument setting

The 3D tunnel advance prediction system uses a simplified network structure for synchronous three-component geophone triggering. The clock synchronization command is sent by the host, and the host system timestamp information is included in the command packet. When the collection substation and the trigger substation receive the command packet, the local timestamp information is recorded. After the clock synchronization is completed, the triggering substation stimulates the artificial seismic wave, and simultaneously triggers the substation to send a data acquisition command, which includes the system timestamp of the seismic wave excitation moment. After receiving the command, the collecting substation calculates the storage address of the collected data with the seismic wave excitation time in the buffer according to the time stamp in the data packet and the current system clock. Taking the address as the starting point, the subsequent data is valid seismic wave data.

Multiple three-component geophone synchronous trigger technology can ensure that different geophones receive seismic wave signals at the same time, which can accurately reflect the propagation time of seismic waves, more accurately locate the specific location of bad geological bodies and improve the accuracy of tunnel advance prediction.

3.4.1.4 Detecting Device of Three-Dimensional Tunnel Geological Advance Prediction

The tunnel environment is relatively harsh. The advanced geological prediction of the tunnel requires simple instrumentation, stable performance, high precision, and high efficiency. The AGI-T3 wireless distributed tunnel advance prediction detecting device comprises a control operation terminal, a wireless data acquisition device, a seismic reflection wave receiving device and a seismic wave excitation device, and specifically comprises a computer terminal, a wireless collector, a three-component geophone, a portable router, and an acceleration geophone and the hammer source (hammer) is shown in Fig. 3.114.

The output end of the seismic wave excitation device and the output end of the seismic reflection wave receiving device are, respectively, connected to the input end of the wireless data acquisition device. The output end of the wireless data acquisition device is connected to the input end of the control operation terminal. The use of wireless distributed instrument (24-bit A/D) and hammer source working mode can be convenient and fast to collect high-precision signals and ensure the safety of work. The three-point receiving side hammering observation system test method can effectively obtain three-dimensional elastic wave field data to carry out high accurate three-dimensional imaging. It also has the advantages of simplified structure, less energy consumption, low production cost, simple arrangement, high detection precision, and high measurement efficiency. Module composition of signal acquisition system is shown in Fig. 3.115.

The number of wireless data collection devices is three, including the first wireless data collection device, the second wireless data collection device, and the third wireless data collection device. The number of seismic reflection wave receiving devices is three, including the first seismic reflection wave receiving device, the second seismic reflected wave receiving device, and the third seismic reflected wave receiving device. The first wireless data collection device is connected to the first seismic reflection wave receiving device. The second wireless data collection device is connected to the second seismic reflection wave receiving device, and the third

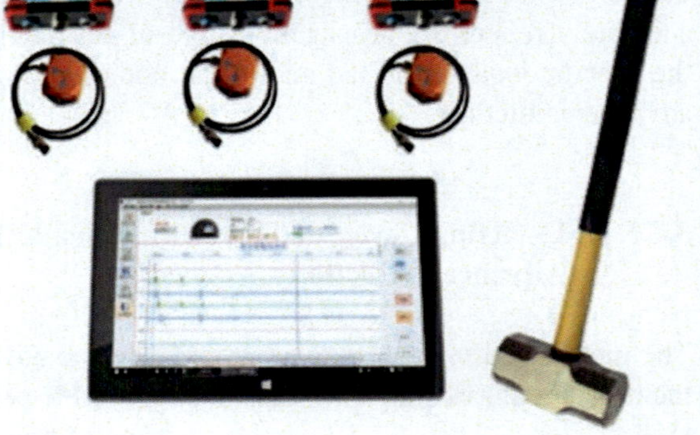

Fig. 3.114 Detecting devices of wireless three-dimensional tunnel advance prediction

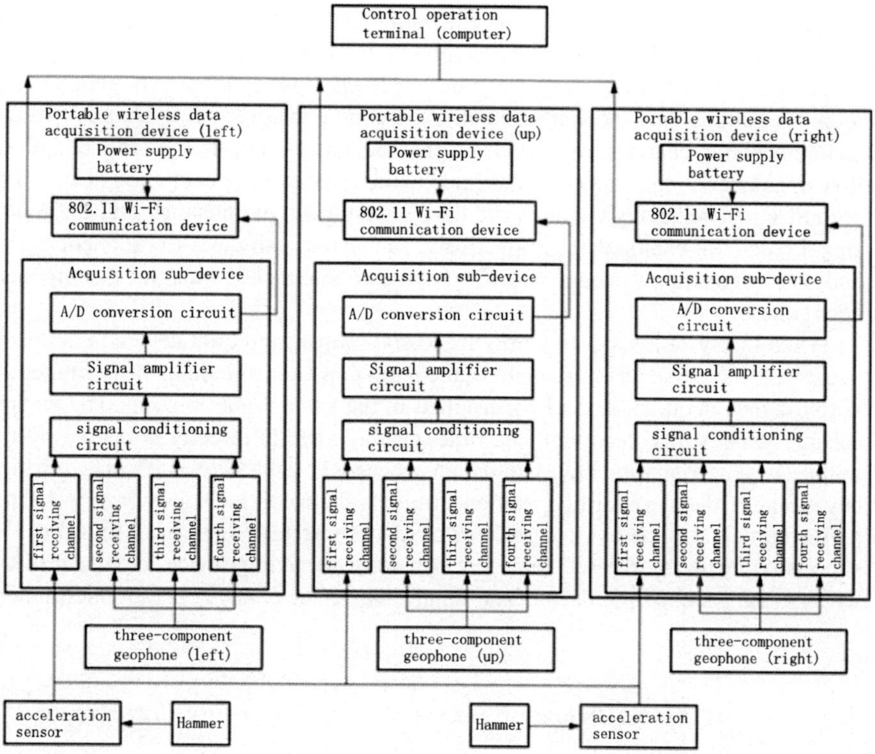

Fig. 3.115 Module composition diagram of signal acquisition instrument

wireless data collection device is connected to the third seismic reflection wave receiving device.

A first installation position, a second installation position, and a third installation position are, respectively, arranged on the left-side wall, the right-side wall, and the top wall of the same tunnel section of the excavated tunnel. The vertical distance between the first installation position and the second installation position from the bottom of the tunnel is 1 m. The third installation location is located at the apex of the tunnel.

There are two sets of seismic source marking points, which are, respectively, set at the same level with a vertical height of 1 m from the bottom of the tunnel on the left and right walls of the tunnel.

The collecting subdevice includes a signal receiving channel, a signal conditioning circuit, a signal amplifying circuit, and an A/D converting circuit. The input end of the signal receiving channel is, respectively, connected with the output end of the seismic reflection wave receiving device and the output end of the seismic wave excitation device. The output end of the signal receiving channel is successively connected with the input end of the A/D conversion circuit after the signal conditioning circuit and the signal amplification circuit. The output end of the A/D conversion circuit is connected with the input end of the communication device.

The seismic reflection wave receiving device is a three-component geophone. It includes a first signal receiving channel, a second signal receiving channel, a third signal receiving channel, and a fourth signal receiving channel. The input end of the first signal receiving channel is connected to the seismic wave excitation device. The second signal receiving channel, the third signal receiving channel, and the fourth signal receiving channel are, respectively, connected with the *X*-component output end, *Y*-component output end, and *Z*-component output end of the three-component geophone.

When the system starts to work, the seismic wave excitation device is first used to excite the seismic waves at each source mark point and simultaneously trigger the wireless data acquisition devices installed in the same tunnel section. The excited seismic wave splits into two. The first seismic wave is directly collected by the wireless data acquisition device and then processed by the wireless data acquisition device and sent to the control operation terminal. The second seismic wave is reflected by the tunnel face and its strong wave impedance interface to generate an echo. The echo is collected by the seismic reflection wave receiving device, then sent to the wireless data acquisition device, and finally sent to the control operation terminal.

3.4.2 Imaging Software of 3D Geological Advance Prediction Technology

(1) **Functional Overview**

The imaging software is designed for the 3D imaging tunnel geological advance prediction instrument system (AGI-T3). The software has the functions of receiving point gather programming, signal spectrum analysis, wavelet analysis, *F-K* two-dimensional filtering, and signal denoising, and for the preprocessed signal, it performs high-precision imaging of the geological structure and the distribution of abnormal bodies in front of the tunnel face based on the 3D imaging data model.

The main processes include:

① Performing 3D wave field separation to filter out the lateral wave and surface wave and retain the front echo.
② Performing longitudinal and transverse wave separation to ensure that the longitudinal and transverse waves are the echoes coming to the front.
③ Performing a scanning analysis of the surrounding rock velocity to determine the velocity distribution of the surrounding rock.
④ Performing migration imaging of geological structures based on the wave velocity of the surrounding rock.

(2) **Operating instructions**

① First inserting the USB dongle, and opening the basic interface of the software AGS (T3) through the [Start] menu, as shown in Fig. 3.116.

② Basic steps for importing raw data (Fig. 3.117).

③ Selecting "Coordinates Arrangement" to confirm whether the data field sending and receiving points are correct. If there is any problem, coordinate editing shall be carried out. The parameters editing is in the right column, including all record, selection record, single-selection record, wave field separation, coordinate proofreading, and point proofreading, as shown in Fig. 3.118.

(a) If the sending and receiving points are correct, the mileage is directly corrected (the purpose: The mileage can be matched with the tunnel construction mileage) and all arranged. Enter the mileage of the receiving point at the "Proofing Miles" and select the forward and reverse (import selection "forward," exit for "reverse"), and "select track number" for "programming classification—all" for separation of transverse wave (WP) and longitudinal wave ($S1$, $S2$) (preferably create a new folder for saving).

(b) If the sending and receiving points are incorrect, coordinate correction needs performing on the sending and receiving points before performing the operation of (b).

④ Reading the data for analysis, as shown in Fig. 3.119.

The signal column on the right side of the data analysis selection includes selection method, processing method (zero calibration, noise removal, waste track processing), and signal enhancement (equalization between tracks, equalization of profile).

⑤ Clicking "Spectrum Analysis," selecting the lithology wave velocity corresponding to the actual rock and soil on the left side, and setting one-dimensional or two-dimensional in the right column. The analysis method includes spectrum

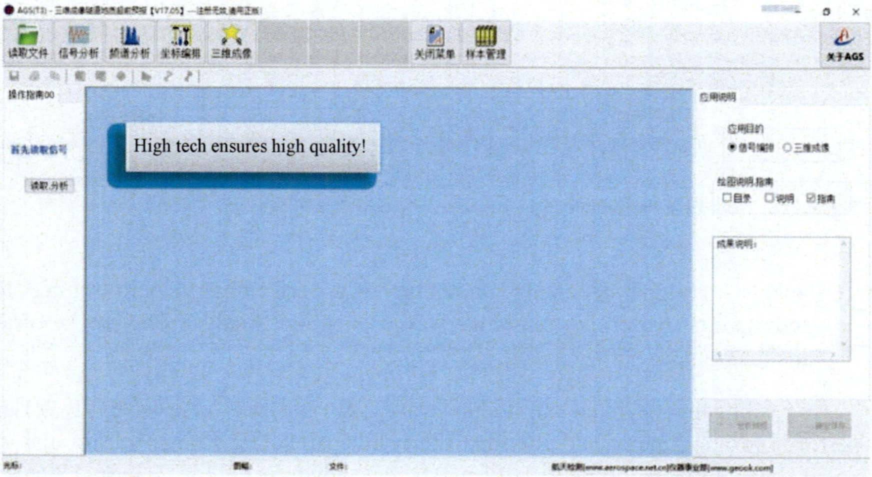

Fig. 3.116 A preview of the basic interface for AGS (T3)

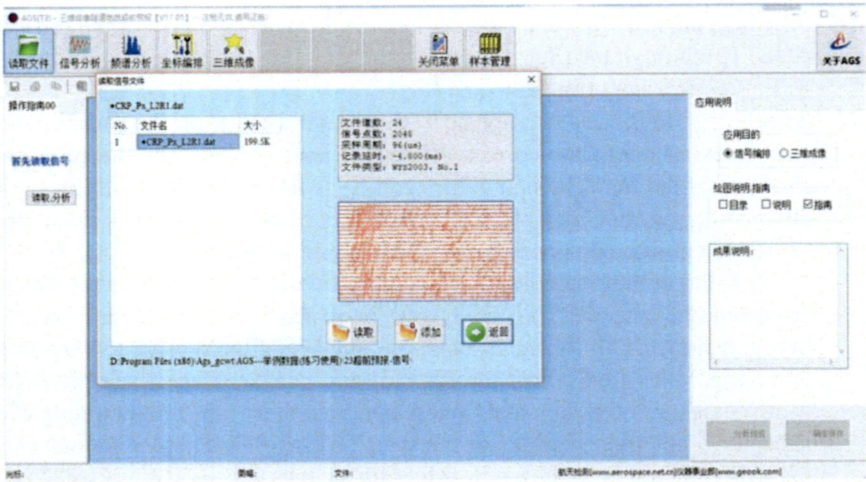

Fig. 3.117 Basic steps for importing data

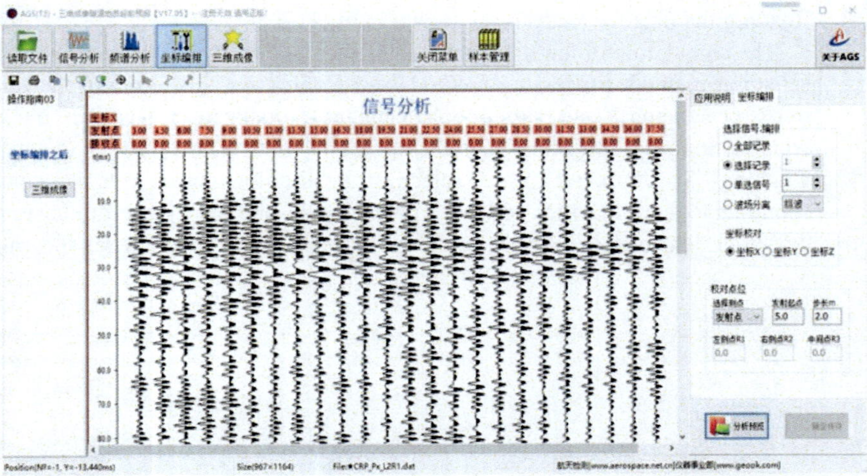

Fig. 3.118 Coordinate arrangement

analysis or signal denoising. Simultaneously, performing velocity settings for longitudinal wave and transverse wave. Clicking "Analyze Preview" to draw the diagram, as shown in Fig. 3.120.

Re-reading the separated longitudinal wave (WP) and transverse wave ($S1$, $S2$) and performing one-dimensional filtering (①one-dimensional $f0$–$f1$—spectrum analysis under the [frequency domain analysis] tab (as shown in Fig. 3.121). Selecting $F0$ and $F1$, respectively. ②clicking "band-pass filtering," confirming the filtering effect,

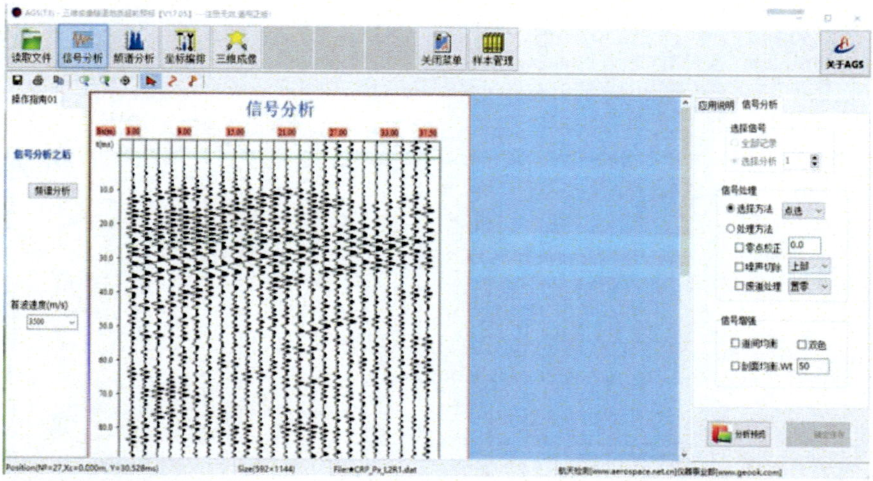

Fig. 3.119 "Analysis" Basic operation steps

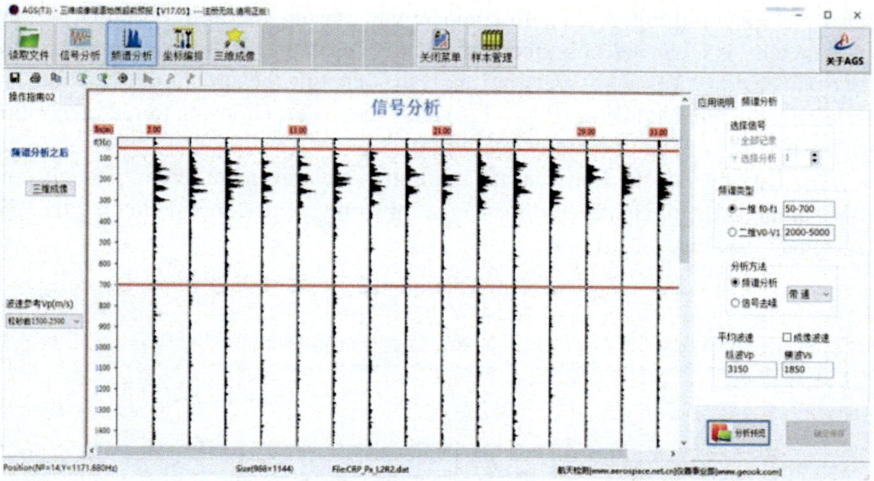

Fig. 3.120 Spectrum analysis

and then clicking "save" when satisfied) and two-dimensional filtering (① two-dimensional $V0$–$V1$—spectrum analysis, selecting $V0$–$V1$, respectively). ②Band-pass filtering: When "signal extraction direct" is selected, the direct wave is reserved, and the reflected wave is filtered. This option is generally used to check the quality of the direct wave of the original signal and calculate the velocity of direct wave, as shown in Fig. 3.122. When "signal extraction reflection" is selected, the direct wave is filtered out in order to retain the reflected wave. This option is generally used to check the quality of the reflected wave and the next step to generate the result

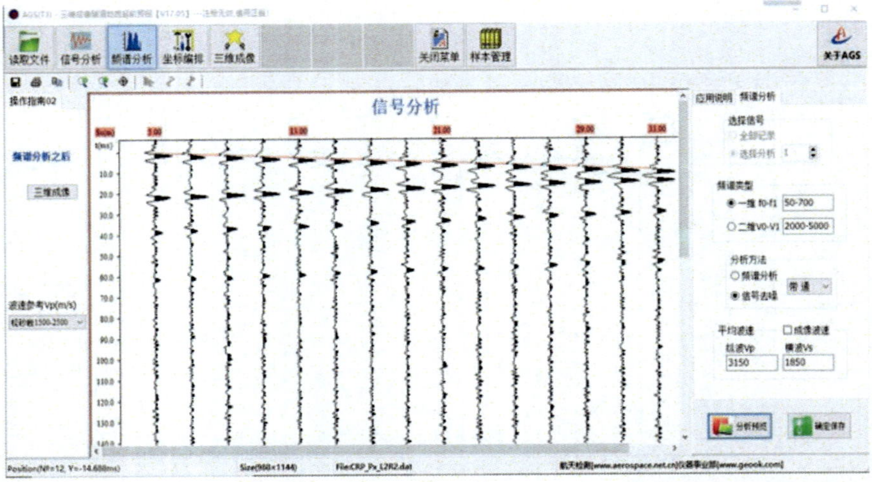

Fig. 3.121 One-dimensional filtering

chart, as shown in Fig. 3.123). To compare the filtering effects of different filtering parameters, you can select a more satisfactory result and click Save. Selecting the two-dimensional *V*0–*V*1—spectrum analysis, selecting the direct wave (Fig. 3.124), selecting the reflected wave (Fig. 3.125) and other tasks, choosing to save according to the filtering effect, or directly overwrite.

The data files of the longitudinal and transverse waves are read separately, and horizontal, vertical, and three-dimensional imaging are performed clicking the "3D imaging" tab (Fig. 3.126).

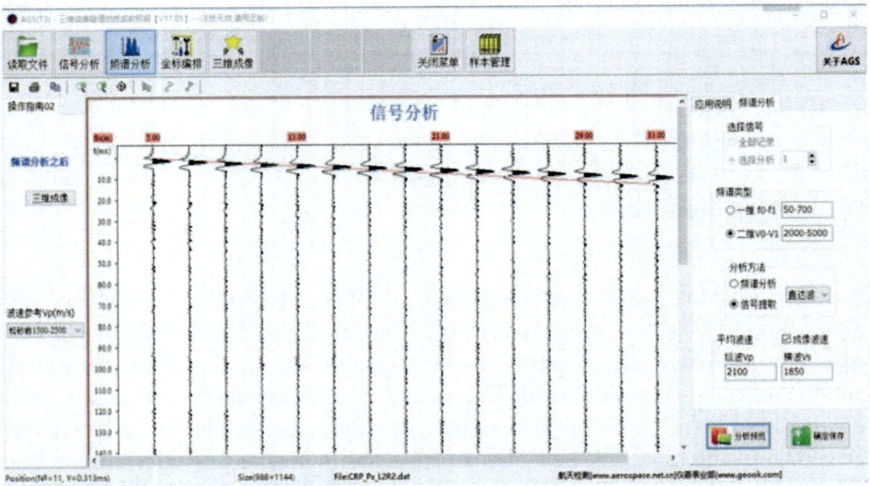

Fig. 3.122 Two-dimensional filtering effect diagram after selecting "direct wave"

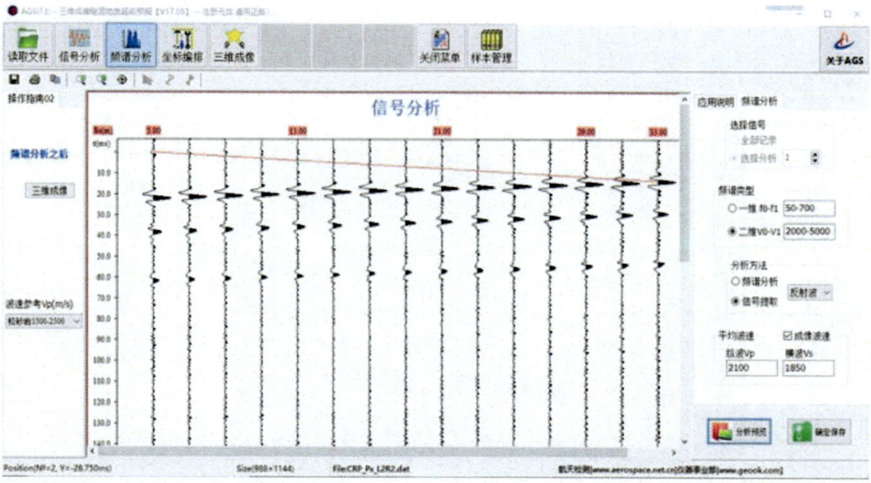

Fig. 3.123 Two-dimensional filtering effect diagram after selecting "reflected wave"

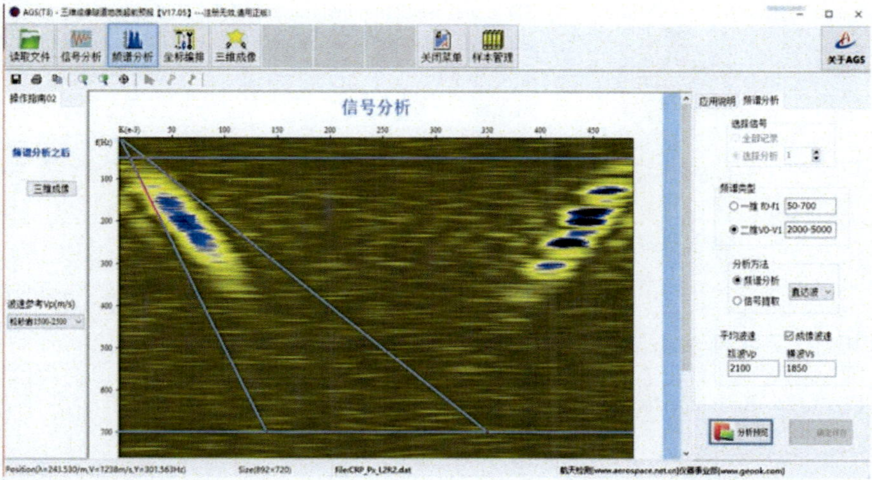

Fig. 3.124 Two-dimensional filtering spectrum analysis after selecting "direct wave"

3.4.3 *Analysis Software of 3D Tunnel Geological Advance Prediction*

3.4.3.1 Design Principle

Three-dimensional visualization of geological bodies is a hot issue that needs to be solved urgently in geological engineering. Computer visualization technology can express geological bodies in the form of three-dimensional visual graphics. Using

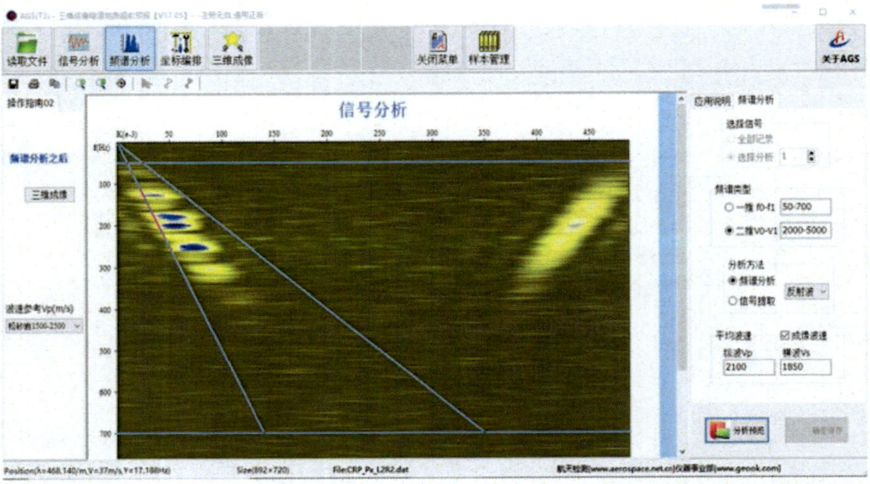

Fig. 3.125 Two-dimensional filtering spectrum analysis after selecting "reflected wave"

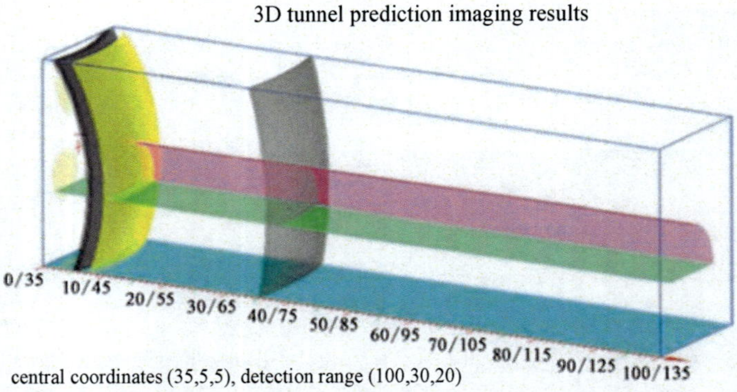

Fig. 3.126 Basic interface of "measuring point coordinates"

the results of 3D visualization, the morphology and structure of the geological body can be directly displayed to geological experts and users, so that they can have an intuitive understanding of the geological distribution in front of the construction site, to propose a reasonable construction plan, reduce the blindness of people's understanding of geological problems, and reduce the huge risk of construction.

(1) **Data reading**

Data reading mainly completes the task of extracting relevant information from the sequence file specified by the user and constructing 3D data, which is divided into three functions: file import, file parsing, and volume data construction. File import imports the contents of the file specified by the user into the memory. The file parsing extracts the useful information in the file one by one according to the custom T3 format and saves them in the memory for each subsequent processing. The volume data structure is to find a sequence file that meets the requirements according to various information extracted from the file and construct a spatial volume data field according to the physical space relationship between the actual pixel and each data point provided in the file.

(2) **2D image browsing**

The resulting three-dimensional image is actually a slice image. Each slice has a certain thickness, and there is also a certain interval between the slices. In general, the analysis and observation of tunnel geology can start from three orthogonal planes of the image, namely the transverse section, the sagittal plane, and the coronal plane. The transverse section is obtained by seismic equipment. The sagittal and coronal planes are obtained after transverse–sectional interpolation and three-dimensional reconstruction. Therefore, image interpolation usually has three constraints:

(1) The interpolated image should be similar to the original slice image.
(2) The similarity between the interpolated image and the two slice images is inversely proportional to the distance between it and the two slices.
(3) The interpolated image should present a gradual process from one original image to another. Common interpolation methods include nearest neighbor interpolation, bilinear interpolation, contour interpolation, and matched interpolation.

(3) **3D reconstruction**

The 3D reconstruction mainly completes the 3D rendering of the read data. The system uses a ray casting method with fast drawing speed and high image quality to realize three-dimensional drawing. The reconstruction process is shown in Fig. 3.127.

It is worth noting here that due to the huge amount of image sequence data collected, it is not convenient for computer processing, so after the data is read, some processing must be done on the original data. The system mainly uses image reduction technology. The principle of the reduction technique is to uniformly select pixels in the original three-dimensional data according to the set parameters and replace the original image data with the selected pixels, thereby achieving the purpose of reducing the amount of data. The vtkImageShrink3D class in VTK implements the image reduction function. The three parameters of the key member functions SetShrinkFactors (int, int, int) represent the reduction factors in the three directions of X, Y, and Z, respectively. After image reduction, the amount of data to be processed by the system is greatly reduced. The program memory occupancy and processing time are reduced, and the number of triangular patches in the reconstructed three-dimensional image is reduced. In addition to increasing the interactivity of three-dimensional images, it is also possible to adjust parameters such as scattered light, ambient light, specular light, light intensity, and sampling distance to get the effect of required three-dimensional reconstruction. At the same time, the corresponding geological anomalies can be stripped out by adjusting the transparency.

(4) **Three-dimensional cutting**

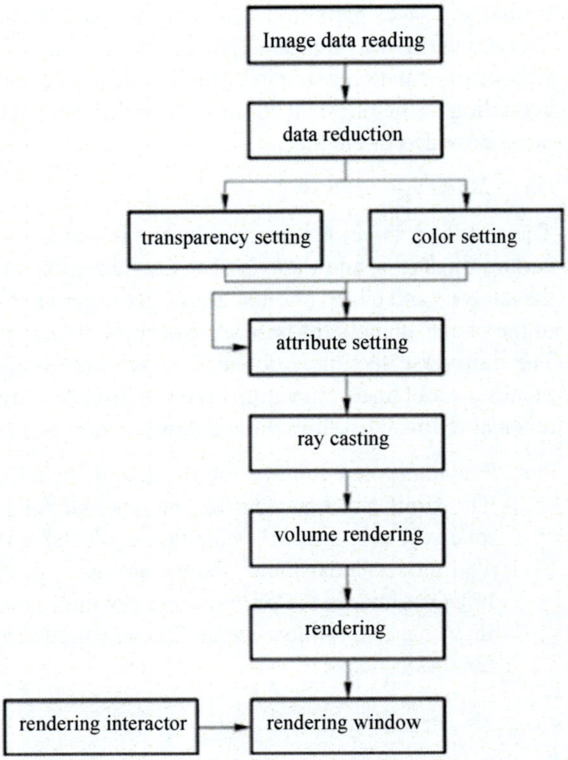

Fig. 3.127 Flowchart of 3D reconstruction

After the 3D reconstruction result is obtained, the simplest and most effective way to see the internal structure of the 3D data field is to cut the volume data field. Cutting can be divided into plane cutting of the data field, isosurface cutting, and segmentation of different ranges. Plane cutting first determines the cutting plane, which can be directly input by the user. For example, a cutting plane is determined by a point and a normal vector. In the same way, two or more planes can be used to slice three-dimensional data. This system mainly studies the plane cutting as follows.

(1) Arbitrary planar cutting. For arbitrary planar cutting, the system uses the point normal vector to determine the plane. Assuming that the cutting plane is always perpendicular to the XOY plane, any two points P_1 $(x_1, y_1, 0)$ and P_2 $(x_2, y_2, 0)$ can be taken on the XOY plane, as shown in the following Fig. 3.128. According to geometric vector operations, the normal vector $\vec{n} = ((y_1 - y_2), (x_2 - x_1), 0)$ of the cutting plane can be obtained.

Using the member functions SetOrigin() and SetNormal() in the vtkPlane class in VTK to set the cutting origin and normal vector. Then adding the cutting plane through AddClippingPlane() in the vtkVolumeRayCastMapper class to achieve planar cutting.

(2) Multi-plane cutting. For multi-plane cutting, it can be achieved in two ways. The first way is to set up several vtkPlane on the basis of arbitrary plane cutting.

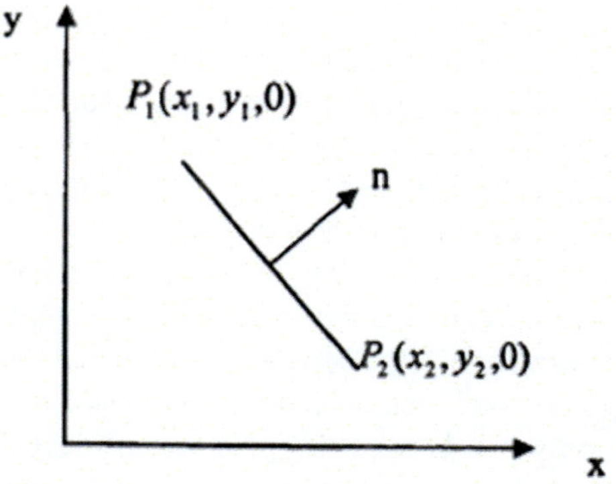

Fig. 3.128 Direction of normal vector

The other way is to use the Crop method provided in VTK. The crop method sets two planes with maximum and minimum values in the *X*-, *Y*-, and *Z*-directions, respectively. The six planes divide a body into 27 small pieces that can be cut as needed. The four most popular crop operations are as follows, as shown in Fig. 3.129.

(5) **Three-dimensional measurement**

3D measurement plays an important role in 3D image analysis systems. At present, conventional methods require manual measurement. But in some cases, such as the measurement of spatial irregular shapes, manual methods are insufficient. However, computer-based 3D measurement can accomplish many tasks that cannot be done

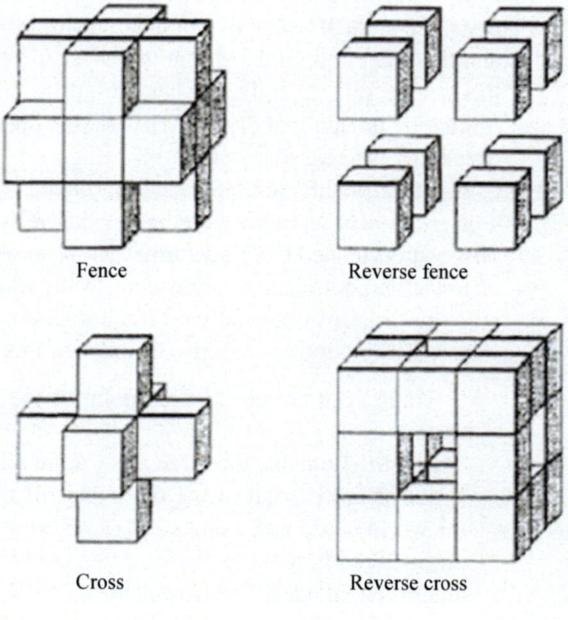

Fig. 3.129 Common crop (cut) operations

manually. For the measurement of the size of anomalous geological bodies, it cannot be achieved manually because of the irregularity of the shape of abnormal geological body. There are two types of measurement of the size of anomalous geological bodies: diameter measurement and volume measurement. The system primarily provides the measurement of the diameter of anomalous geological bodies. To measure the distance between any two points of anomalous geological body, it is necessary to observe the anomalous geological body from any angle. However, the two-dimensional screen of the computer can only display the two-dimensional projection of one side of the measured object. Calculating the distance between any two points requires the two-dimensional coordinates of the two points, so it can only be obtained by three-dimensional coordinate picking and three-dimensional coordinate conversion.

① Three-dimensional coordinate picking. To achieve the measurement of the diameter of abnormal geological bodies, the most important thing is to obtain the spatial coordinates of any two points, which can be obtained by picking up the three-dimensional coordinates. The three-dimensional coordinate picking is to convert the two-dimensional coordinates obtained by clicking the mouse on the computer screen into the three-dimensional coordinate values of the measured object. The system adopts a selective picking mechanism of "pinprick method" under parallel projection to complete the three-dimensional coordinate picking of the measured object.

The principle of the "pinprick method" is that when observing a non-transparent object, one always sees the point closest to the observer on the object. To put it bluntly, if a needle whose length can be freely increased or decreased is inserted into the object from the direction parallel to the observer's eye, the needle, and the object have n ($n > 0$) intersections. Among the n intersections, the point at which the tip first touches is the point closest to the observer, and the point on the object that the observer can actually see. The points that this needle subsequently touches are all invisible points. This method is not limited by the amount of data. The picking speed is fast, and the smallest unit of picking is a real space point.

② Conversion of three-dimensional coordinate. The world coordinates must be converted to observation coordinates before the object is projected onto the observation plane. The transformation of an object from world coordinate to observation coordinate is equivalent to the transformation of the observation reference system onto the world coordinate system using a basic geometric translation rotation operation. The order of this transformation is:

① Translating the observation point to the origin of the world coordinate system.

② Rotating to make the three axes of the observation coordinate system X, Y, and Z correspond to the three axes of the world coordinate system. If the world coordinate point (x, y, z) is specified as the observation reference origin, the origin of the world coordinate system of this point is obtained by the following matrix transformation.

$$T = \begin{bmatrix} 1 & 0 & 0 & -x_0 \\ 0 & 1 & 0 & -y_0 \\ 0 & 0 & 1 & -z_0 \\ 0 & 0 & 0 & 1 \end{bmatrix} \tag{3.59}$$

The rotation sequence that causes the observation axis to change to the world axis depends on the direction of the selected N. In general, if N does not coincide with any of the coordinate axes, the observation coordinate system and the world coordinate system can be coincident with the transformation series Rz·Ry·Rx. That is, first rotating around the X-axis of the world coordinate system to make Zy on the XZ plane of the world coordinate system. Then, rotating around the Y-axis of the world coordinate system to make the Z-axis of the world coordinate coincide with the Zy-axis. Finally, rotating around the Z-axis of the world coordinate system so that the Y-axis of the world coordinate system and the Yy-axis are aligned. If the viewing reference frame is left-handed, reflection from one of the viewing axes is also required. The composite transformation matrix is then applied to the world coordinate description of the object to transform them into the observation coordinate system.

The transformation matrix from the complete world coordinate system to the observation coordinate system is obtained by matrix product:

$$M = R \cdot T \tag{3.60}$$

Once the world coordinate description of the object in the scene is converted to the observed coordinates, the three-dimensional object can be projected onto the n-dimensional observation coordinate. There are currently two basic projection methods: parallel projection and perspective projection. In parallel projection, the coordinate positions are transformed along the parallel line onto the viewing plane. If the three-dimensional coordinates of the entire object have minimum and maximum values in the X-, Y-, and Z-directions, respectively xmin, xmax, ymin, ymax, zmin, zmax, Left = -max(|xmin|, |xmax|), Right = Max(|xmin|, |xmax|), Bottom = −max(|ymin|, |ymax|), Near = -max(|zmin|, |zmax|), Far = max(|xmin|, |xmax|), then the model of parallel projection is shown in Fig. 3.130.

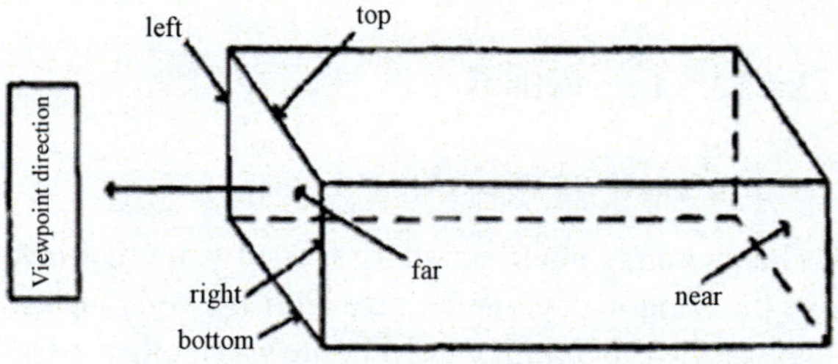

Fig. 3.130 Parallel projection

Parallel projection keeps the relevant scale of the object unchanged. This is the method of generating scale drawing in 3D object drawing. The precise view of each side of the object is obtained by parallel projection.

③ Three-dimensional measurement method. The system mainly calculates the distance between any two points of the abnormal geological body. The user can arbitrarily select two points on the screen to obtain the two-dimensional coordinate values on the screen. According to the three-dimensional picking algorithm to obtain the three-dimensional coordinates of two points, the precise distance between any two points of the abnormal geological body can be obtained. Its calculation formula is as follows:

$$D = \mathrm{SQRT}((x_2 - x_1) \times 2 + (y_2 - y_1) \times 2 + (z_2 - z_1) \times 2) \tag{3.61}$$

where D—the true distance between two points.

3.4.3.2 Functional Overview

The 3D analysis software of tunnel geological advance prediction is the same as the software of other current systems. It needs to meet the requirements of real time, interactivity, scalability, and portability. The main functions of this system are as follows:

① Selection and display of any slice of X-axis, Y-axis, and Z-axis.
② Zoom, rotation, and movement of images of the X-axis slice, the Y-axis slice, and the Z-axis slice.
③ Arbitrary adjustment of window width/window level.
④ Three-dimensional rendering of tunnel geological image data.
⑤ Cutting of any part of the three-dimensional image.
⑥ Choice of multiple styles for 3D rendering effects.
⑦ Zoom, rotation, and movement of the 3D drawing images.
⑧ Arbitrary switching of "1/4" of the display window.
⑨ Storage and printing of images in any window.

The main functional module is shown in Fig. 3.131.

3.4.3.3 Instructions

(1) Operating environment

The operating environment of the software requires the computer operating system to be Windows XP and above Chinese version, with .NET Framework 3.5 SP1 or above, CPU Pentium IV, clock rate greater than 1.6 GHz, memory of 1 GB or greater, free space of hard disk greater than 5 GB, display resolution of 1024 × 768 or higher. It can be normal operation on ordinary desktop or laptop computers.

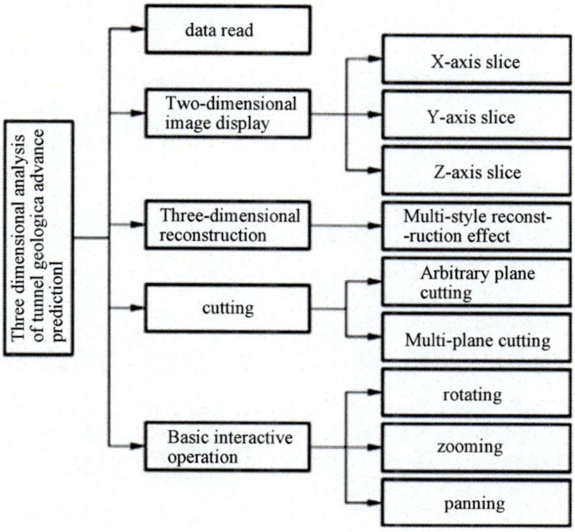

Fig. 3.131 Module of system functions

(2) Software operation

① Interface of the system. The main interface of the system is shown in Fig. 3.132.

All displays and operations are performed in the main interface. The "Toolbox" and "Configuration" windows are hidden in the coordinates of the main interface, which can be opened when needed.

Figures 3.133, 3.134, 3.135, and 3.136 are menu items, respectively

Figure 3.137 shows the contents of the Toolbox, which can be used to set the proportions of the length, width, and height of the 3D image, the position setting of the 3D image slice, the position selection of the 3D image slice, and the color selection.

Fig. 3.132 Main interface of the system

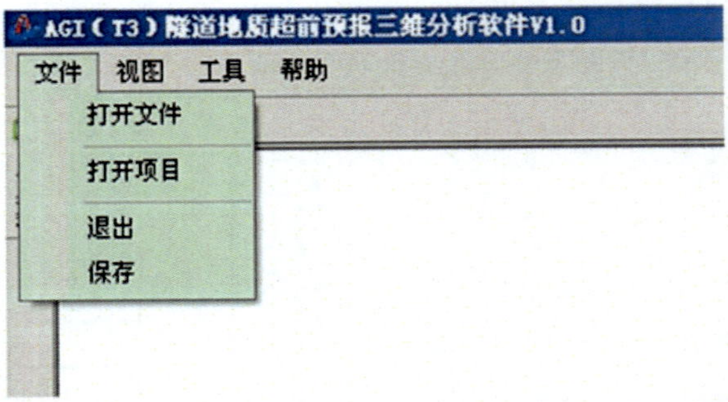

Fig. 3.133 File menu

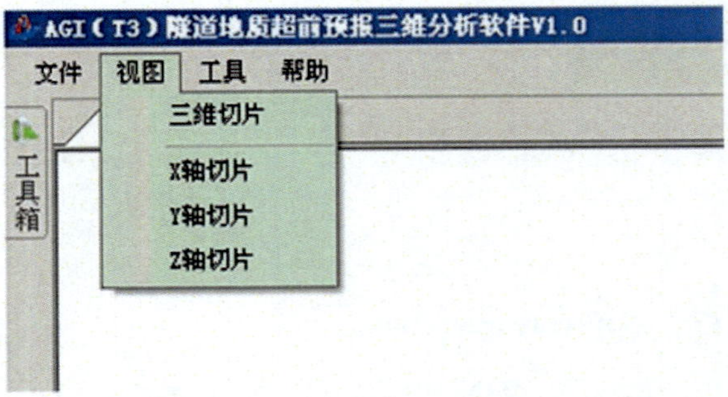

Fig. 3.134 View menu

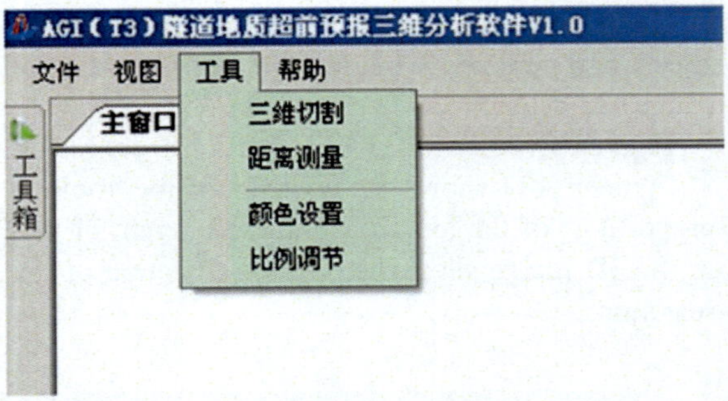

Fig. 3.135 Tool menu

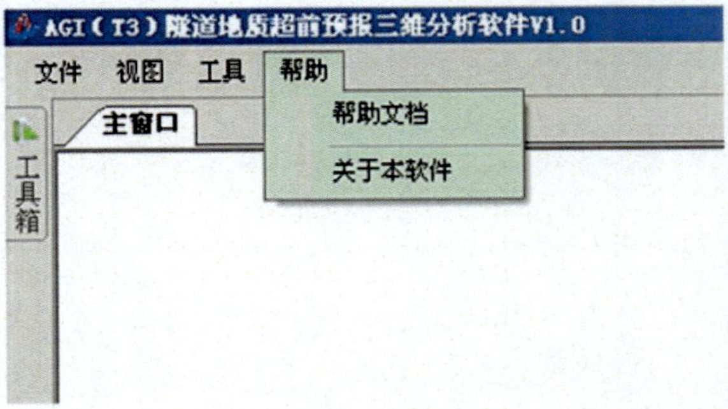

Fig. 3.136 Help menu

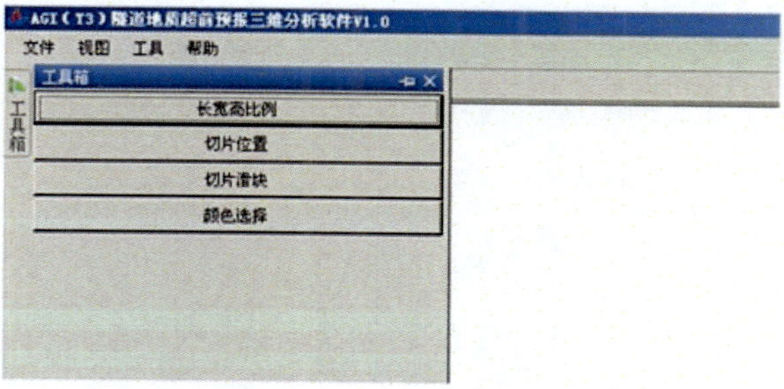

Fig. 3.137 Contents of Toolbox

② Basic interaction. In this system, the user can use the various functional buttons provided in the system interface to achieve a fixed scale of zoom and rotation. In addition, the mouse and shortcut keys are also can be used to interact with the system as follows:

(a) When the left button of mouse is pressed, the 3D image is rotated around its coordinate origin.
(b) When the < shift > + left button or the middle button of mouse is pressed, the panning operation of the 3D image can be realized.
(c) When the right button of mouse is pressed, the 3D image can be zoomed.

③ Data reading. The system can read RAW format data and custom T3 data, where RAW format files are standard unprocessed and uncompressed data, and T3 data is a system-defined tunnel geological 3D data format. The "Open File" menu in the "File" menu can be used to pop up the dialog box shown in Fig. 3.138 to select the corresponding data for subsequent processing.

If "T3_Vertical Section.T3" is opened, the interface shown in Fig. 3.139 can be obtained.

④ Display and operation of three-dimensional image. Opening the 3D data file (*.T3) can visually see the 3D geological conditions as shown in Fig. 3.140.

3D images can be zoomed in, zoomed out, rotated, and panned. These operations can refer to basic interactions and slicing operations. For slicing operations, by selecting the transverse section, coronal plane, and sagittal plane in the view menu, slicing in the corresponding direction can be performed. Then the selecting of corresponding slice position can be realized by using the slice position and slice slider in the Toolbox.

⑤ Ratio setting of length, width, and height. If the default display of 3D geological conditions in the original scale is not convenient for observation, the display ratio of length, width, and height can be adjusted. It can be set in the "Ratio of Length, Height and Height" of the "Toolbox," as shown in Fig. 3.141 .The result are shown in Fig. 3.142.

⑥ Selection of color card. If the default display color of the image is not suitable for observation, a different color scheme can be selected. The appropriate color scheme can be selected in the "Color Table" of the "Toolbox," as shown in Fig. 3.143.

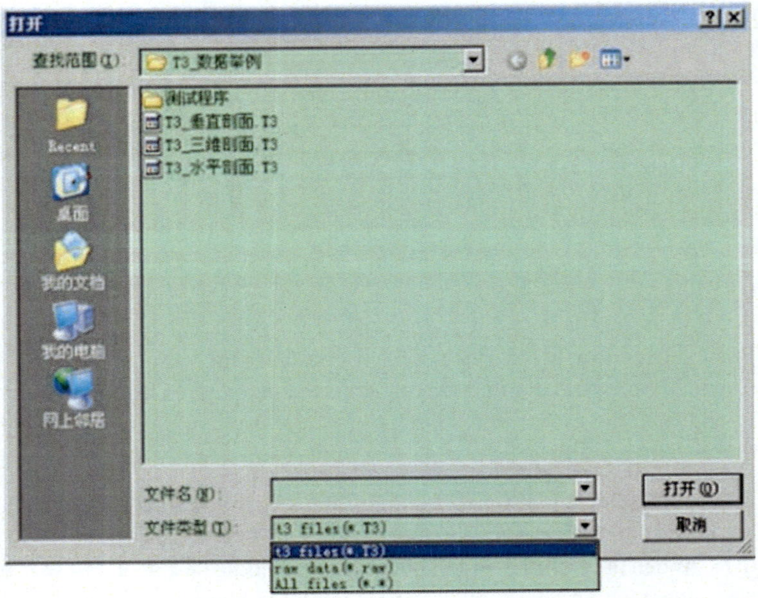

Fig. 3.138 Data open dialog

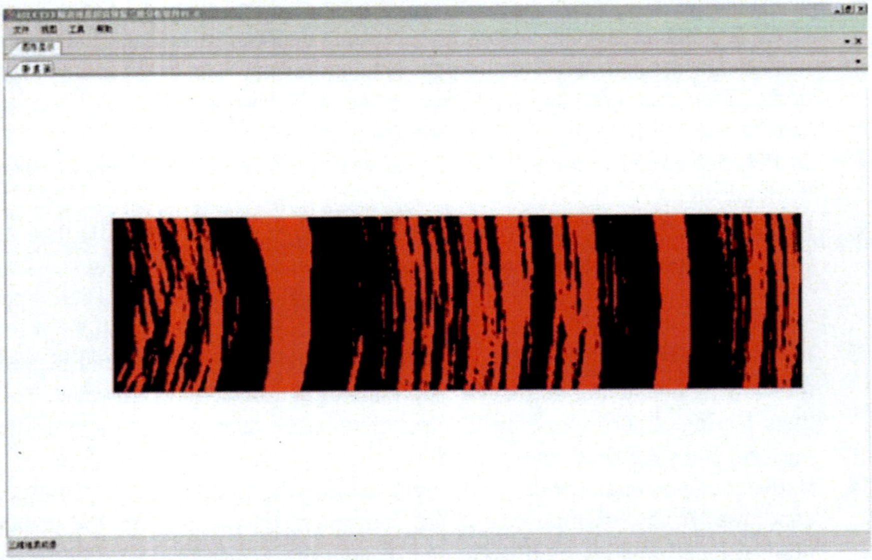

Fig. 3.139 Figure corresponding to "T3_Vertical Section.T3"

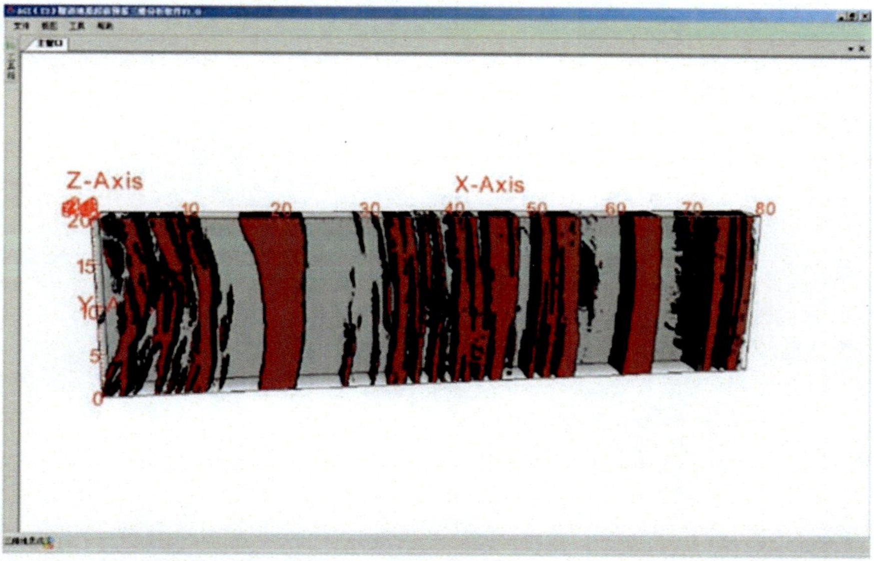

Fig. 3.140 3D reconstructed image

Fig. 3.141 Ratio setting of length, width, and height

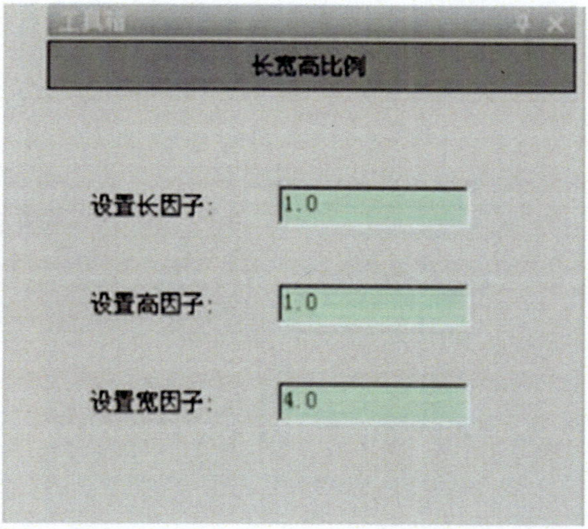

⑦ Three-dimensional distance measurement. In the 3D image, if the straight-line distance from one point to another point is to be measured, it is needed to click "Distance Measurement" in "Tools," then select the starting position in the 3D image by clicking the right button of mouse and selecting the "Set as starting point" in the right-click menu, as shown in Fig. 3.144.

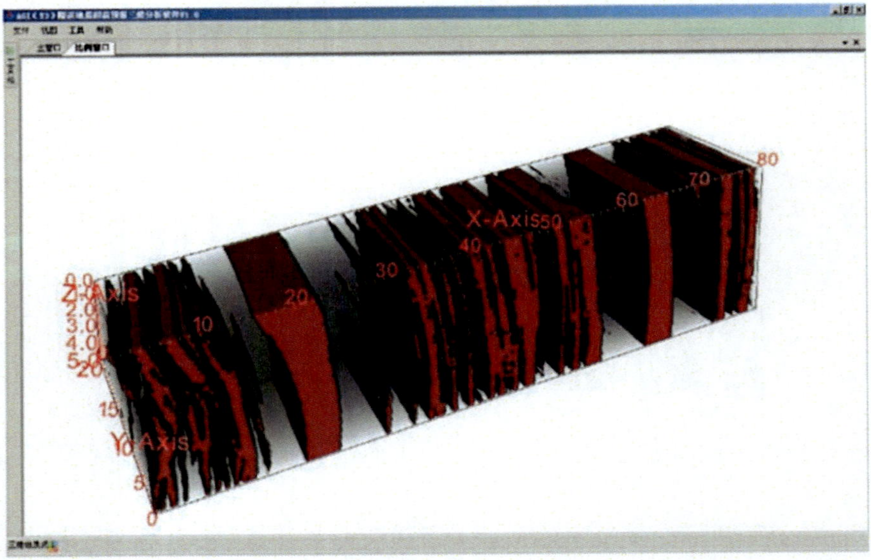

Fig. 3.142 Effect of setting the ratio of length, width, and height

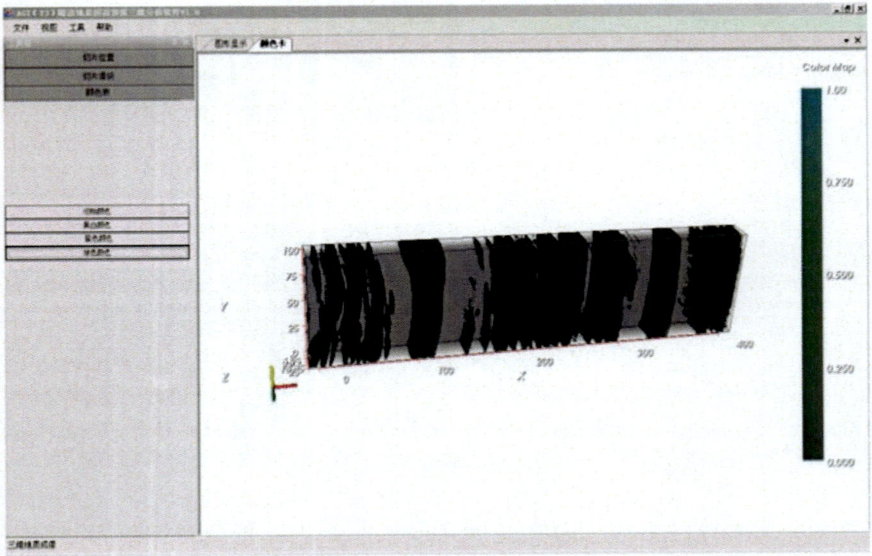

Fig. 3.143 Selection of color scheme

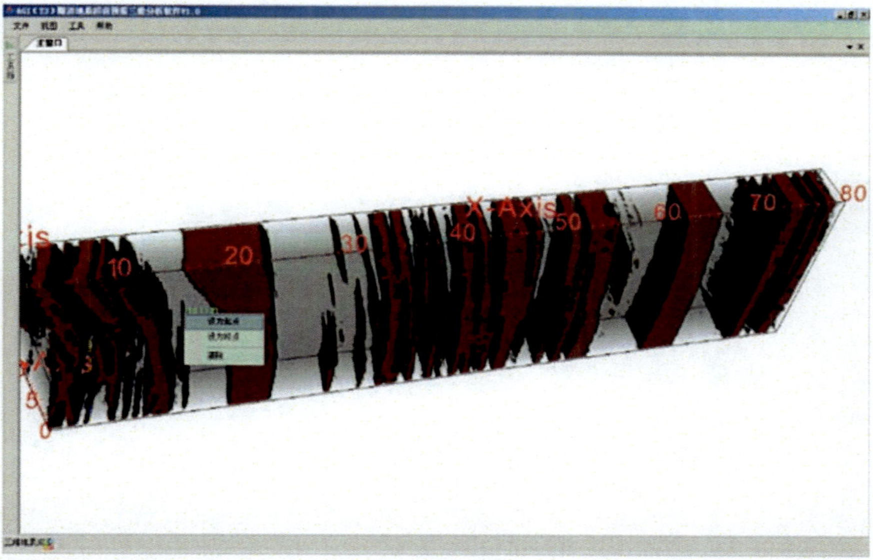

Fig. 3.144 Starting point selection for 3D ranging

Moving the mouse to the end position, clicking the right button of mouse and clicking the "Set as end point" in the right-click menu, then the actual distance between the two point can be obtained, as shown in Fig. 3.145.

⑧ Three-dimensional cutting. In the 3D image, if a certain area is to be cut out and zoom in, it is needed to click "Distance Measurement" in "Tools," then select

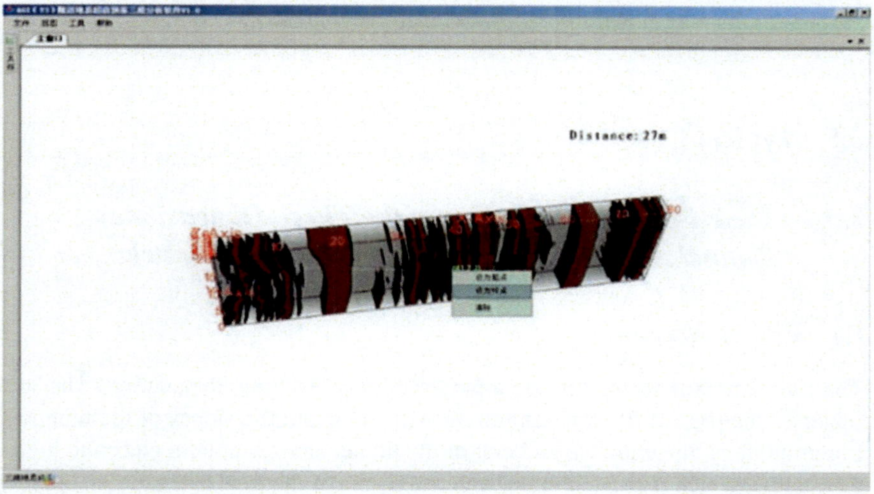

Fig. 3.145 End point selection and results of 3D ranging

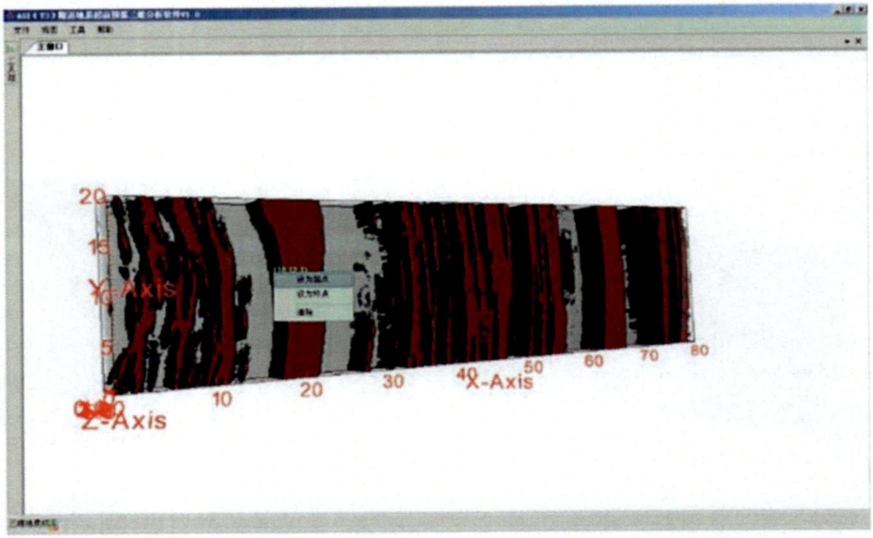

Fig. 3.146 Starting point selection for 3D cutting

the starting position in the 3D image by clicking the right button of mouse and selecting the "Set as starting point" in the right-click menu, as shown in Fig. 3.146.

Moving the mouse to select the area to be cut, then clicking the right button of mouse and selecting the "Set as end point" in the right-click menu, as shown in Fig. 3.147.

The area after cutting is obtained, as shown in Fig. 3.148.

The results of the final three-dimensional imaging are shown in Fig. 3.149.

3.5 Typical Case

3.5.1 Case 1 of the Application of the Three-Dimensional Tunnel Advance Prediction System in the Tunnel

(1) **Project overview**

The tunnel area is located in the alpine section of tectonic denudation. The route traverses the pass of the Baimang Snow Mountain and the slopes of the mountain. Longitudinally, the entrance and exit of the tunnel are low at both ends and high in the middle, which is located in the watershed area of the pass of the Baimang Snow Mountain. In the lateral direction, the two sides are high, and the middle is low, with

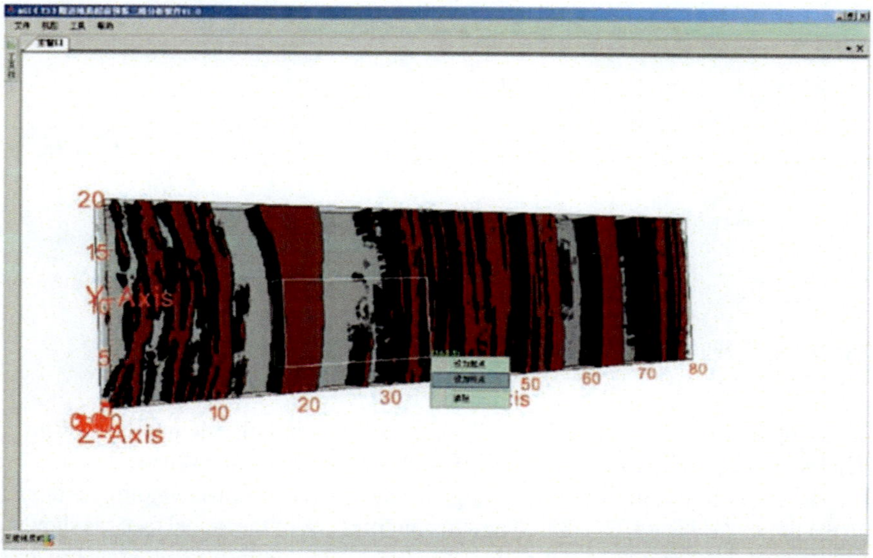

Fig. 3.147 End point selection for 3D cutting

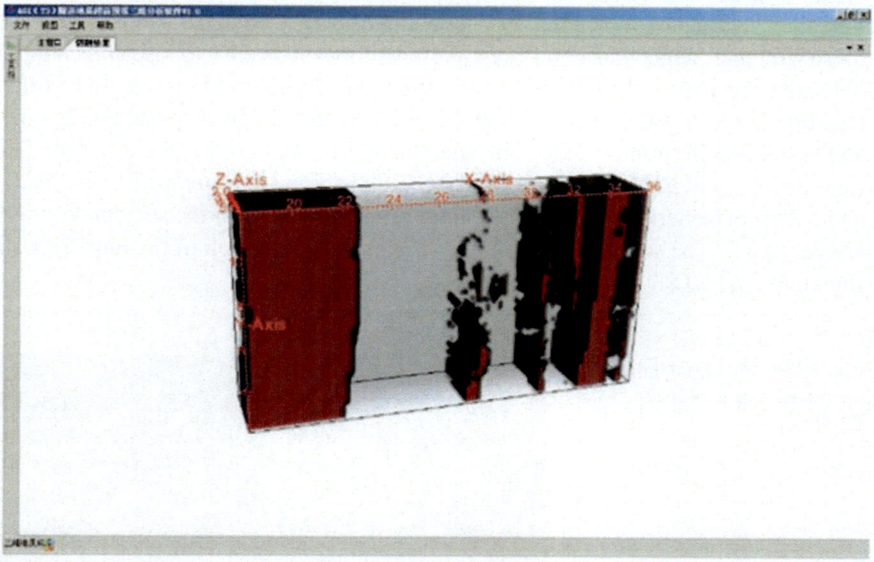

Fig. 3.148 Result after 3D cutting

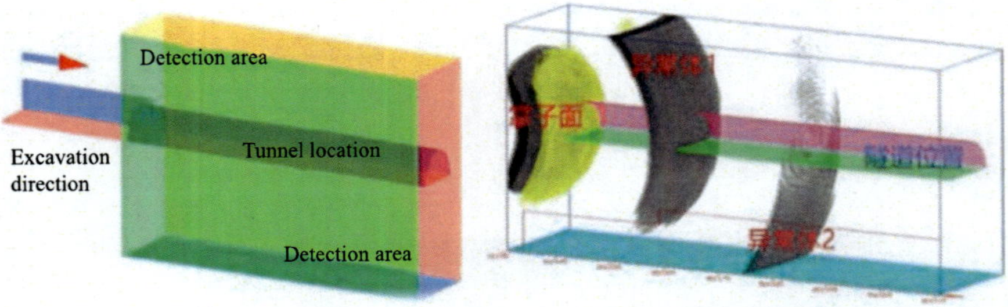

(a) :Observation system diagram (b) Imaging result diagram

Fig. 3.149 Results of 3D imaging

an altitude of 4075.65–4363.34 m. The starting and ending pile numbers of the 1# tunnel are K124 + 650 to K129 + 830, with a total length of 5180 m.

The three-dimensional tunnel geological advance prediction technology is applied in the tunnel face at K127 + 225. The surrounding rock of the tunnel face (as shown in Fig. 3.150) is the Quaternary moraine. The lithology is cohesive soil, gray, brownish yellow, dense, accounting for 40%, with pebbles, gravel, 5% of the particle size of 30–60 cm, 30% of 10–30 cm, and 25% of those less than 10 cm. The rocks are granodiorite, diabase, siliceous rock, sandstone. As a whole, the tunnel face has a loose structure, which has good water permeability and is easy to lose and collapse. Secondly, the part of the tunnel crossing is covered by glacial moraine. In addition, the surface water source is adequately recharged, and the groundwater is abundant and gushes out in strands. Long-term soaking of the rock mass causes softening. The surrounding rock is very unstable. And the vault is prone to water inrush, mud burst, and collapse. Generally speaking, the integrity and stability of the surrounding rock are very poor. The schematic diagram of the geological sketch of the tunnel face is shown in Fig. 3.151.

Fig. 3.150 Photograph of the tunnel face at mileage K127 + 225

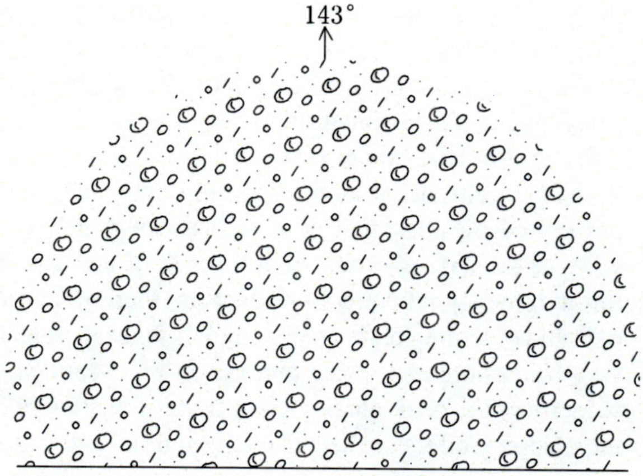

Fig. 3.151 Schematic diagram of the geological sketch of the tunnel face at mileage K127 + 225

(2) **Layout of survey lines and data collection**

According to the on-site test conditions, the data collection location is selected at K127 + 190 at the entrance of the tunnel 35 m away from the tunnel face to collect data. The layout of survey lines is shown in Fig. 3.152. Three three-component geophones and three wireless acquisition modules are, respectively, fixed to the vault, the left side wall (about 2 m vertical distance from the invert), and the right-side wall (about 2 m vertical distance from the invert) of the tunnel. The offset is 4 m. At the same height of the left and right-side walls of the tunnel, a hammer point is placed every 2 m, and the left- and right-side walls are hammered 11 times each to stimulate the vibration wave. The seismic wave is reflected by the tunnel face and the wave impedance interface. The generated echo is received by the geophone, processed by the acquisition module, and sent to the computer terminal.

(3) **3D data imaging and result interpretation**

The zero calibration of the collected data is performed, and the clutter in the signal is removed by noise processing to reduce the interference of the interference wave on

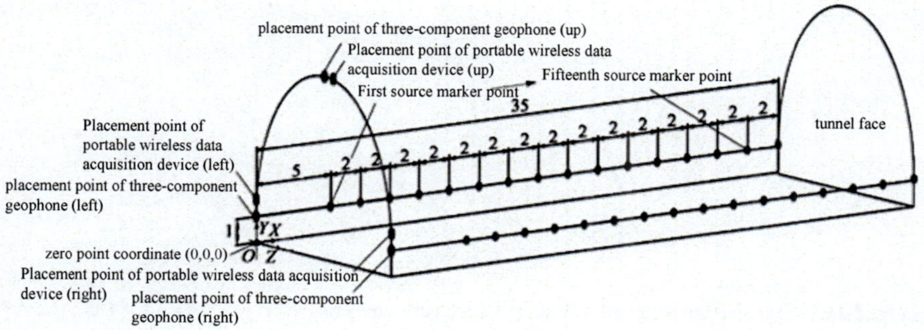

Fig. 3.152 Engineering layout (unit: m)

the target reflection signal. Through a series of signal analysis and processing steps such as spectrum analysis, extraction of effective reflection wave, the recognition of target signal is improved. The results of effective data are shown in Fig. 3.153. The range of data acquisition includes the reflection data of geological bodies within 100 m behind the tunnel face.

After preliminary analysis of the data, two-dimensional spectrum analysis is performed on all signal records, as shown in Fig. 3.153. The effective reflected wave of the data is extracted as a result of the two-dimensional filtering of the data through the signal extraction function, thereby improving the quality of the overall data and obtaining high-quality signals that can reflect the target body. The data of the two-dimensional spectrum analysis is first scanned by velocity to extract the wave velocity of the geological body in front of the tunnel face, and then the parameters of the mapping effect are set. The longitudinal and transverse waves with better effects in the signal are selected for mapping, and the longitudinal wave and the transverse wave are used for combined imaging. Meanwhile, according to the actual situation, the three-dimensional imaging of the same side, the opposite side or all the data in the option of selection arrangement is obtained to obtain the effect diagram of 3D profile imaging, as shown in Figs. 3.154, 3.155, and 3.156.

According to the detection results of the new generation of 3D imaging tunnel geological advance prediction system, the results of the physical and mechanical indexes of the rock mass at the entrance of the main tunnel of the tunnel at K127 + 225 to K127 + 325 are summarized in Table 3.6.

Combined with the comprehensive analysis of the geological conditions of the tunnel face, there is strong reflection at the mileage segment K127 + 225 to K127 + 235. It is speculated that the surrounding rock mass is very fragmented due to

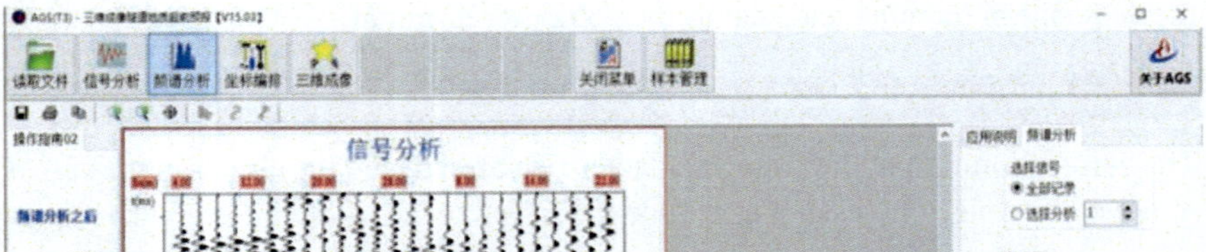

Fig. 3.153 Signal after zero calibration and spectrum analysis

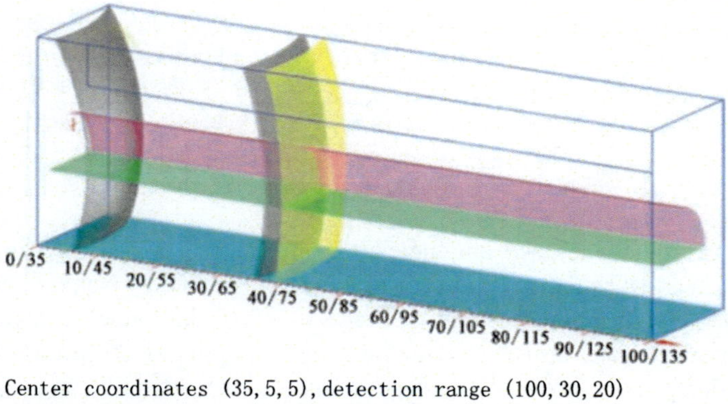

Fig. 3.154 Three-dimensional imaging of longitudinal wave

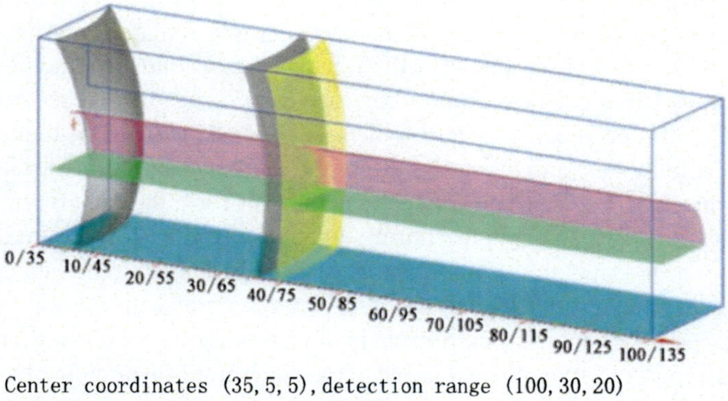

Fig. 3.155 Three-dimensional imaging of transverse wave

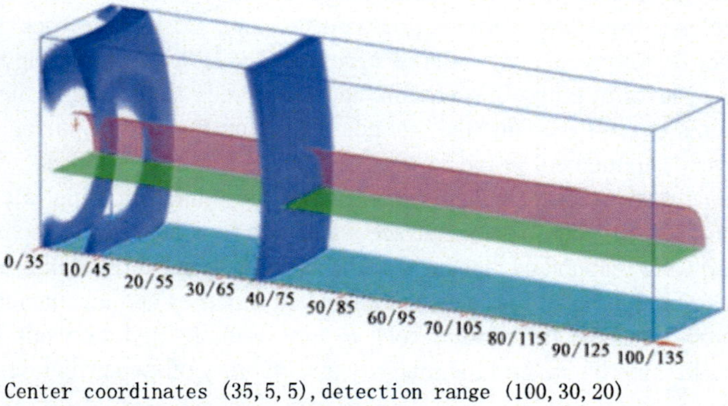

Fig. 3.156 Combined imaging of longitudinal wave and transverse wave

Table 3.6 Summary of physical and mechanical indicators of rock mass of the tunnel

The name of the rock	Mileage (m)	Average P-wave velocity V_p (m/s)	Average S-wave velocity V_s (m/s)
Cohesive soil	K127 + 225 to K127 + 325	2945	2360

blasting near the tunnel face, and there are many cracks, showing a loose, and scattered structure. During the support construction, the secondary support construction should be strengthened according to Grade V. When the tunnel is excavated, the vault is prone to water inrush, mud outburst, and collapse. It is required to follow the construction principles of short footage, weak blasting, strengthening support, and closely following the second lining.

Longitudinal wave imaging, transverse wave imaging, and joint imaging of longitudinal, and transverse waves show that K127 + 261 has strong reflection. Combined with the geological characteristics of the tunnel, it is speculated that there is a fault at K127 + 261, which is the surface F1 fault. The secondary structure traverses this location. The secondary support construction is strengthened according to the Grade V for the surrounding rock near the section K127 + 261. Due to the influence of $F1$ large fault zone and secondary structural fault, as well as the development of groundwater, which soaked and softened surrounding rock, the vault is prone to water inrush, mud outburst and collapse during tunnel excavation. It is required to follow the construction principles of short footage, weak blasting, strengthening support, and closely following the second lining. For the section with fissure water in the surrounding rock, drainage is carried out by means of advanced drilling, etc., and the advanced support is strengthened to make the waterproof layer and reduce the time for water to soak and soften the surrounding rock.

After the actual excavation, the surrounding rock at the tunnel face of K127 + 261 is the Quaternary moraine. The lithology is cohesive soil, accounting for 40%, with pebbles, gravel. The rocks are granodiorite, diabase, siliceous rock, sandstone. The tunnel face represents a scattered structure. The lower surrounding rock is the interbedded layer of purple-red phyllite and gray sandy slate. Due to the influence of the F1 major fault zone and the secondary structural faults, the surrounding rock mass is very fragmented, in the form of thin slices and loose bodies, all of which are extremely soft rocks. Secondly, the part of the tunnel crossing is covered by glacial moraine. In addition, the surface water source is adequately recharged, and the groundwater is abundant and gushes out in strands. Long-term soaking of the rock mass causes softening and reduces the physical properties and mechanical strength of the rock mass. The surrounding rock is very unstable, and the vault is prone to water inrush, mud burst. And the integrity and stability of the surrounding rock are very poor.

Through the physical engineering three-dimensional tunnel geological advance prediction, the geological features of the unfavorable geological body in front of the tunnel face are explained accurately, clearly, and intuitively, which provides a timely and effective basis for the tunnel construction. Through the actual excavation verification, the imaging result of three-dimensional tunnel geological advance prediction technology are in good agreement with the actual geological excavation, which proves that the method is accurate and effective. The combined imaging of longitudinal and longitudinal waves reflects the actual geological conditions more clearly than the separate imaging of longitudinal and transverse waves, improves the resolution and accuracy of 3D tunnel advanced geological forecast, provides a more accurate basis for tunnel construction under complex geological conditions, and guarantees the safety of the tunnel construction.

3.5.2 Case 2 of Application of 3D Tunnel Advance Prediction System in Tunnel

(1) **Project overview**

The tunnel is located in a highway in Yunnan. The length of the road is 16.658 km. The surrounding rock on the face is gray-black, gray-white strong-fully weathered granite, and kaolin fillings. The rock mass is broken, joints, and fissures are developed, showing a fragmented structure. Affected by geological structure, structural fissures are relatively developed. The surrounding rock is in the shape of sandy breccia with crushed stones, with a soft structure. The surrounding rock is easy to slide and fall along the layer, and it is easy to collapse. The structural surface of the rock mass is poorly bonded and wet. The integrity of the rock mass and the self-stability of surrounding rock are poor.

The design grade of the surrounding rock in this section is Grade V1. According to the classification of the surrounding rock in Sect. 3.6 of the "Code for Design of Highway Tunnels" (JTGD70-2004), it is determined that the geological conditions of the tunnel face are in accordance with the conditions of the surrounding rock of V2. The schematic diagram of the geological sketch of the tunnel face is shown in Fig. 3.157. The photograph of the surrounding rock of the tunnel face is shown in Fig. 3.158.

(2) **Survey line layout and data acquisition**

As shown in Fig. 3.159, 10–15 seismic focuses (about 1.5–2.0 m apart) are set on the left and right walls of the tunnel starting from the tunnel face under ideal conditions according to the requirements of the principle of 3D imaging advance prediction. If it is a hammer source, a three-component geophone is placed about 5 m away from the first seismic focus. On the contrary, if the explosive source is used, a three-component geophone is placed at about 10–15 m away from the first seismic focus to ensure safety. One geophone (3 in total) is installed at the height of about 1.0 m on the left and right walls and at the vault. On the other hand, the instrument application takes into account the complex diversity of field conditions. The detection method has great flexibility at the same time. The seismic focus can be set unilaterally on

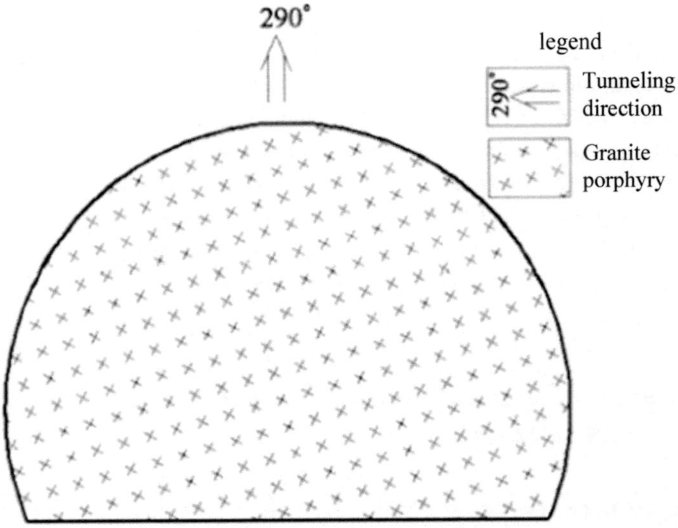

Fig. 3.157 Geological sketch map of the tunnel face at K281 + 904

Fig. 3.158 Photograph of the tunnel face at K281 + 904

the left or right walls of the tunnel, and 1–3 geophones can be selected according to the conditions.

According to the condition on site, one geophone (3 in total) is installed at the height of about 1.0 m on the left and right walls and at the vault. The geophone is installed 3 m away from the first hammering point. A total of 8 explosion points are

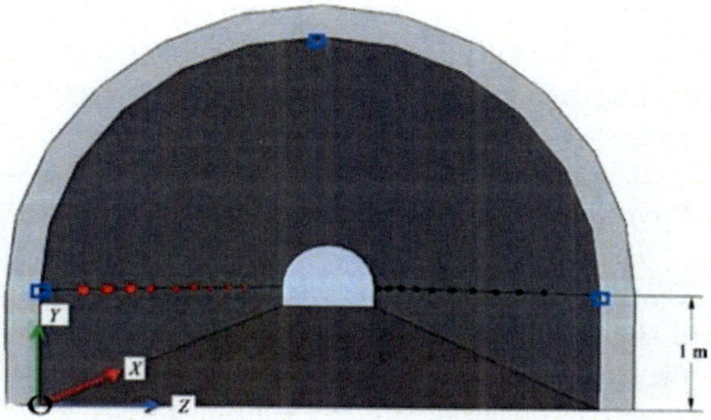

Fig. 3.159 Schematic diagram of the work arrangement

arranged at a height of about 1.0 m on the left wall (with a pitch of 1.5 m), and a total of 6 explosion points are arranged at a height of about 1.0 m on the right wall (with a pitch of 1.5 m). The original data collected by the instrument is processed by professional software. The detection range is K281 + 904 to K282 + 985 (81 m). The tunnel axis is the baseline. The detection area is 80 m in the axial direction, and the circumference is 20 m wide and 30 m high. In the interpretation of the results, the longitudinal wave data is mainly used to divide the rock layers, and the geological phenomena of transverse wave data are combined for interpretation.

(3) **3D data imaging and result interpretation**

Figure 3.160 shows the three-dimensional result image of the tunnel geological structure. The average velocity V_P of K281 + 904 to K281 + 957 is 2000 m/s. The lithology is mainly gray-black, gray-white strong-fully weathered granite. Joint fissures are developed and there are fillings. The rock mass is broken, and the integrity and stability of the surrounding rock are poor.

In the range of K281 + 910 to K281 + 925, there is an irregular reflection interface. It is speculated that the integrity and stability of the surrounding rock are poor, and the structural fracture is developed.

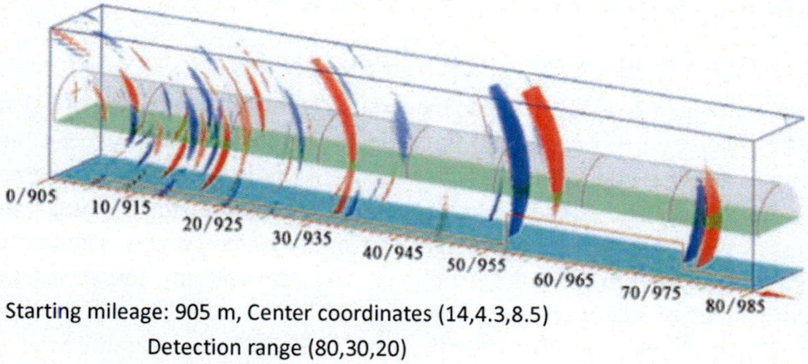

Fig. 3.160 3D imaging result

At the position K281 + 937, there is a clear reflection interface. It is speculated that the rock mass in this section is affected by the geological structure and the structural fissures are developed, and there may be fractured rock mass area or water-bearing area.

It is inferred that the grade of surrounding rock in this section is Grade V2. Protective measures should be strengthened during construction to ensure safety.

The average speed V_P of K281 + 957 to K281 + 985 is 2800 m/s, and the lithology is mainly gray-black, gray-white strong-fully weathered granite. The wave velocity is slightly improved. The bonding between the surrounding rock layers is poor, and the integrity and stability of the surrounding rock are poor.

At the two positions K281 + 957 and K281 + 977, there are clear reflection interfaces. It is speculated that the rock mass in this section is affected by the geological structure and the structural fissures are developed, and there may be fractured rock mass area or water-bearing area.

It is inferred that the grade of surrounding rock in this section is Grade V2. Protective measures should be strengthened during construction to ensure safety.

3.5.3 Case 3 of Application of 3D Tunnel Advance Prediction System in Tunnel

(1) **Project overview**

The collapse occurred in section K103 + 616 to 574 of the right hole of the Tunnel. The section of the tunnel was completely blocked by the collapse materials. 284 m behind the collapsed section is a formed tunnel. The landslide section is in fault fracture zone. The surrounding rock is the Jurassic Zhongtong strong-fully weathered andesite, partially interposed with a small amount of dolomite, which is a soft rock. The joint fissures of rock mass are extremely developed, and the fissure surfaces are mostly filled and cemented by fully weathered calcium clay material. The surrounding rocks are mostly loose and fragmented. The integrity and stability are extremely poor, and they are prone to collapse.

(2) **Survey line layout and data acquisition**

As shown in Fig. 3.161, according to the requirements of the principle of 3D imaging advance prediction and combined with the conditions on site, a three-component detector is installed on the left and right-side walls of the K103 + 635 Section (1.0 m above the ground) and on the vault, respectively. Twelve seismic focuses are placed on the left wall of the tunnel at a height of 1 m above the ground. The first seismic focus is offset by 3 m from the geophone, and a hammering source is used. The detection range of the observation system is K103 + 616 to K103 + 536 (80 m). The tunnel axis is the baseline. The detection area is 80 m in the axial direction, and the circumference is 50 m wide and 50 m high.

(3) **3D data imaging and result interpretation**

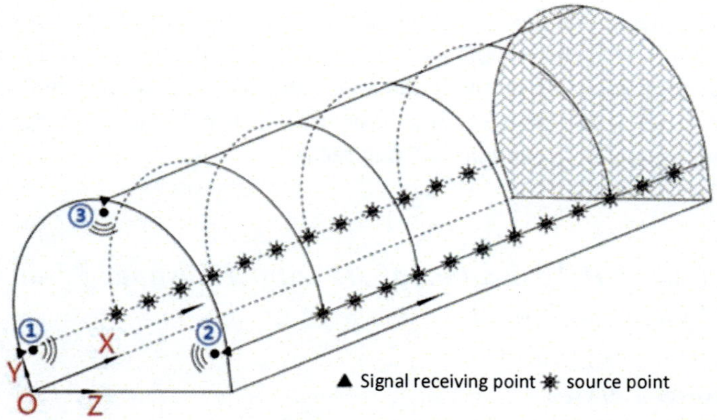

Fig. 3.161 Schematic diagram of 3D observation mode of AGI-T3 tunnel geological detection system

The seismic three-dimensional imaging detection range is K103 + 616 to K103 + 536(80 m). The tunnel axis is the baseline. The detection area is 80 m in the axial direction, and the circumference is 50 m wide and 50 m high. The detection result is shown in Fig. 3.162.

As can be seen from Fig. 3.162:

On the whole, K103 + 616 to K103 + 576 (40 m) has strong reflection and is disorderly, which indicates that the surrounding rock of the tunnel is broken and belongs to the easy collapse section. The reflection interface of K103 + 576 to K103 + 536 is not obvious, and a small amount of weak reflection occurs locally. It is speculated that the surrounding rock of this section is relatively good.

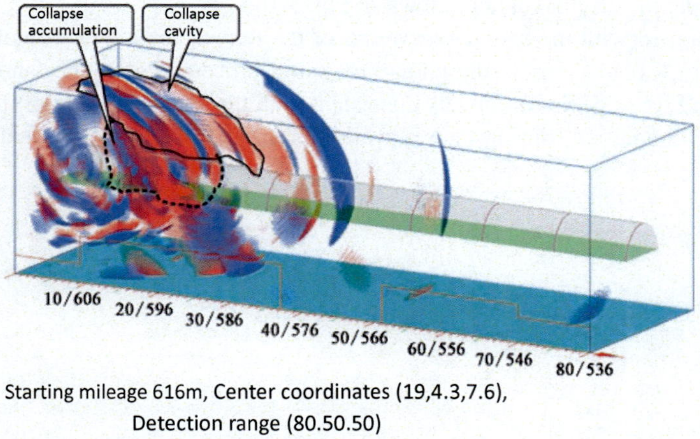

Fig. 3.162 Result of three-dimensional seismic imaging detection of the right tunnel (section K103 + 616 to K103 + 536)

It is speculated that the cavitation cavity is located in front and above the accumulation body of K103 + 609, as shown in the solid line range in the figure. The cavity range is about 23 m above the vault, and the longitudinal mileage is K103 + 609 to K103 + 585. This result is basically consistent with the imaging result of three-dimensional of ground penetrating radar.

3.5.4 Case 4 of Application of 3D Tunnel Advance Prediction System in Tunnel

(1) **Project overview**

A tunnel passes through the middle mountainous area outside the Erhai Fault Depression Basin. The topography of this area is deep cut, steep terrain, dense vegetation, abundant rainfall, complex topography, and vertical and horizontal valleys. According to geological survey and drilling data, the left and right sections of the tunnel are in the same geomorphic unit.

The groundwater type in the tunnel area is mainly Quaternary loose rock pore water, bedrock fissure water, tectonic fissure water, and karst water. The water richness of the rock layer is generally good. Due to the large buried depth of the tunnel, the large water catchment area, and the abundant groundwater in some parts, the structural fracture zone or karst water development zone during the rainy season is relatively rich in water. In general, the hydrogeological conditions of the tunnel are poor, which is unfavorable for tunnel excavation and operation.

(2) **3D data imaging and result interpretation**

According to the detection depth of 60 m, the three-dimensional result after processing the original data is shown in Fig. 3.163.

Combined with the actual excavation of the tunnel, the mileage segment K121 + 542 to K121 + 565 is the paragraph being processed after the landslide. The result in Fig. 3.163 is completely consistent with the actual mileage of the collapse section. In addition, during the excavation process, it is also found that the entire

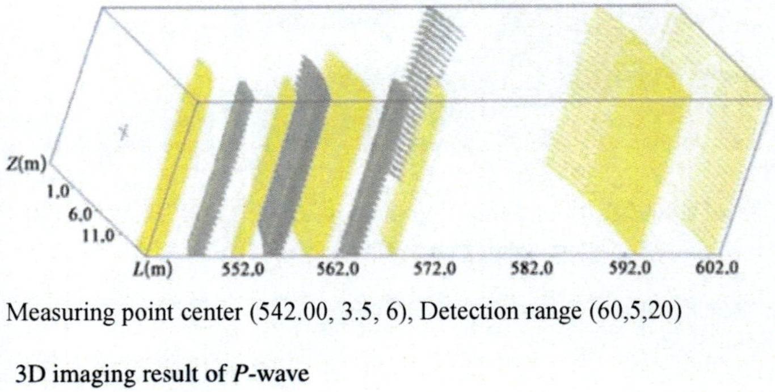

Measuring point center (542.00, 3.5, 6), Detection range (60,5,20)

Fig. 3.163 3D imaging result of *P*-wave

collapsed mileage section is all broken and loosely accumulated landslide soil, and the predicted result is consistent with the actual situation.

It can be inferred from Fig. 3.163 that there are strong reflection interfaces at the mileage points K121 + 594 and K121 + 600. It is speculated that there are abnormal geological conditions, which are later verified to be water-rich areas after excavation. The prediction of the abnormal area is consistent with the actual situation.

Chapter 4
Three-Dimensional Detection Technique for Groundwater-Containing Potential of Tunnel Surrounding Rocks

The geological conditions in tunnel construction are unpredictable. However, unfavorable geological conditions such as fault fracture zones, weak surrounding rocks, and karst caves often appear during the construction process (especially in the karst areas of southwest China). Rush construction is very easy to cause landslides, water inrush, and other unfavorable geological diseases, especially in karst and fault fracture zone areas with high incidence and high risk of water inrush, which is a major key problem in tunnel construction. Therefore, more and more experts, scholars, and technicians pay attention to advanced water exploration in tunnel. The advanced geological prediction of tunnel geological structure has been described in detail in Chap. 3. The advanced geological prediction methods widely used in the detection of the hidden water hazards of the surrounding rock in tunnel include the ground penetrating radar method and the electrical method. High-density resistivity method is the most commonly used method. The author and his team have improved the high-density resistivity method in terms of power supply, cables, and terrain correction, forming a set of high-power electrical sounding 3D imaging technology suitable for deep buried long tunnels.

4.1 Evaluation and Analysis on Advance Water Exploration Technology of Tunnels at Home and Abroad

Groundwater is the water stored in the pores of the formation below the vadose zone, including water in rock pores, fissures, and caves. The amount of groundwater is huge. According to estimates, the total amount of groundwater in the world is as much as 150 million km^2, which accounts for almost 1/10 of the total amount of water on the earth. Hydrogeological surveys often need to combine geology,

G. Cao and H. Liu, *Three-Dimensional Exploration Technology of Tunnel Geology*, Key Technologies for Tunnel Construction under Complex Geological and Environmental Conditions, https://doi.org/10.1007/978-981-16-9225-3_4

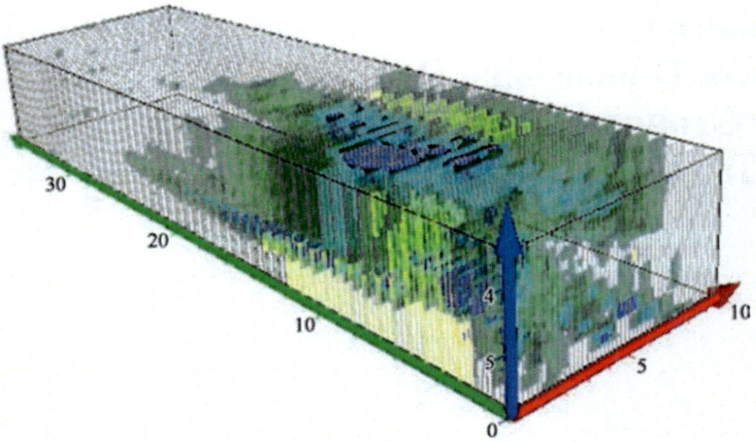

Fig. 4.1 Ground penetrating radar three-dimensional water exploration technology

hydrology, geophysical prospecting, and other methods for comprehensive analysis. Firstly, the storage conditions of groundwater are analyzed by topography, structure, and hydrology to determine the occurrence area of groundwater, and then the detailed distribution of groundwater is detected by geophysical methods.

Geophysical prospecting for water is mainly based on the difference in physical properties between the aquifer and the surrounding rock. For example, the electrical method is based on the difference in resistivity (polarizability) between the aquifer and the surrounding rock. The sandstone aquifer generally has a resistivity greater than that of the surrounding rock such as mudstone. Structural water is generally where the resistivity of the water-bearing position is lower than that of the surrounding rock. Commonly used geophysical methods to find water include ground nuclear magnetic resonance method, high-density resistivity method, infrared detection method in the cave, and ground penetrating radar method (Fig. 4.1). In addition, remote sensing technology can also be used for large-scale groundwater detection.

In recent years, foreign companies represented by Seismo Electronics LLC of the USA have tried to use the seismic effect to find water and achieved good results.

4.2 High-Power Electric Sounding Three-Dimensional Advance Forecasting Technology

4.2.1 High-Density Resistivity Method

(1) **Principle**

The high-density resistivity method is an electrodetection system developed according to the actual needs of hydrological, engineering, and environmental geological surveys. The system includes data acquisition and data processing. In the field measurement, it is only necessary to lay all the electrodes on the measuring points at a certain interval (the density of the measuring points is much larger than the conventional resistivity method, generally 1–10 m), and then connect them to the program-controlled multi-channel electrode transfer switch with multi-core cable. The electrode transfer switch is an automatic transfer device controlled by a single-chip microcomputer. It can automatically carry out the conversion of electrode device form, pole distance, and measuring point according to needs. The measurement signal is sent to the electrical measuring instrument through the electrode transfer switch, and the measurement results are sequentially stored. After the data is played back and sent to the microcomputer, the original data can be processed according to the given program.

Because the high-density resistivity method can realize the rapid acquisition and computer processing of field data, it changes the traditional working mode of resistivity exploration, greatly improves work efficiency, reduces labor intensity, and makes the degree of intelligence of electrical exploration a step forward. In summary, the high-density resistivity method has the following characteristics compared with the conventional electrical method:

① The arrangement of the electrodes is completed at one time, which not only reduces the malfunction and interference caused by the electrode setting but also provides conditions for the rapid automatic acquisition of field data.
② The density of observation is large and scanning measurements of various electrode arrangements can be performed, so that a wealth of information about the structural characteristics of the geoelectric section can be obtained.
③ Field data collection is automated or semiautomated, which not only speeds up acquisition but also avoids errors caused by manual operations.
④ The data can be preprocessed and the profile curve shape can be displayed. After offline processing, various result maps can be automatically drawn and printed.
⑤ Compared with the traditional resistivity method, it has obvious advantages such as low cost, high efficiency, rich information acquisition and convenient interpretation, and its exploration capability is significantly improved.

The basic principle of high-density resistivity method is the same as that of ordinary electrical method. The difference between the two is only the instrument and the way they work.

(2) Arrangement format of collected data

The arrangement of the data in the data collection process is related to the selected device form, and the specific data arrangement is as follows:

The Pole-Pole device is shown in Fig. 4.2. The data patterns are arranged in a trapezoidal shape.

The P-Dpole device is shown in Fig. 4.3. The data patterns are arranged in a rectangular.

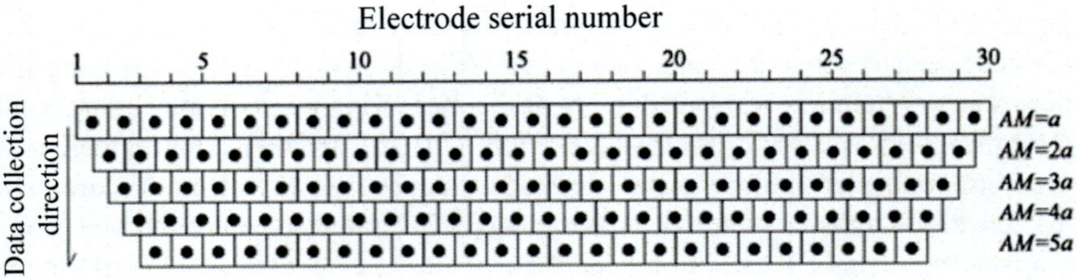

Fig. 4.2 Pole-Pole device

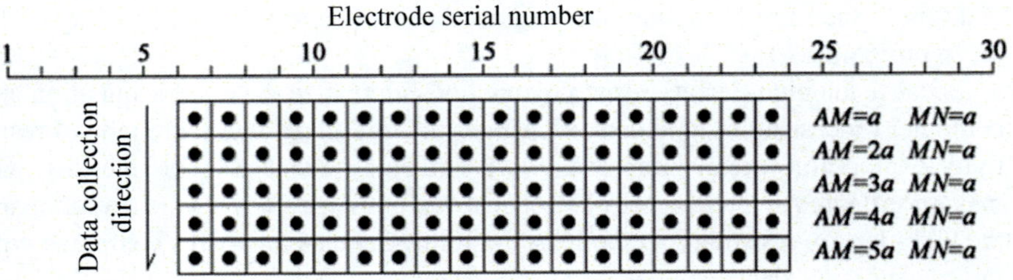

Fig. 4.3 P-Dpole device

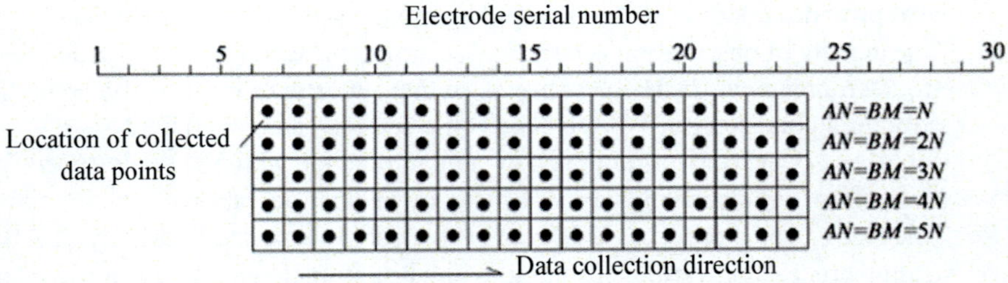

Fig. 4.4 Schlumberger device

The Schlumberger device is shown in Fig. 4.4, and the data patterns are arranged in a rectangular.

The Wenner device is shown in Fig. 4.5. The data patters are arranged in a trapezoidal shape.

4.2.2 Principle and Working Method of High-Power Electrical Sounding Three-Dimensional Detection

The high-power electric sounding three-dimensional detection technology is an improved method based on the high-density resistivity method. Compared with the traditional high-density resistivity method, its single array length can reach 4 km,

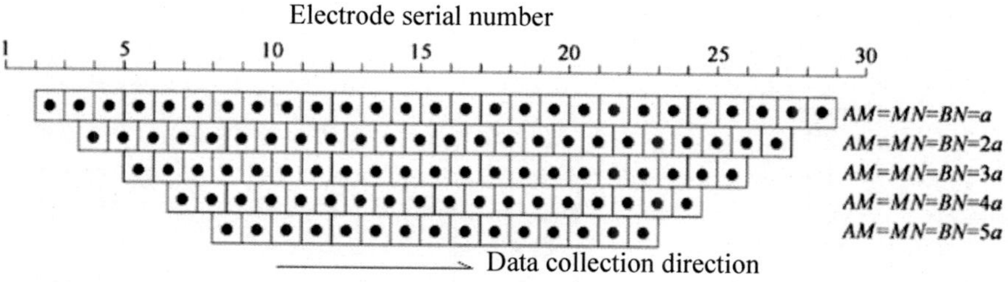

Fig. 4.5 Winner device

and the power supply voltage can reach up to 1200 V. In addition, it introduces new terrain correction technology and 3D imaging technology. The improvement of the above aspects effectively improves the depth of detection and the accuracy of detection, making the method suitable for the special geological survey of deep buried long tunnels.

4.2.2.1 Terrain Correction

The fluctuation of the topography has a major influence on the distribution of the electric field, and thus, the measurement result of the apparent resistivity also causes a large error. The topography and resistivity correction of the measurement results of the high-density resistivity method is necessary to improve the accuracy and reliability of the exploration results, especially for mountainous areas in western China.

The distance and depth of the electrical survey are measured by the electrode spacing a. In the measurement, the supply electrode AB and the electrode for measuring the voltage are collectively referred to as measuring electrodes, and their positions and distances are varied. There are many ways to arrange the form without loss of generality. Taking the most commonly used Wenner observation method as an example, the arrangement of measuring electrodes is AMNB, and the distance between adjacent measuring electrodes is equal, which is called measuring pole distance. In the measurement of the first layer, one electrode distance is taken: AM = MN = NB = a. In the measurement of the second layer, two electrode distances are taken: AM = Mn = NB = $2a$, and so on. The horizontal coordinate of the apparent resistivity data for each measurement is the midpoint of the four measuring electrodes, and the depth coordinate is the pole pitch of the measuring electrode. That is to say, the depth of the first layer is a, $2a$ depth the second layer, and so on. Obviously, this layered data arrangement can reflect the geological structure characteristics of flat terrain. The data and surface morphology are horizontal. If the terrain fluctuates greatly, the connected surface between the measured polar distances is inclined, and the elevation position also varies greatly and then the coordinates of the apparent resistivity measuring points are no longer the positions assumed by the model, so the coordinates of the ground surface and the apparent resistivity measuring points

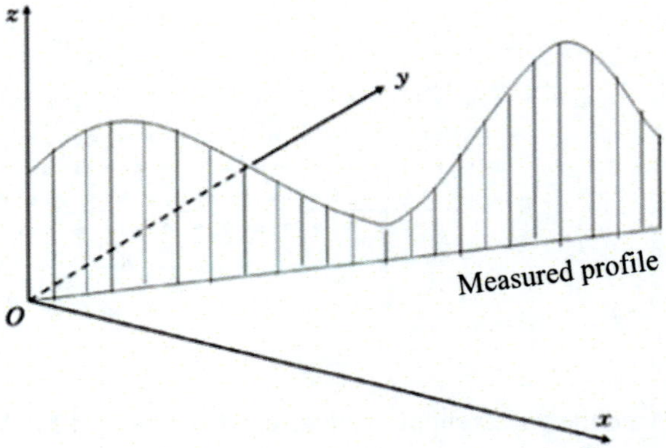

Fig. 4.6 3D to 2D schematic diagram of high-density resistivity measurement profile

must be corrected. The early terrain correction only adds or subtracts the elevation coordinates, which will change the true occurrence of the geological structure, and would not get the true position of the apparent resistivity point. The method in this chapter is to introduce a coordinate transformation to correct the true position of the apparent resistivity point. The transformation formula is as follows:

$$\begin{pmatrix} x \\ y \end{pmatrix} = \begin{bmatrix} \cos\theta & \sin\theta \\ -\sin\theta & \cos\theta \end{bmatrix} \begin{pmatrix} X \\ Y \end{pmatrix} \tag{4.1}$$

where

- θ the inclination of the formation, positive upwards;
- X, Y the coordinates of the apparent resistivity point given by the polar distance position and the layer depth;
- x, y the true coordinates of the apparent resistivity point after correction.

The three-dimensional to two-dimensional schematic diagram of a high-density resistivity measurement profile is shown in Fig. 4.6. The schematic diagram of correction of apparent resistivity point coordinates is shown in Fig. 4.7.

4.2.2.2 Apparent Resistance Correction

The high-density resistivity method is based on the difference in apparent resistivity between subsurface media to infer geological structural features. If the apparent resistivity value is not true, the interpretation will inevitably lead to misleading. Conventional calculation of apparent resistivity assumes that the ground is horizontal, and the apparent resistivity can be calculated according to the supply current between AB, the measuring voltage between MN, and the device coefficient of the observation system. Equation (4.2) is the apparent resistivity of the Wenner observation mode, and Eq. (4.3) is the apparent resistivity of the dipole observation mode:

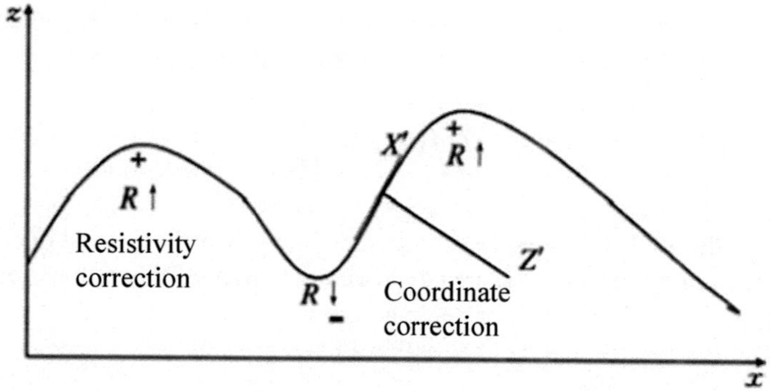

Fig. 4.7 Schematic diagram of correction of apparent resistivity point coordinates

$$\rho_\alpha = 2\pi a \frac{\Delta V_{\text{MN}}}{I_{\text{AB}}} \tag{4.2}$$

$$\rho_\beta = 6\pi a \frac{\Delta V_{\text{MN}}}{I_{\text{AB}}} \tag{4.3}$$

where

I_{AB}	the supply current of the supply electrode AB;
ΔV_{MN}	the voltage between the measuring electrodes MN;
a	electrode spacing.

When the land is flat, the distribution of the current field is relatively uniform. When the terrain fluctuates greatly, the distribution of electric field deviates greatly from that of flat terrain. The current density of the mountain beam is small, and the voltage ΔV_{MN} between the measurement points MN is small, and then the calculated apparent resistivity of the mountain beam is small, which should be corrected to improve the apparent resistivity. For the valley topography effect, the opposite is true. For the valley topography effect, the opposite is true. The current density of the valley is high, and the measured voltage ΔV_{MN} is large, and then the calculated apparent resistivity is large, which should be reduced to correct. Resistivity correction can be calculated as follows:

$$\rho_\alpha = 2\pi a f(\theta) \frac{\Delta V_{\text{MN}}}{I_{\text{AB}}} \tag{4.4}$$

$$\rho_\beta = 6\pi a f(\theta) \frac{\Delta V_{\text{MN}}}{I_{\text{AB}}} \tag{4.5}$$

where $f(\theta)$—the correction factor, which is a function of the angle of the undulation of the terrain, and has different forms for the two-dimensional and three-dimensional topographical functions.

In the two-dimensional case, the simplest form of the function is taken as the following formula, and as the research progresses, it can have a more precise form:

$$f(\theta) = 1 - \frac{\Delta\theta}{180} \tag{4.6}$$

where $\Delta\theta$—the angle between the topographic line and the horizontal line. The angle of the mountain girder terrain is negative, and the angle of the valley topography is positive.

4.2.2.3 Three-Dimensional Imaging

High-power electrical sounding three-dimensional detection technology can realize three-dimensional imaging of unfavorable geology such as groundwater and faults. The imaging procedure is as follows (Figs. 4.8, 4.9, and 4.10):

Step 1: Convert multiple 2D resistivity profile data to 3D space according to the coordinates through 3D modeling software.
Step 2: Mesh the data in 3D space.
Step 3: Use 3D imaging software to reproduce the underground low resistivity distribution in the ground and realize 3D imaging.

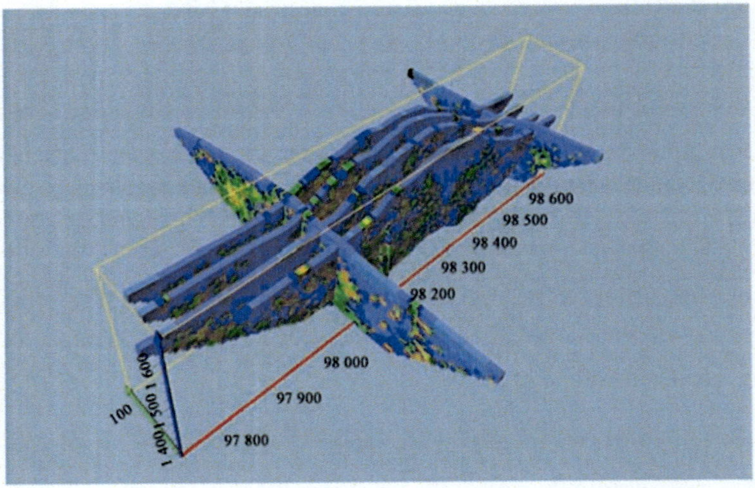

Fig. 4.8 Spatial distribution of multiple 2D sections

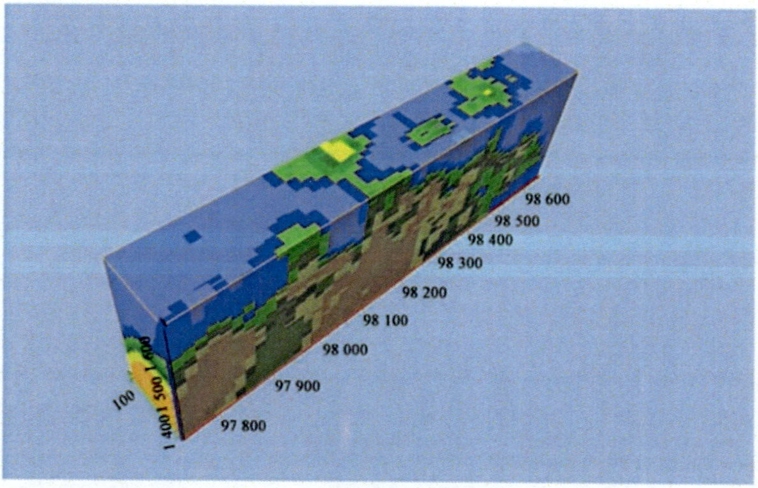

Fig. 4.9 Meshing the resistivity data into a three-dimensional block of data

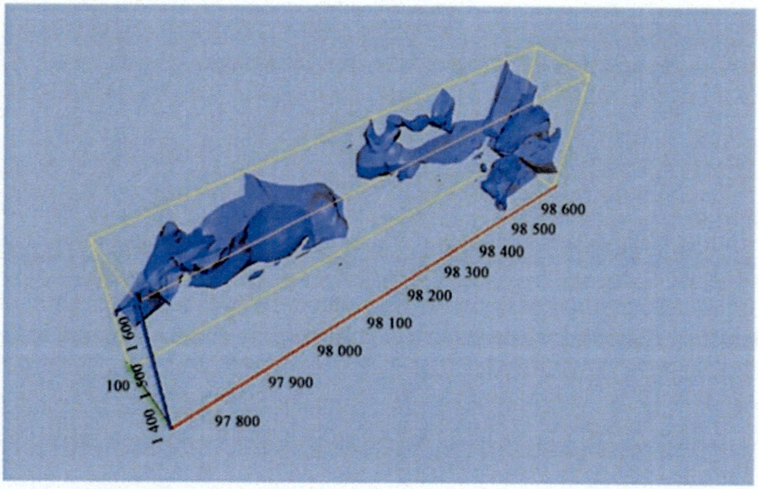

Fig. 4.10 Three-dimensional imaging of the underground low resistivity

4.2.2.4 Practical Application

(1) **Application examples of terrain correction technology in tunnel high-density resistivity method advance prediction**

Figure 4.11 shows the high-density resistivity diagram before terrain correction. The high-density resistivity diagram that is really needed in the work is obtained after processing by the terrain correction software, as shown in Fig. 4.12.

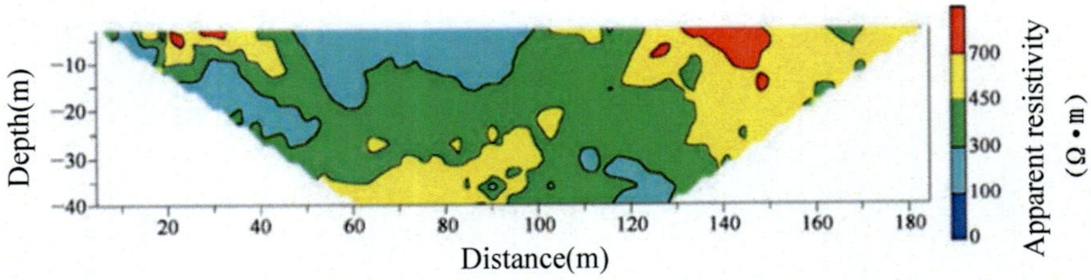

Fig. 4.11 High-density resistivity diagram before terrain correction

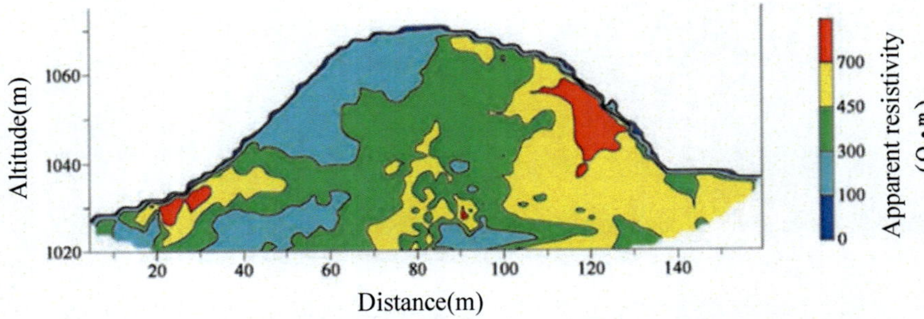

Fig. 4.12 High-density resistivity diagram after terrain correction

The distribution of underground resistivity is shown in different colors. Red and yellow indicate areas of high resistivity, representing complete and dry bedrock. The geological conditions are good. Blue and green indicate areas of low resistivity, representing water-bearing, fractured zones. These areas are bad geological bodies.

When using the high-density resistivity method for tunnel advance prediction, it is often detected in areas with uneven terrain. However, the terrain will affect the detection results, causing the detected rock strata and the abnormal shape and position of apparent resistivity caused by the geological structures to be distorted and displaced, and even meaningful anomalies are concealed. Therefore, when making geological interpretations of electrical observation data, terrain impact correction should be carried out to ensure the effect of advance prediction. In this chapter, the terrain and resistivity correction software are improved to solve this problem.

(2) **Identification mode of water-bearing zone**

The water in the rock mass is easy to dissolve the minerals in the rock and the soil. It contains cations such as potassium, sodium, calcium and magnesium, and anions such as sulfate, chloride, and bicarbonate. It has good electrical conductivity. Therefore, the water-bearing zone appears as a low-resistance body in the high-density resistivity image, and the lower the electrical resistivity, the better the water-containing potential. The morphology of the low-resistance body is closely related to the shape and scale of the water-bearing zone.

The mechanical properties of aquifers and dry rocks are quite different. The cohesion and friction coefficient of the water-bearing rock mass are greatly reduced, and

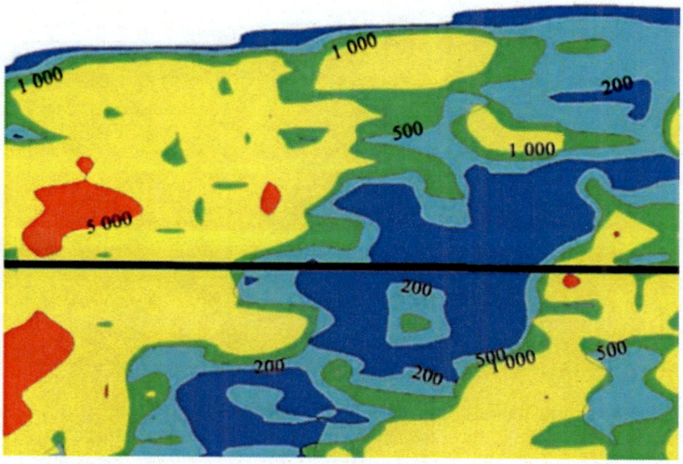

Fig. 4.13 Fractured water-rich zone

Fig. 4.14 Water content change contact zone

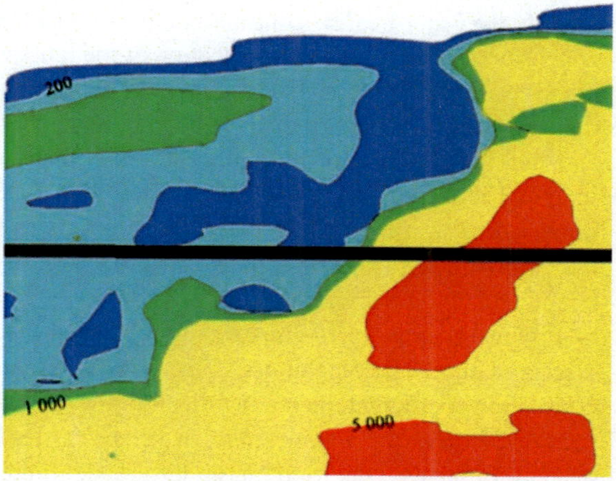

the modulus is reduced. It is extremely easy to deform and the stability is poor. It is a location prone to geological disasters. There are two types of water-bearing zones that should be noted. One is the fracture aquifer and the other is the contact zone aquifer. The former is characterized by a high-resistance zone on both sides and a low-resistance zone in the middle, with a width of several meters to several tens of meters, which is the most dangerous. The latter one is changed from high resistance to low resistance. It is the contact zone between the intact rock mass and the fractured water-rich rock mass. The rock mass at the contact boundary is relatively broken, which is a dangerous section. For large areas of low-resistance areas in the

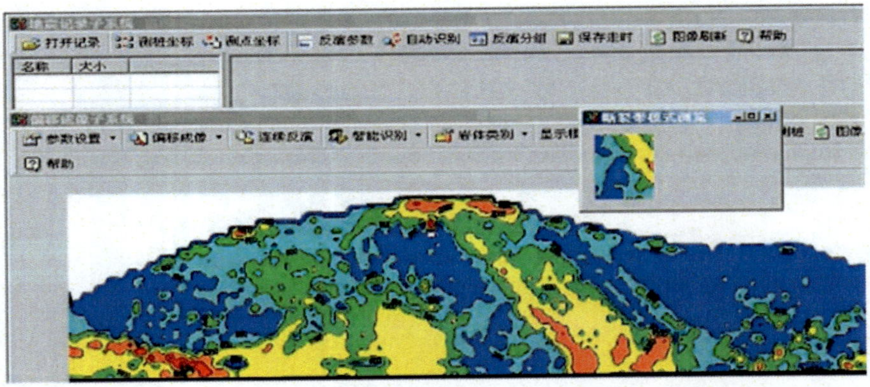

Fig. 4.15 Representation and discrimination mode of resistivity image

image, most of them are caused by mudstone or coal-bearing strata, which is easy to distinguish. Figures 4.13 and 4.14 represent the two modes described.

The intelligent identification system of tunnel advance prediction can be used to identify the water-bearing zone automatically. The identification of the water-bearing zone is very similar to the identification of the fault zone. The correlation coefficient identification method is also used to scan with the mode image. The difference is that the identification of the water-bearing zone is scanned on the resistivity image, and the identification of the fault zone is scanned on the reflected seismic offset image. The resistivity image is converted into a BMP bitmap file, and the three-dimensional array is composed of the coordinates (X, Y) of the pixel points and the color values of the pixel points. The purpose of this is to apply the correlation coefficient identification method.

Figures 4.14 and 4.15 are a section of tunnel resistivity profile and the result of discrimination. The blue area in Fig. 4.16 represents the part with a large positive correlation coefficient, indicating that it is a water-rich section. Similarly, the identification of the water-bearing zone can also be carried out by ground penetrating radar detection images. Due to the large difference in dielectric constant between water and surrounding rock media, when the electromagnetic wave emitted by the radar encounters the water-bearing zone, strong reflected signal will appear on the radar scanning image due to the different electrical differences of the medium. A strong reflected signal appears on the radar scanned image. Therefore, by discriminating the radar scan image, it can be determined that the strong reflection areas in the image can be judged as water-rich zones.

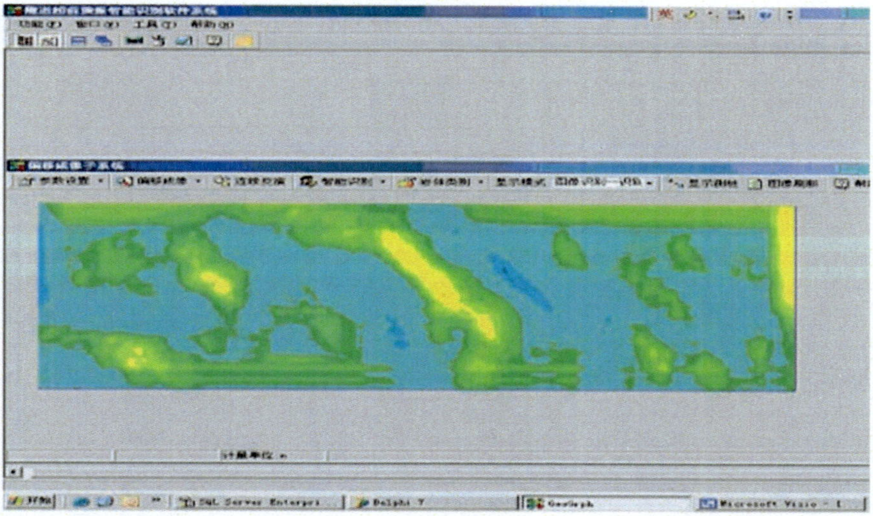

Fig. 4.16 Intelligent recognition result mode of resistivity image

4.3 Ground Penetrating Radar Three-Dimensional Detection Technology

Ground penetrating radar has been widely used in the field of engineering geophysical exploration because of its fast, efficient, and informative features, and its market application field is also very broad. At present, ground penetrating radars at home and abroad use low-frequency antennas to achieve detection in the depth range of 30 m.

Ground penetrating radar +100 MHz shielded antenna has become one of the commonly used methods for tunnel geological prediction. This method is especially sensitive to water and can identify the water damage in front of the tunnel face. However, the current prediction results of ground penetrating radar are two-dimensional. In order to show the distribution of water in front of the tunnel face more clearly and intuitively, the ground penetrating radar three-dimensional geological advance prediction technology is researched and developed based on the theory and method of ground penetrating radar.

4.3.1 Detection Principle of Ground Penetrating Radar

According to Sect. 2.3.3, the basic detection principle of ground penetrating radar is shown in Fig. 4.17.

The equation of electromagnetic wave propagation is

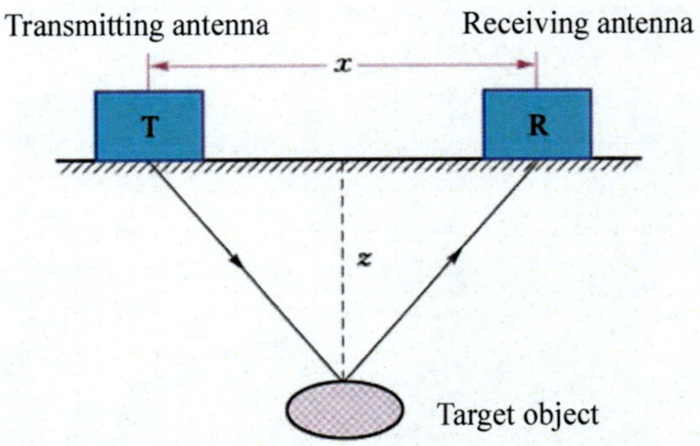

Fig. 4.17 Schematic diagram of detection principle of ground penetrating radar

$$t = \sqrt{4z^2 + x^2}/v \tag{4.7}$$

where

t travel time of pulse wave (ns);
z depth of reflector (m);
x the distance between T and R (m);
v the velocity of the radar pulse (m/ns).

For a shielded integrated antenna, x can be taken as 0, and the equation is simplified to

$$t = \frac{2z}{v} \tag{4.8}$$

Compared with the high-frequency electromagnetic pulse used in ground penetrating radar, the medium encountered in engineering exploration and detection is usually a low-loss medium mainly composed of displacement current. In this type of medium, the reflection coefficient and the wave velocity mainly depend on dielectric constant:

$$r = \frac{\sqrt{\varepsilon_1} - \sqrt{\varepsilon_2}}{\sqrt{\varepsilon_1} + \sqrt{\varepsilon_2}} \tag{4.9}$$

$$v = \frac{c}{\sqrt{q}} \tag{4.10}$$

where

r the reflection coefficient;
v velocity;
ε relative dielectric constant;
c the velocity of light.

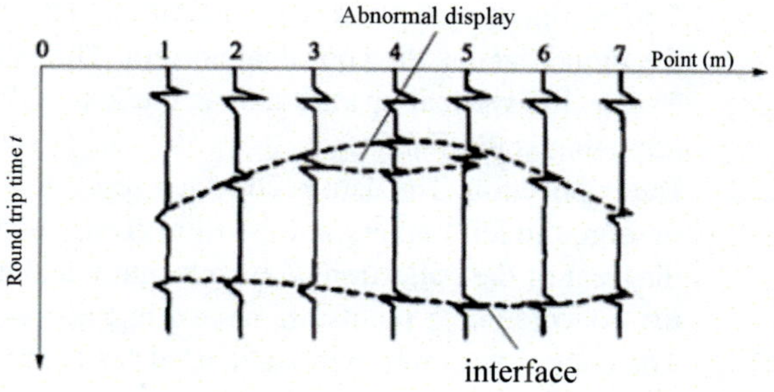

Fig. 4.18 Ground penetrating radar profile

The subscripts "1" and "2" indicate the upper and lower media, respectively. The relative dielectric constant of air is 1, the smallest. The relative dielectric constant of water is 81, the largest.

The receiving antenna receives the reflected wave signal and digitizes it, which is then recorded by the computer in the form of reflected wave. The signal waveform and the cross-sectional image can be displayed in real time on the spot, and the data is automatically saved, as shown in Fig. 4.18. After the corresponding data is processed, the spatial position, structure, and distribution of the underground target can be recognized according to the travel time, amplitude, and waveform of the reflected wave. Ground penetrating radar is based on the analysis of the characteristics of the reflected waveform to identify the underground target body, so its detection effect mainly depends on the electrical difference between the underground target and the surrounding medium, the degree of attenuation of the electromagnetic wave, the buried depth of the target body, and the strength of external interference. Among them, the greater the electrical difference between the target and the medium, the clearer the interface between the two, which is shown in the radar profile that the in-phase axis is discontinuous. It can be said that the electrical difference between the target and the surrounding medium is the basic condition for ground penetrating radar detection. The ground penetrating radar detection method has the characteristics of non-destructive high efficiency, continuous, and intuitive result profile, and thus has been widely used and plays an important role in engineering construction.

4.3.2 Ground Penetrating Radar Three-Dimensional Detection Observation System and Method

Using the characteristics of the ground penetrating radar antenna to control the directional radiation of the radar wave, through the three-dimensional modeling and inversion, the three-dimensional detection of the hydrogeology of the surrounding rock in front of the tunnel and 30 m surrounding area is completed. The implementation steps of the ground penetrating radar three-dimensional detection method are as follows:

① Three-dimensional coordinate definition. The midpoint of the bottom side of the tunnel face is the coordinate origin. The horizontal axis to the right is the X-axis. the vertical upward axis is the Z-axis. The axis along the direction of tunneling is the Y-axis.

② Data collection. The data is collected separately at the front, the periphery, the bottom, the left corner, and the right corner of the tunnel face. The schematic diagram of the collection of each location and the three-dimensional model of the collected data points are shown in Figs. 4.19, 4.20, 4.21, 4.22, and 4.23. The continuous mining mode is used for the front survey lines. Only the start and end coordinates of each survey line are recorded. For spot measurements at several other locations, one file is stored for each location. The initial state and the ultimate state of the 100 MHz antenna is upright and close to the tunnel face. Gradually falling down (M times) at each place on the bottom. Gradually lying on your back (M times) around each place. The corners are performed M times until they fall down, and then rotating a certain angle to the side wall and repeating for a total of N times.

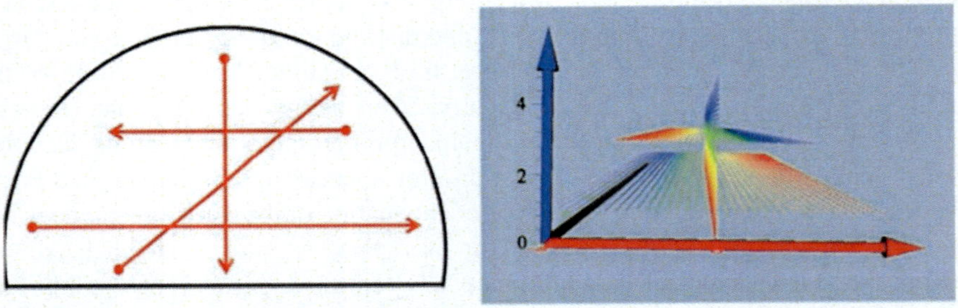

(a) Survey line layout drawing (continuous collection, one folder for each survey line)
(b) Spatial layout drawing of measured data

Fig. 4.19 Layout of the survey lines in front of the tunnel face and the spatial distribution of the measured data

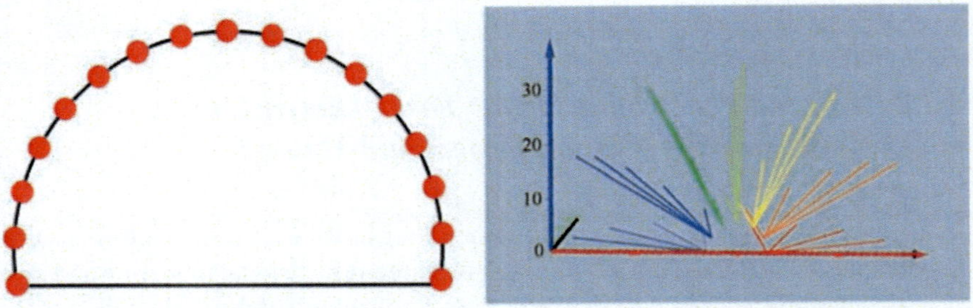

(a) Survey line layout drawing (Spot measurement, one file)
(b) Spatial layout drawing of measured data

Fig. 4.20 Layout of the tunnel lines around the tunnel face and the spatial distribution of the measured data

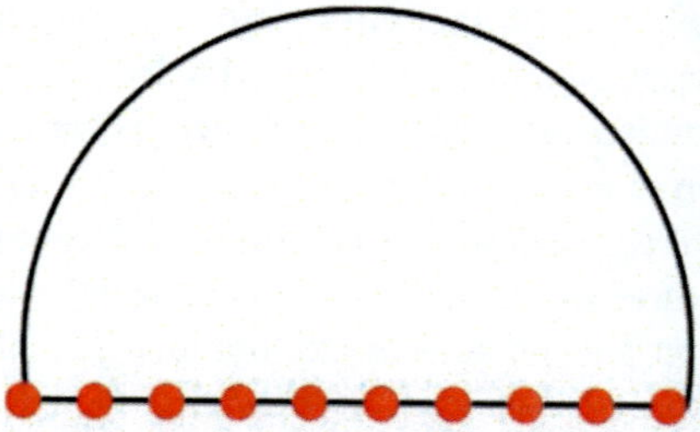

(a) Survey line layout drawing
(Spot measurement, one file)

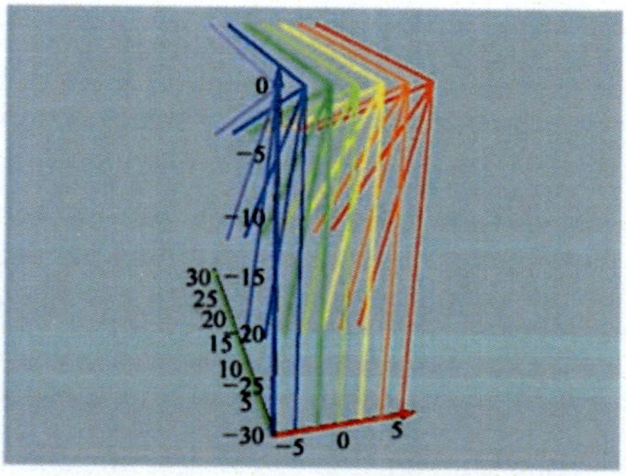

(b) Spatial layout drawing of measured data

Fig. 4.21 Layout of the survey lines at the bottom of the tunnel face and the spatial distribution of the measured data

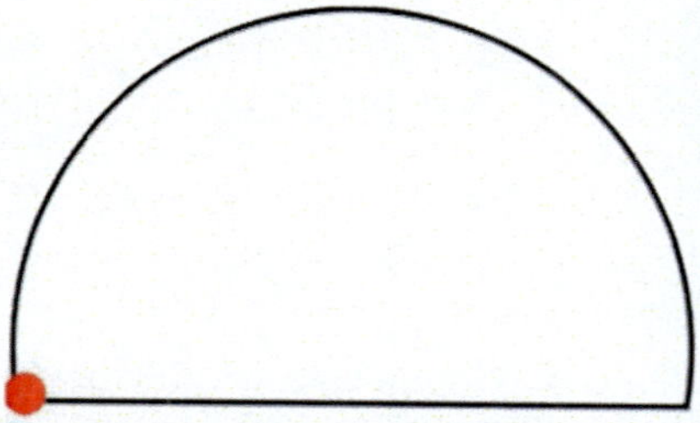

(a) Survey line layout drawing
(Spot measurement, one file)

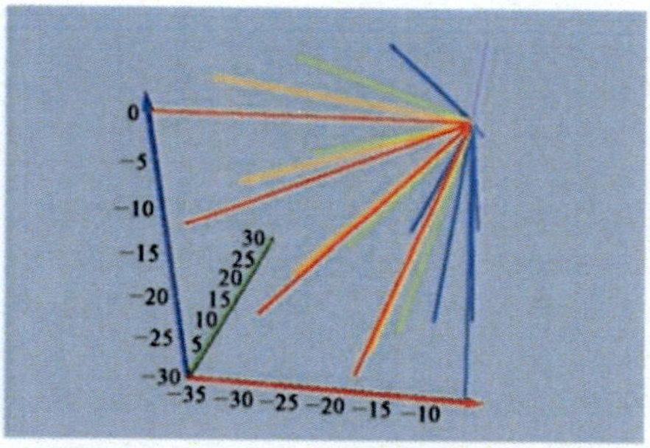

(b) Spatial layout drawing of measured data

Fig. 4.22 Layout of the survey lines of the left wall of the tunnel face and the spatial distribution of the measured data

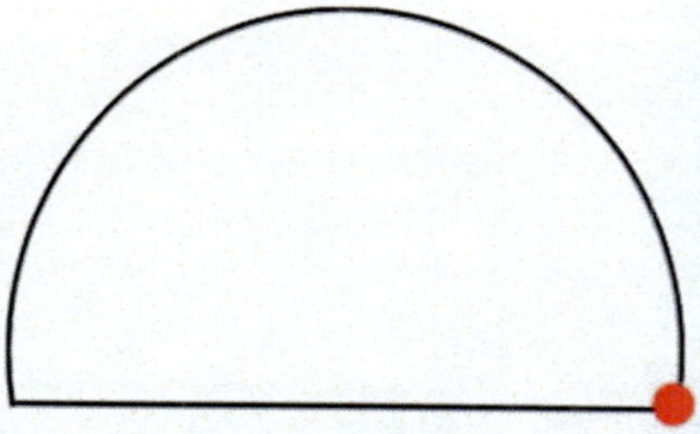

(a) Survey line layout drawing
(Spot measurement, one file)

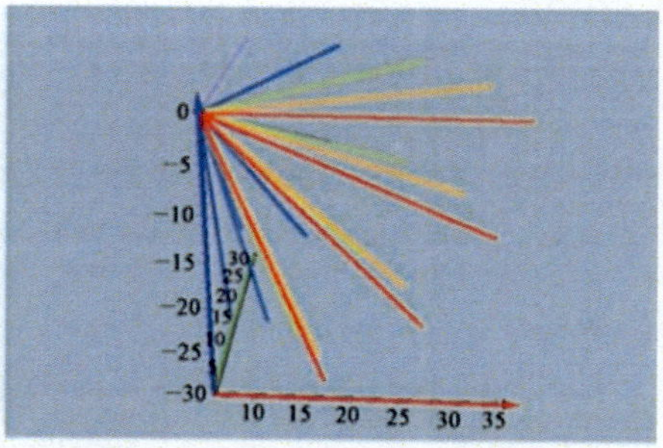

(b) Spatial layout drawing of measured data

Fig. 4.23 Layout of the survey lines of the right wall of the tunnel face and the spatial distribution of the measured data

③ The GPR-3D modeling software is used to transform the coordinates of each sampling point on each survey line of the GPR into the defined three-dimensional coordinate system, as shown in Figs. 4.24 and 4.25.

④ The data transformed into the three-dimensional space is meshed to form a three-dimensional data block, as shown in Fig. 4.26.

⑤ Using the three-dimensional data of GPR, the professional three-dimensional imaging software is used to realize three-dimensional mapping, which can realize fine three-dimensional detection and geological exploration. The 3D GPR detection result of underground space is obtained, as shown in Fig. 4.27.

Electromagnetic waves entering the rock or soil from the air will have strong reflections (corresponding to the ground, and because the electromagnetic waves in the air travel faster, the ground at this time corresponds to a negative phase). Likewise, when electromagnetic waves propagate from the soil layer to the rock layer or other

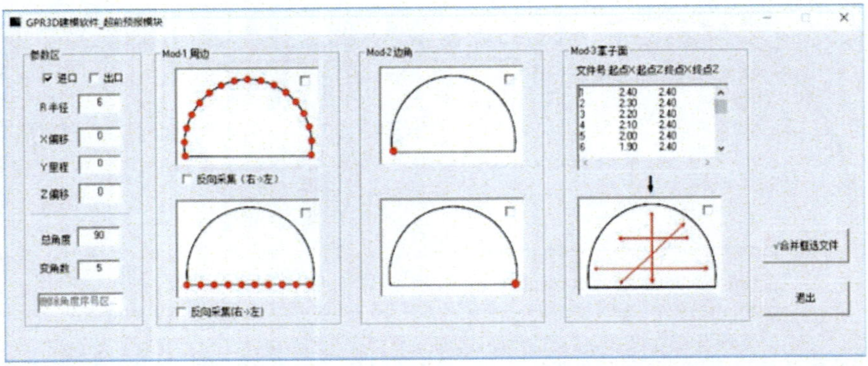

Fig. 4.24 Interface of GPR-3D modeling software

Fig. 4.25 Spatial distribution map of each sampling point of the ground penetrating radar three-dimensional detection

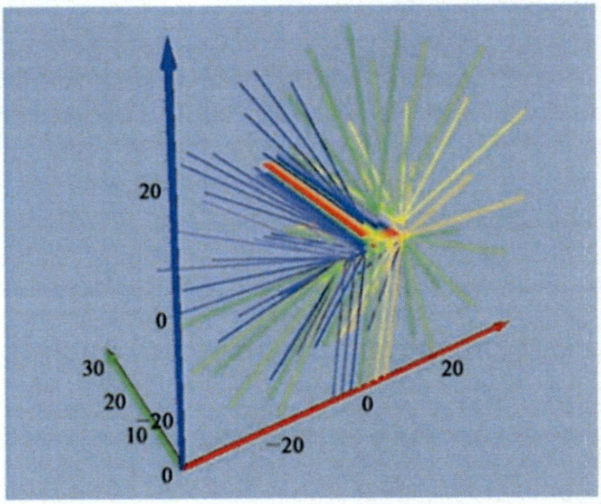

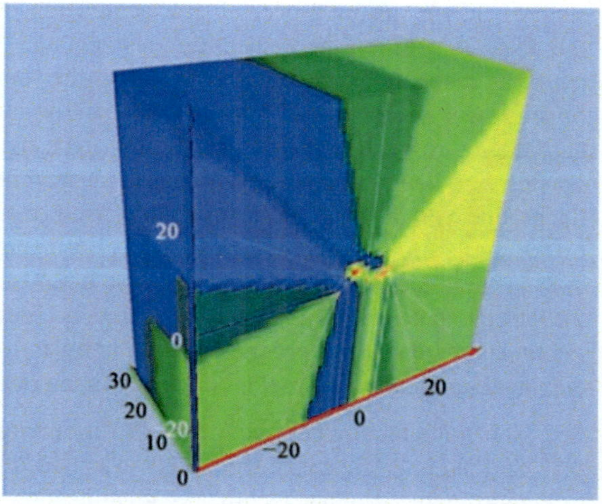

Fig. 4.26 Three-dimensional data block after meshing

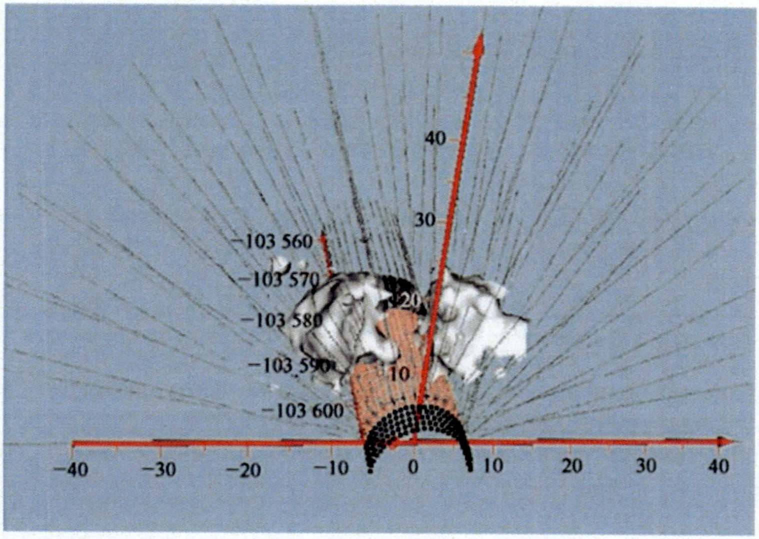

Fig. 4.27 Detection results of 3D GPR

layers, that is, when they propagate on the layers of different media, if the junction is not well-fitted or there are gaps, it will also cause the phase and amplitude of the radar profile to change. The received signal on the host screen will fluctuate, so that the thickness of the layer can be determined. Therefore, in the process of electromagnetic wave propagation, the host will receive the electromagnetic waves emitted by the antenna to judge the approximate thickness of the soil layer and rock formation, the size of the cave and the underground cavern, the location of the fault and the fissure water, so as to predict for the construction. In view of the current state of the art, in the aspect of data acquisition, the radar antenna usually fixes the distance between the transmitting antenna and the receiving antenna. The two antennas move along the measuring line at the same time, and the collected (co) offset data is obtained. Although this acquisition method can quickly obtain radar data, there are various noises in the GDR data, including random noise from the outside world, regular interference caused by the reflection of ground and objects on the ground, and the noise of the instrument itself (the noise of the instrument appears on the single-offset profile as a periodically occurring event axis parallel to the event axis of the direct wave), which masks the effective weak signal and brings great difficulty to data processing and interpretation. To further improve the accuracy of geological prediction of ground penetrating radar, it is necessary to solve three key technical problems of current ground penetrating radar in tunnel advance prediction:

① The contradiction between detection distance and resolution.
② Serious interference of multiple waves and other clutter, low signal-to-noise ratio of the original record, the identification of the effective wave, and the interpretation of the result.
③ Too little amount of spatial information of the detected object, true recognition of explain ambiguity of its data results.

4.4 Typical Cases

4.4.1 Application of High-Power Electric Sounding Three-Dimensional Detection Technology in a Tunnel

(1) **Project overview**

The project exploration area is located in a highway tunnel in Yunnan Province, which belongs to the high mountain landform of structural denudation and cutting. The tunnel passes through an alpine mountain with an altitude of 2000–3000 m and a relative height difference of 800–1000 m, as shown in Fig. 4.28. The tunnel area is crisscross with ravines with water system developed, and the valley is deep. The average buried depth of the tunnel is 765 m.

Rivers and streams in the tunnel area are very developed, and there are reservoirs, lakes and fishponds distributed along the tunnel area. Vegetation develops, and trees and shrubs flourish. When the tunnel is excavated, it is very easy to leak and collapse, which brings great difficulties to the tunnel construction.

The geological ages of the strata in the area where the tunnel crosses are mainly sedimentary rocks of the Paleozoic Cambrian (∈3b), Ordovician (O_1, O_2, O_3), and part of the Silurian (S_3) and Devonian (D_1). The lithology includes silty clay rock, fine sand, coarse sand, limestone, mudstone, shale, etc. The lithology is broken in part of the mileage.

There are many fault zones in the tunnel area, including the Ashizhai fault, the ticketing dam fault, the Yancheng fault, the Hongmuling dry-twisted fault, and the

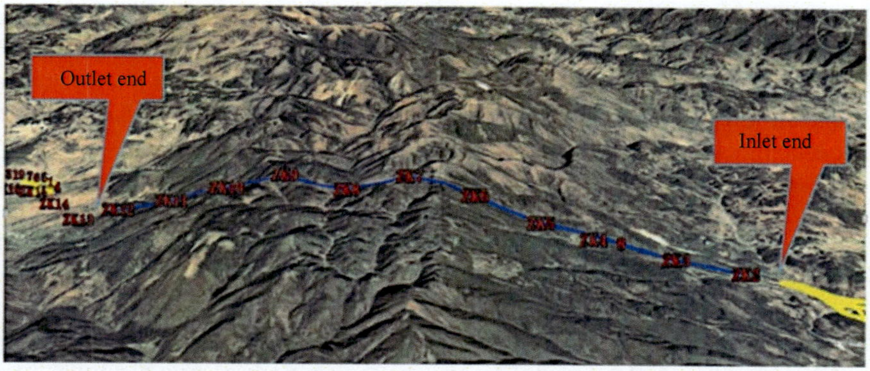

Fig. 4.28 Topographical features of the tunnel area

Zhushitun twist-torque fault, which are almost orthogonal to the tunnel. There are as many as 16 secondary faults derived from the main fault, which greatly affects the safety of tunnel construction.

(2) 3D data imaging and result interpretation

Data collection is performed on the tunnel surface, and the collected data is processed and inverted. Combined with the geological conditions, a tunnel resistivity sectional view is obtained, as shown in Fig. 4.29.

As shown in the figure, the axial resistivity of the tunnel is divided into two layers. The upper part has a low resistivity of about 50–300 Ω•m, and the buried depth is about 50 m underground. There are five low-resistance areas in the tunnel, which are K6+200, K7+100 to K7+300, K8+100 to K8+300, K9+800, and K8+700 to K9+500, and the average resistivity value is about 70–240 Ω·m. They are the geological hidden areas during tunnel construction.

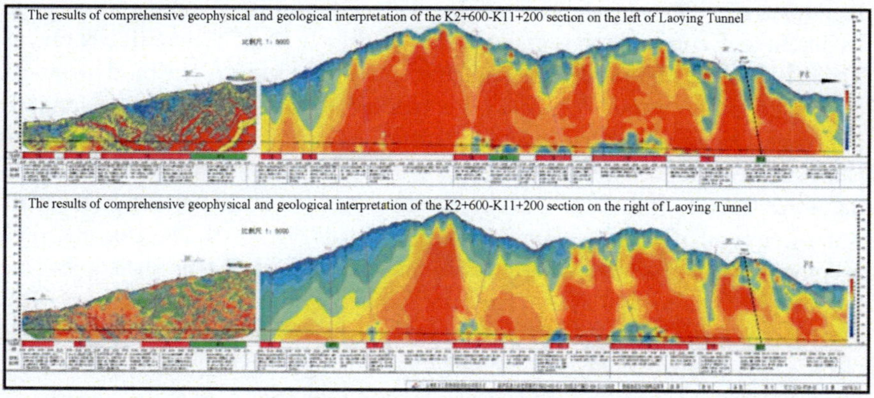

Fig. 4.29 Sectional view of the resistivity at the mileage section K2+600 to K11+200 of the tunnel

Combined with the characteristics of surface fracture, it is concluded that there are eight faults in the tunnel axis from F8 to F15. The five faults F10, F11 1, F12, F13, and F14 correspond to the low-resistance zone, and the intersections of the fault and the tunnel are geological disease points. The fault F10 is located at K6+200, where the lithology is calcareous siltstone, fine sandstone, and limestone. It is speculated that the rock at the crossing point of the tunnel is broken. This fault has poor integrity and abundant groundwater. It is a hidden geological point during tunnel construction. The fault F11-1 is located at K7+100 to K7+300, where the lithology is calcareous siltstone and fine sandstone. It is estimated that the fault is affected by the fault and the lithology is broken, which is prone to water leakage and collapse. The fault F12 is located at K8+100 to K8+300, where the lithology is limestone, calcareous siltstone, and fine sandstone, which is also the geological disease point during tunnel excavation. The F13 fault is located at K8+700 to K9+500, where the lithology is calcareous siltstone, fine sandstone and argillaceous siltstone. The area is 800 m long. When the tunnel is drilled here, the advanced geological prediction should be done, and the support should be strengthened. The F14 fault is located at K9+800 and the tunnel location are a high-resistance response. At present, the tunnel is still under construction, and the excavated section is consistent with the detection results.

4.4.2 Application of Ground Penetrating Radar Three-Dimensional Detection Technology in a Tunnel

(1) **Project overview**

The research result has been applied in the mileage section P1K2+192 to P1K2+220 of a tunnel guide hole. The section of the rock formation consists of the hail layer, the third member of the Upper Triassic Jiazhao Formation, and the Upper Permian. The hail layer is composed of glacial melted sedimentary sand, clay, and gravel. Its thickness varies with the topography, with certain sorting and rhythm changes. Its parent rock composition is granodiorite, sandstone, SLATE, andesitic basalt, etc. The hail layer is rich in water and has sufficient surface supply. The third member of the Upper Triassic Jiajila Formation is composed of purple-red silty mudstone with purple-gray intermediate layered clastic sandstone and glutenite lens. The rock mass is broken, and the rock mass is sand-like and fragmented. The rock mass is soft and the water stability is poor. It is easy to weather and crack. The joint fissures with a water-rich level of weak-medium is very developed (Fig. 4.30). The Upper Permian is composed of gray Anshan basalt, slate, metamorphic sandstone slate, gray-green rhyolite, Ying'an rock, diabase. The joint fissures with a water-rich level of weak-medium are developed, and the rock mass is relatively broken. The contact surface of the same overburden layer is not integrated in angle, and the water permeability and water richness near the contact surface are strong (Fig. 4.31).

(2) **3D data acquisition and analysis**

Fig. 4.30 Lithology of the third member of the Upper Triassic Jiajila Formation

Fig. 4.31 Permian upper lithology pole contact unconformity

The instrument adopts SIR 3000 mainframe and is equipped with 100 MHz antenna. For the purpose of large-scale deep exploration, 64 times of superposition point measurement method is adopted, and the time window is set to 1500 ns. According to the site conditions of the tunnel face, the antenna is manually lifted, and the radiation angle of the antenna is controlled to achieve the purpose of directional radiation of radar waves. Radar data collection is performed on the surrounding rock in front of the tunnel face excavation and the surrounding rock.

By analyzing the amplitude spectrum of the standard reflected signal of the radar wave, the frequency response spectrum of the polarization of the electromagnetic wave by the target medium such as rock, air and water is separated (Fig. 4.32). According to the configuration of the ground penetrating radar 100 MHz antenna, the characteristics of frequency response lines of the target medium are calibrated

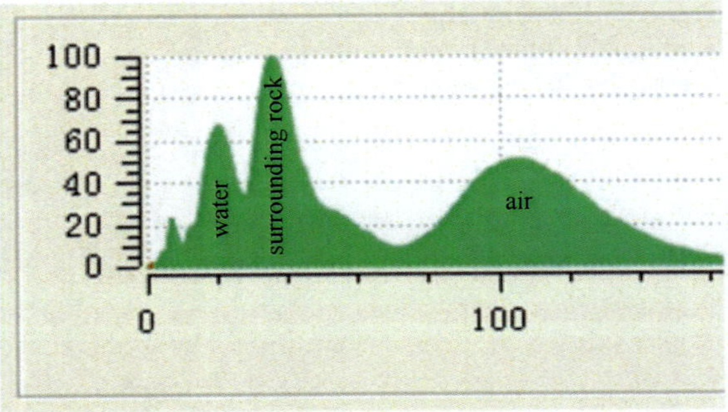

Fig. 4.32 Frequency response spectrum of the target medium to electromagnetic wave polarization

separately, identified, and extracted. According to the calculation formula, the spectral energy intensity of air and water is calculated. The integrity of the surrounding rock mass in front of the tunnel face, the distribution range and scale of groundwater, and other factors that affect construction safe are evaluated, and suggestions for tunnel construction are put forward to ensure the safety of tunnel construction (Table 4.1).

(3) **Interpretation of the results**

As shown in Fig. 4.33, the 3D image reflects that the surrounding rock in front of and 30 m around the tunnel face is rich in water (green marked part). The upper arch is in a loose and dry state (white marked part). Therefore, the following interpretation results of advance prediction are given:

(1) The tunnel face of Pilot tunnel (P1K2+192) has entered the water-rich area of the stratum structure, and the vault of the construction working surface is in the shape of stream and water. It is rich in water.
(2) At P1K2+192 to P1K2+220 (28 m) in front of the tunnel face, the upper part presents a loose and dry area and the lower part presents a water-rich area. It is inferred that the stratum structure is rich in water area. Due to the presence of drainage channels above the construction work surface, the upper part in front of the tunnel tunneling is a dry state and the lower part is in a water-bearing state, which is easy to cause roof fall and collapse.

In order to verify the forecast results, the lithological characteristics and geological conditions of the section were revealed by continuous real-time tracking and recording of the site, indicating the water-rich features and distribution in this section, as shown in Fig. 4.34.

Table 4.1 Frequency response characteristics of different media for 100 MHz antenna calibration

Target media	Dominant frequency response (MHz)	Bandwidth (MHz)
Air	103	20
Water	16	10

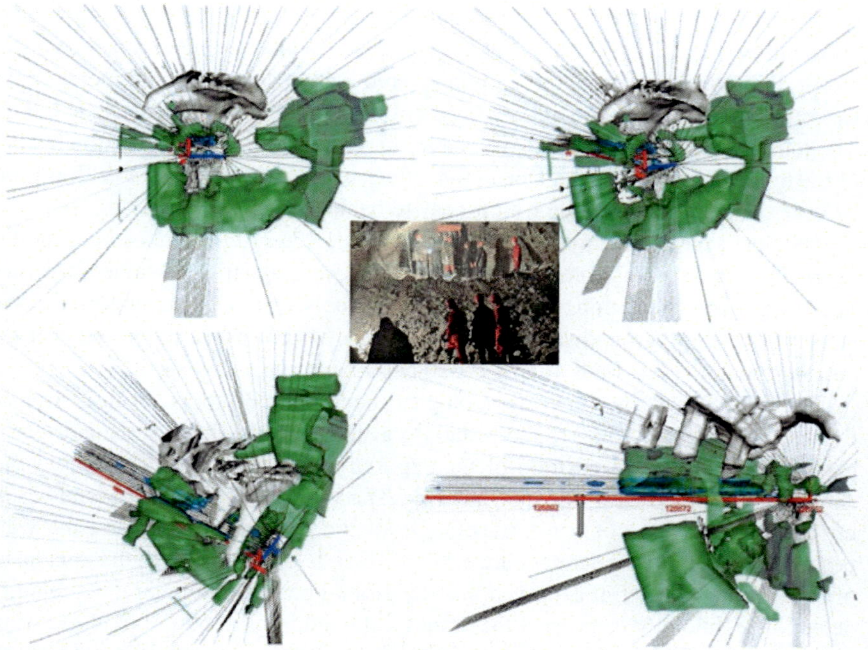

Fig. 4.33 Effect drawing of 3D geological radar tunnel geological advance prediction

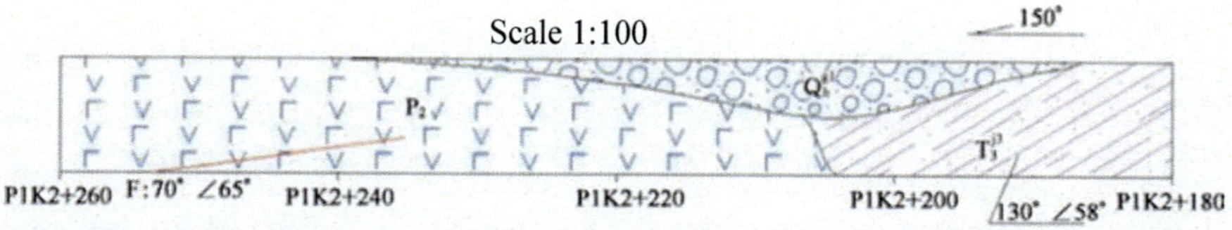

The surrounding rock stratum of this section is composed of Quaternary moraine(Qhq1), the third member of the Upper Triassic Jiajila Formation(T3j3), purplish red mudstone mixed with sandstone, Upper Permian, andesite basalt and Slate

Fig. 4.34 Sketch map for real-time tracking of surrounding rock stratum geology on site excavation

The surrounding rocks in the range P1K2+190 to P1K2+205 are composed of purple-red strong weathered silty mudstone with purple-gray intermediate layered clastic sandstone. The joint fissures develop and are not integrated with the underlying strata. The rock mass is wet and seepage, and the part is in the shape of being drenched by rain. The rock mass is broken and is sand-like and fragmented. The rock is soft. The water stability is poor, and it is easily weathered and cracked. After five days of excavation, collapsed cavities appeared above the tunnel face of P1K2+195 to P1K2+189 in the Pingdao Hole. The collapsed material is mainly pebbling with confluent water.

The mileage of P1K2+205 to P1K2+220 is composed of hailstone pebble layer and drifting stone sand. The lower part is composed of gray strong weathered Anshan basalt, light gray green thin to medium layered pebbly sandstone and metamorphic sandstone. The rock mass is in the shape of block and pick, and it is easy to fall off. The joint fissures develop. There is a large surge of water in the arch on the left side of the tunnel face P1K2+205. Due to the influence of groundwater, the vault is prone to large collapse, and the side wall is easy to be unstable.

The upper part of the mileage P1K2+220 to P1K2+240 is composed of the hail layer. The lower part is composed of brown-yellow, gray strong weathered Anshan basalt, light gray green thin to medium layered pebbly sandstone, metamorphic sandstone. Affected by groundwater, there are risks of water inrush, slabs, and roof fall. Among them, the left vault of P1K2+220 has a large surge of water.

The forecast results show the geological conditions of about 30 m in front of the tunnel face. According to the actual excavation results, the loose and dry areas are consistent with the position of the exposed Quaternary hail layer Q_h^{al}, which should be the riverbed of the ancient river. The water-rich part corresponds to the unconformity of the contact surface angle of the overlying strata, and the water permeability and water richness are strong, which is consistent with the conclusion of excavation tracking. It is proved that the ground penetrating radar 3D geological detection technology can detect the geological conditions about 30 m ahead of the tunnel face, with high accuracy and intuitive imaging. It is an efficient and accurate method for tunnel advance prediction.

Chapter 5
Three-Dimensional Detection Technology for the Side Slope of the Tunnel

With the rapid development of China's social economy and infrastructure construction, more and more railways, highways and hydropower stations will be built in the mountainous areas, and tunnel engineering is one of the necessary structures. Due to the erosion of surface water, the development of weathering fissures and the excavation of rock–soil mass, the stability of the side slope around the tunnel entrance is poor. In the mountainous area, due to the restriction of topographical conditions and line shape, the location of the tunnel entrance must be selected at the position of "possibly unstable slope." Under the unloading action of the rock–soil mass, the failure form of landslide or collapse may occur, which may cause economic loss and casualties.

This chapter mainly discusses the use of comprehensive geophysical methods to survey the side slope around the tunnel entrance to understand the factors affecting its stability and provide guarantee for the safe construction of the tunnel.

5.1 Characteristics of the Side Slope of the Tunnel

5.1.1 Overview of the Characteristics of the Side Slope of the Tunnel

The portal section of the tunnel is generally on a slope that is severely eroded by surface water and has relatively developed weathering fissures. In addition, the buried depth of the tunnel at portal section is often shallow, and the rock–soil mass in the upper part of the structure is difficult to form a load-bearing arch. Therefore, the surface slope of the side slope of the tunnel is easy to be pulled and cracked. If the surface water penetrates into the crack, its stability is difficult to guarantee. Therefore, the portal section of the tunnel is often one of the sections causing collapse and

G. Cao and H. Liu, *Three-Dimensional Exploration Technology of Tunnel Geology*, Key Technologies for Tunnel Construction under Complex Geological and Environmental Conditions, https://doi.org/10.1007/978-981-16-9225-3_5

cracking damage, so that the stability of the side slope around the tunnel entrance is a problem that must be taken seriously during tunnel design and construction.

When the deformation of the side slope around the tunnel entrance reaches a certain level, the stability of the side slope will be destroyed. According to the failure mode of the side slope, its failure mode can be classified into plane failure, wedge failure, rotation failure, dumping failure, and collapse failure. In engineering practice, the main failure mode is wedge failure.

5.1.2 Factors Affecting the Stability of the Side Slope of the Tunnel

The basic factors affecting the stability of the side slope are various, and the stability of the rock slope is temporary and conditional. Slopes are always in an unstable–stable–unstable cycle. If the range of slope stress changes within the allowable strength of the rock slope, the stress adjustment will not cause the slope failure; otherwise, it will lead to deformation of the slope.

There are many factors that change or affect the stress state of the slope rock mass, mainly including:

① The influence of topography. Slopes with unfavorable form and scale often produce tensile stress at the top of the slope, causing tensile cracks on the top of the slope, strong shear stress at the foot of the slope, and shear failure zone, which reduces the stability of the slope.

② The influence of rock mass properties. The mechanical properties of the rock mass determine the way of slope failure. For soft rocks, such as shale, tuff, marl, and other rocks with a layered structure, it is easy to be softened by soaking in water, and the strength is reduced, forming a weak layer, which is prone to landslides. In general, the better the engineering geological properties of the rock–soil mass, the higher the stability of the slope.

③ The influence of rock mass structure. Rock mass failure is not only a material failure, but also a structural instability. The rock mass structural factors affecting the stability of the slope are mainly the number of structural planes, the dip angle and inclination of the structural plane, the continuity of the structural plane, and the surface properties of the structural surface. The deformation of slope of different rock mass structures is different.

④ The influence of geological structures. It is manifested by the development degree, scale, connectivity, filling degree of the structural plane and the influence of the filling state and the output state of the structural plane on the slope stability. In areas with complex regional structures and relatively active neotectonic movements, the stability of slopes is poor. The development mode and density of unstable slopes are closely related to the direction and position of the structural line.

⑤ The influence of other factors, such as atmospheric precipitation, groundwater, climate and weather changes, vegetation and time.

The stability of the side slope of the tunnel affects the success or failure of the entire tunnel project, and the tunnel entrance is also the most complicated and most prone to safety accidents in the entire tunnel project. Therefore, how to ensure the safety and stability of the side slope during tunnel construction is still a concern of tunnel construction. Therefore, an effective method is adopted to understand the engineering geological conditions of the rock–soil mass around the tunnel entrance, which is of great help to the safe construction of the tunnel construction.

5.2 Typical Case

5.2.1 Application of High-Density Resistivity Method in a Tunnel

(1) **Project overview**

A highway tunnel in Yunnan is a two-hole, one-way, two-lane tunnel with separated left and right sides. The length of the right tunnel is 1213 m. The length of the left tunnel is 1236.5 m. The maximum buried depth is about 204 m.

The surrounding rock of the tunnel is dominated by gneiss. Due to the influence of regional geological structure, the joint fissures are developed, weathered and broken, and weathered deeply. The overlying residual slope contains gravel subclay and gravel soil. The entrance section is thick, which is composed of landslide stacked subclay with a thickness of 15.7 m. The surrounding rock is very unstable.

During the construction of the tunnel, large-scale collapse occurred in the entrance section, and the exit of the left tunnel was completely buried. Nearly, ten large cracks appeared in the mountain above the tunnel, and the width of the crack was up to 2.5 m (Fig. 5.1). A few days later, tow collapses occurred. The right tunnel entrance collapsed. It is concluded that the tunnel disease is caused by a landslide.

In order to ascertain the engineering geological and hydrogeological conditions of the tunnel, analyze the cause of tunnel collapse, find out the scope, thickness and scale of the landslide body, and provide scientific and reliable information for the next step, the entrance section of the tunnel was specially surveyed for geological diseases.

(2) **Engineering geological conditions**

① Topography. The tunnel is located in the middle of the north bank of the Red River Valley, which belongs to the mountainous landform of the structural erosion canyon. The tunnel has developed lateral valleys, with deep cuts, steep terrain, and steep slopes ranging from 20° to 45°, with a maximum of 65°. The vegetation is relatively developed. Many of the trunks are inclined downwards

Fig. 5.1 Collapse of the tunnel entrance

due to the downward collapse of the surface rock. Between the left and right tunnels, there is a paddy field in the projection range of the left tunnel. Due to long-term water storage, the water content of the rock and soil underneath has changed greatly.

② Composition of the strata.

(a) Subclay. It is brown, brown-yellow, and it is the loose accumulation of gravel-bearing debris in the Quaternary of the Cenozoic. The entrance of the tunnel is thick with a thickness of 15.7 m, which is a landslide deposit. As the slope collapses and slips downwards, many tensile cracks appeared on the surface, with maximum width 2.5 m, maximum depth 10 m, visible depth 4 m. The tensile cracks run east west and are perpendicular to the direction of the slope. In addition, there is a group of shear fissures whose occurrence is the same as that of a group of weak structural planes 75°∠55° of bedrock. The cracks appear in the middle of the left and right entrance and develop to the upper right for about 3 m. There is obvious scratch between the cracks. On the right upper slope of the entrance of the right tunnel, shear cracks in the form of landslide steps can be seen, which run along the slope and is about 20 m long.

(b) Sericite gneiss. It is gray, and it is the metamorphic rocks of the Yaoshan Group in the Proterozoi, with joints, foliation and fissures develop, and the morphological occurrence is 75°∠55°. Fully weathered to strong weathering, the rock mass is relatively broken to broken, with loose structure. According to the Code for Design of Highway Tunnels, the surrounding rock is classified into Grade V and $B_Q \leq 250$.

(c) Geological structure. The tunnel is about 3 km east of the Honghe deep fault. The fault is active for several periods. The geological structure has a large impact range, and the joint fissures of rocks develop. The fracture group is mainly 225°–300°∠58°–85°.

(d) Hydrogeological conditions. The type of groundwater in this area is weathered bedrock fissure water with weak to medium water abundance. Due to the large terrain cutting, the groundwater is buried deep. No groundwater is found under the floor of the tunnel. Atmospheric precipitation in the rainy season is used for replenishment, and the water seeps into the deep, which is predicted to have little impact on tunnel construction.

(3) Program layout

As shown in Figs. 5.2 and 5.3, a total of 12 detection sections of high-density resistivity method were arranged for this work.

Six sections of high-density resistivity method were placed on the surface of K44+950 to K45+450 along the line direction, numbered EZ1, EZ2, EZ3, EZ4, EZ5, and EZ6, with a line spacing of 30 m. Among them, EZ3 and EZ5 were arranged directly above the left tunnel and the right tunnel, respectively. Each survey line is about 500 m long. 55–75 layers are collected, and the sounding depth is about 120 m.

Six sections of high-density resistivity method were placed perpendicular to the line direction, numbered EH1, EH2, EH3, EH4, EH5, and EH6, with a line spacing of

Fig. 5.2 Schematic diagram of the whole project and geophysical plane layout

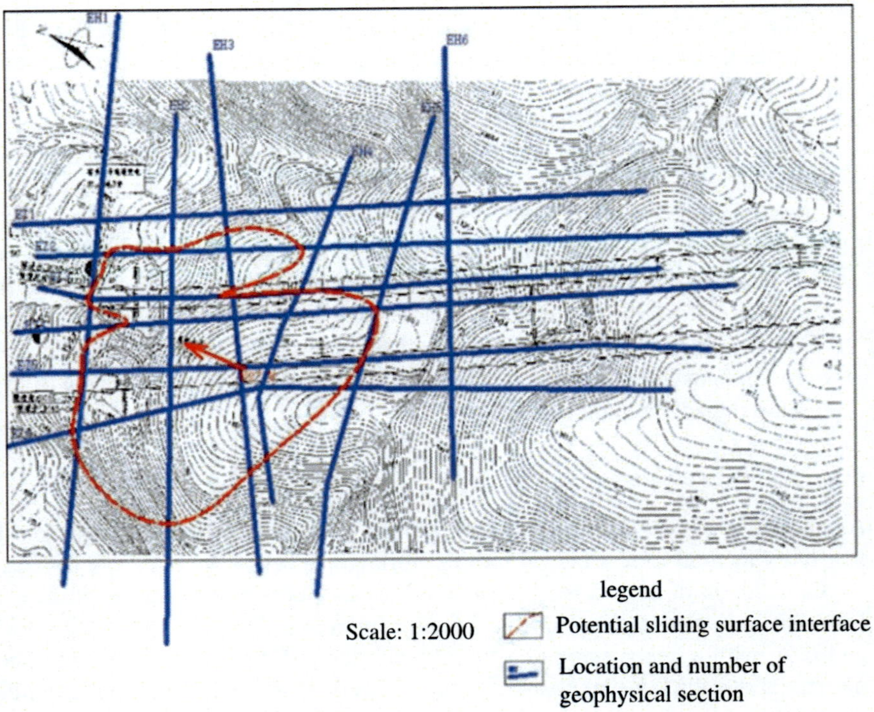

Fig. 5.3 Flat picture of comprehensive results

about 50 m. The six profiles cross the sections with pile numbers K45+000, K45+056, K45+100, K45+150, K45+200, and K45+250 of the left line. The electrode spacing of each survey line is 2 m. Each line is about 400 m long. Fifty layers are collected, and the sounding depth is 100 m.

The length of each section and the number of sounding points are shown in Table 5.1.

(4) **Detection result**

By performing data conversion, dead pixel culling, and terrain correction on the collected raw data, 12 apparent resistivity images are obtained. Figures 5.4 and 5.5 are schematic diagrams showing the spatial position of the apparent resistivity in the longitudinal and transverse sections.

The red and yellow parts of the apparent resistivity distribution image represent areas of high apparent resistivity, and the dark blue and light blue parts represent areas of low apparent resistivity. The red dotted line in the resistivity image represents the inferred potential sliding surface, and the parts circled by black dotted line represent the collapsed loose areas.

Figures 5.6, 5.7, 5.8, 5.9, 5.10, and 5.11 are apparent resistivity images corresponding to longitudinal sections of EZ1 to EZ6 (along the direction of the tunnel). The six images are now explained as follows.

Table 5.1 Table of physical workloads of tunnel geological disease survey

The work items		Unit	The workload	Note
Measurement	1:1000 topographic profile measurement geological survey	Bar km^2	12 (5544 m) 0.5	
Geophysical prospecting	Resistivity imaging detection profile EH1	Length (m)/number of sounding points	430/216	Pass through K45+000
	Resistivity imaging detection profile EH2	Length (m)/number of sounding points	406/204	Pass through K45+056
	Resistivity imaging detection profile EH3	Length (m)/number of sounding points	406/204	Pass through K45+100
	Resistivity imaging detection profile EH4	Length (m)/number of sounding points	286/144	Pass through K45+150
	Resistivity imaging detection profile EH5	Length (m)/number of sounding points	358/180	Pass through K45+200
	Resistivity imaging detection profile EH5	Length (m)/number of sounding points	310/156	Pass through K45+250
	Resistivity imaging detection profile EZ1	Length (m)/number of sounding points	550/276	
	Resistivity imaging detection profile EZ2	Length (m)/number of sounding points	550/276	
	Resistivity imaging detection profile EZ3	Length (m)/number of sounding points	526/264	
	Resistivity imaging detection profile EZ4	Length (m)/number of sounding points	574/288	
	Resistivity imaging detection profile EZ5	Length (m)/number of sounding points	574/288	
	Resistivity imaging detection profile EZ6	Length (m)/number of sounding points	574/288	
Total		Length (m)/number of sounding points	5544/2784	

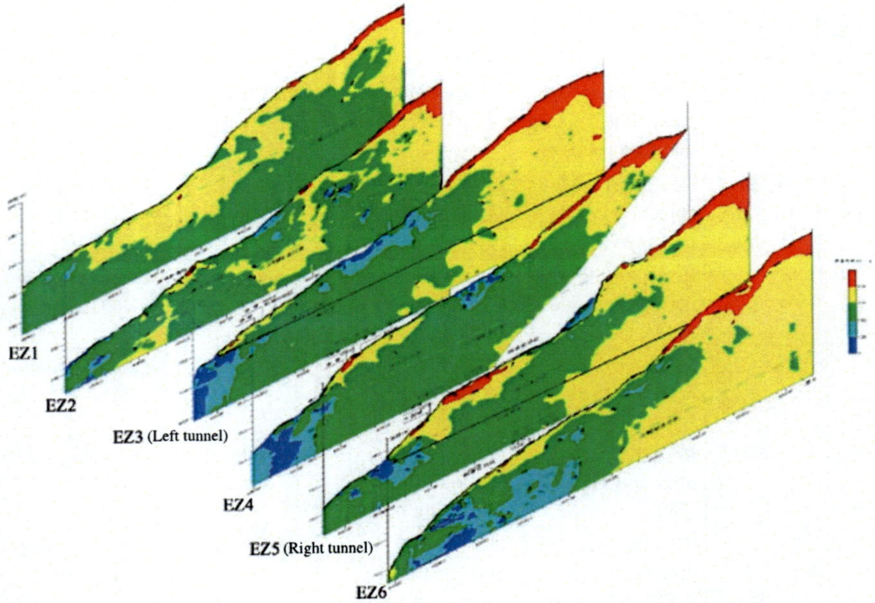

Fig. 5.4 Schematic diagram of the spatial position of the longitudinal section of the apparent resistivity

Figure 5.6 is the apparent resistivity image corresponding to the section EZ1. This section is placed 60 m to the left of the left tunnel (from a small mileage to a large mileage). The distribution of apparent resistivity in the figure is uniform, and there is no obvious apparent resistivity anomaly, indicating that the lithology of the section is stable and there is no collapsed and loose area.

Figure 5.7 shows the apparent resistivity image corresponding to the section EZ2. The section is placed 30 m to the left of the left tunnel (from small mileage to large mileage). The figure shows that there is a high-impedance anomaly area in the section K45+075 to K45+140, which is represented by the parts circled by black dotted line in the image. It is inferred that the anomaly area is a loose area caused by collapse. The area is 65 m long with an average thickness of 15 m.

Figure 5.8 shows the apparent resistivity image corresponding to the section EZ3. The section is placed directly above the central axis of the left tunnel. The figure shows that there is a high-impedance anomaly area in the section K45+000 to K45+090, which is represented by the parts circled by black dotted line in the image. It is found that there are many ground fissures in the area. It is inferred that the anomaly area is a loose area caused by collapse. The area is 90 m long with an average thickness of 15 m. The position of the potential sliding surface is indicated by a red dotted line in the figure. It can be seen from the figure that the anomaly area is shallow and only affects a small section of the tunnel entrance, indicating that the tunnel has not been completely buried.

Figure 5.9 shows the apparent resistivity image corresponding to the section EZ4. The section is placed between the left and right tunnels. The figure shows that there

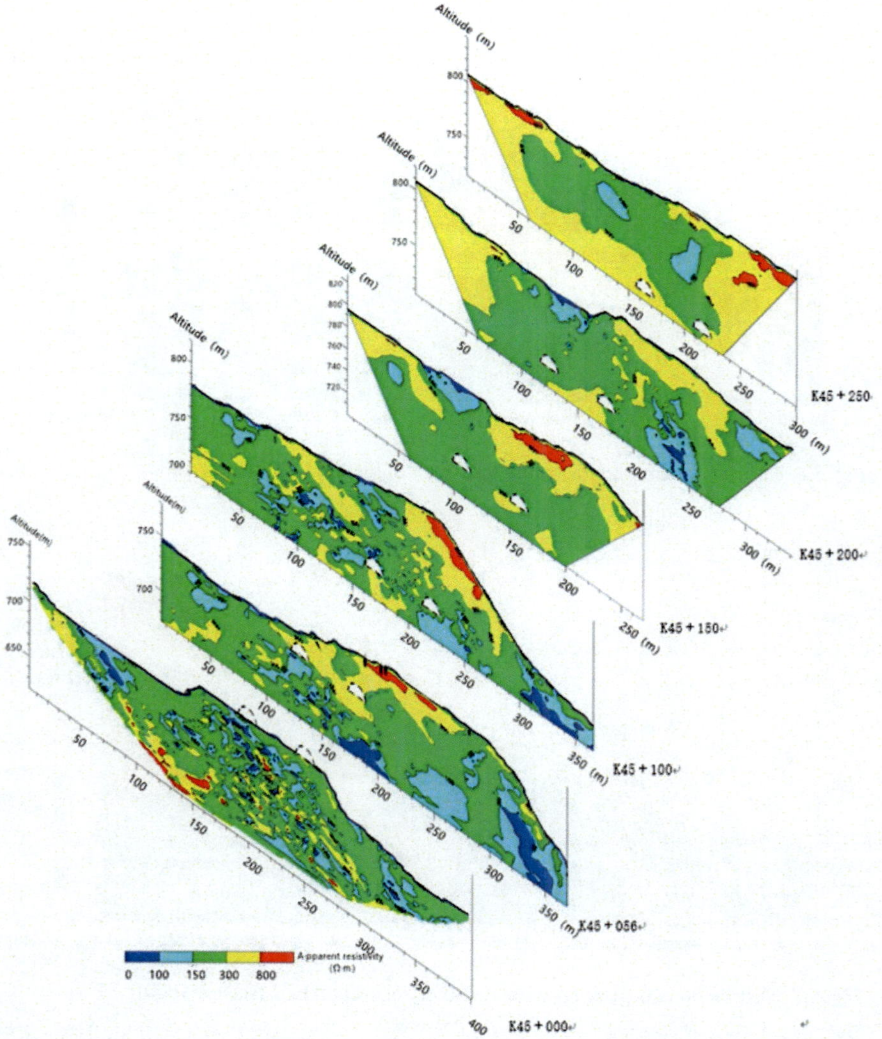

Fig. 5.5 Schematic diagram of the spatial position of the transverse section of the apparent resistivity

is a high-impedance anomaly area in the section K45+030 to K45+150, which is represented by the parts circled by black dotted line in the image. It is found that there are many ground fissures in the area. It is inferred that the anomaly area is a loose area caused by collapse. The area is 120 m long with an average thickness of 25 m.

Figure 5.10 shows the apparent resistivity image corresponding to the section EZ5. The section is placed directly above the central axis of the right tunnel. The figure shows that there is a high-resistance anomaly area in the section K45+030 to

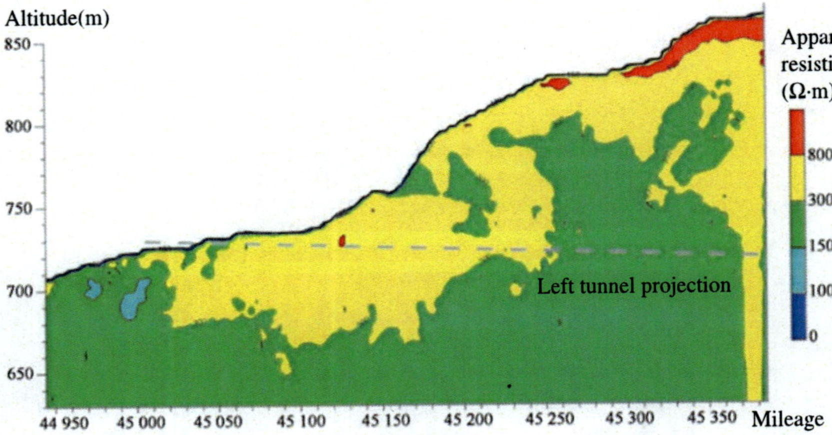

Fig. 5.6 Distribution image of apparent resistivity of section EZ1 (scale 1:1000)

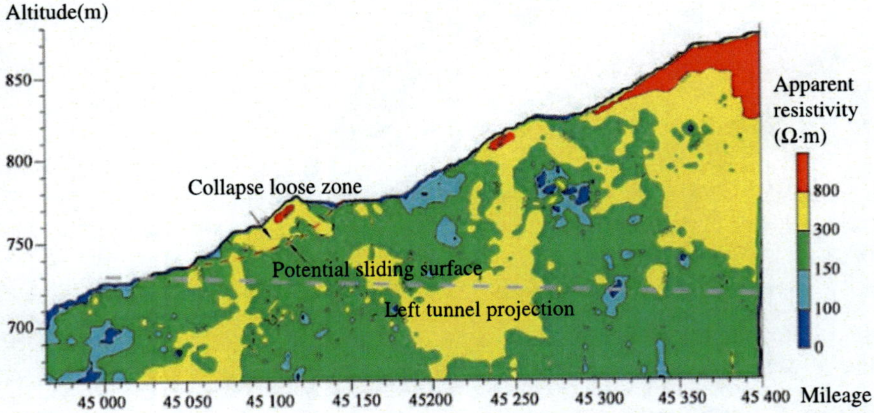

Fig. 5.7 Distribution image of apparent resistivity of section EZ2 (scale 1:1000)

K45+260, which is represented by the parts circled by black dotted line in the image. It is found that there are many wide ground fissures in the area. It is inferred that the anomaly area is a loose area caused by collapse. The area is 230 m long with an average thickness of 40 m. The position of the potential sliding surface is indicated by a red dolled line in the figure. The loose area in the section K45+100 to K45+120 has extended to the tunnel position, indicating that the collapse of the right tunnel is serious.

Figure 5.11 shows the apparent resistivity image corresponding to the section EZ6. The section is placed 30 m to the right of the right tunnel (from small mileage to large mileage). The figure shows that there is a high-impedance anomaly area in the section K45+020 to K45+240, which is represented by the parts circled by black

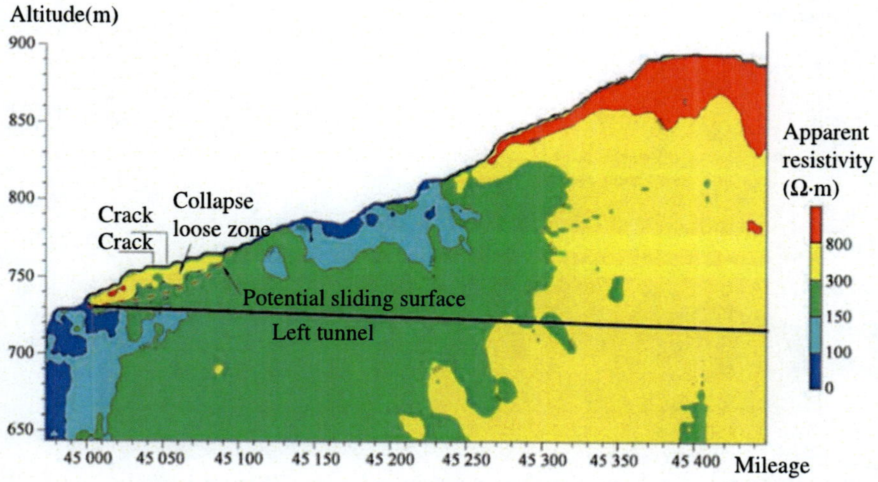

Fig. 5.8 Distribution image of apparent resistivity of section EZ3 (scale 1:1000)

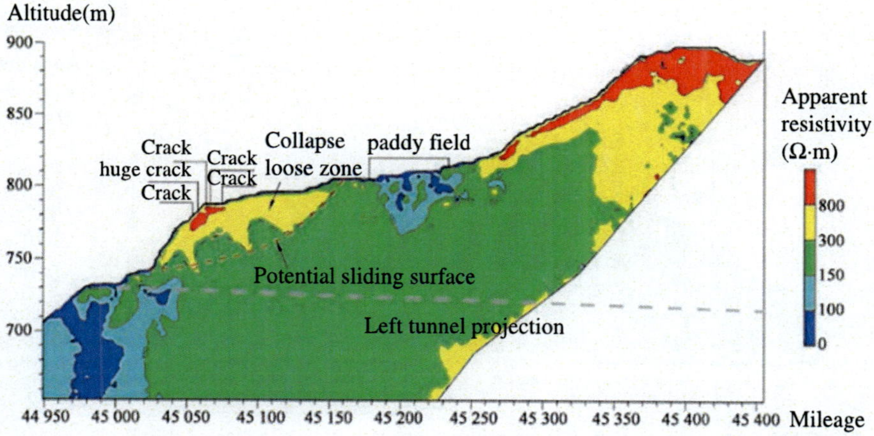

Fig. 5.9 Distribution image of apparent resistivity of section EZ4 (scale 1:1000)

dotted line in the image. It is found that there are many ground fissures in the area. It is inferred that the anomaly area is a loose area caused by collapse. The area is 220 m long with an average thickness of 20 m.

Figures 5.12, 5.13, 5.14, 5.15, 5.16, and 5.17 are apparent resistivity images corresponding to transverse sections (perpendicular to the direction of the tunnel) of EH1–EH6. The collapsed loose areas estimated on the longitudinal sections are also reflected in the two images EH2 and EH4. The apparent resistivity image corresponding to the section EH2 shows that there is a high-resistance anomaly area, which is represented by the parts circled by black dotted line in the image. It is inferred that the anomaly area is a loose area caused by collapse. The area is 180 m long

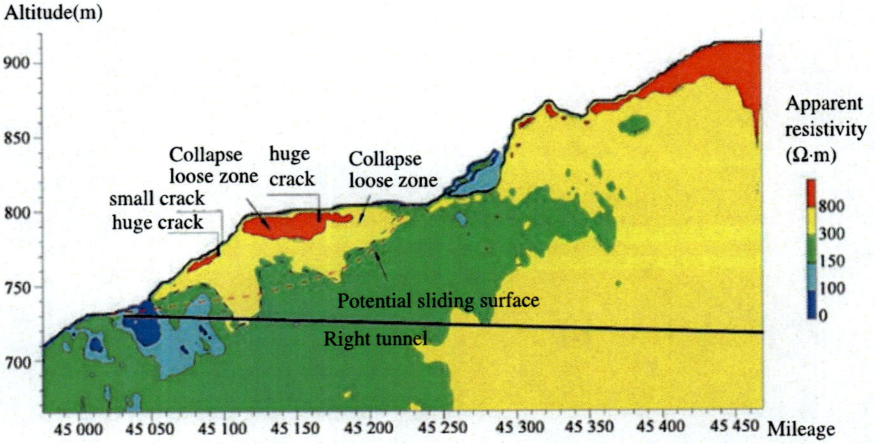

Fig. 5.10 Distribution image of apparent resistivity of section EZ5 (scale 1:1000)

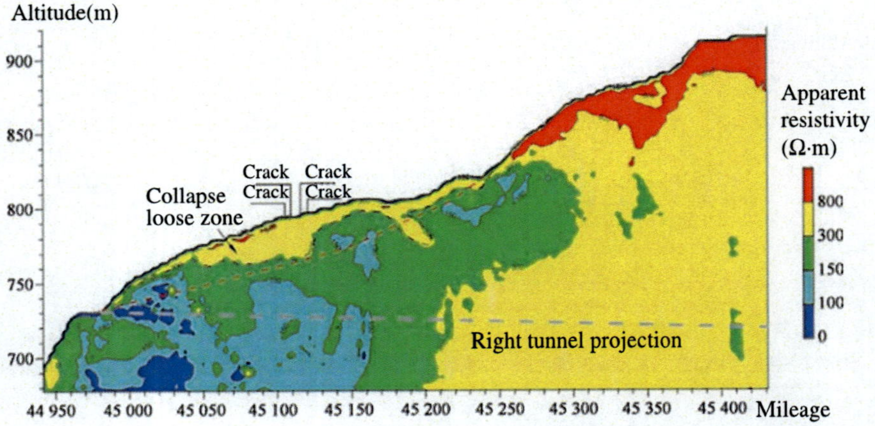

Fig. 5.11 Distribution image of apparent resistivity of section EZ6 (scale 1:1000)

with an average thickness of 30 m. It can be seen from the figure that this area has a greater influence on the right tunnel. The apparent resistivity image corresponding to the section EH24 shows that there is a high-resistance anomaly area, which is represented by the parts circled by black dotted line in the image. It is inferred that the anomaly area is a loose area caused by collapse. The area is 100 m long with an average thickness of 30 m. Since this section crosses K45+150 perpendicular to the direction of the tunnel, and the buried depth of the tunnel is deep, it has little effect on the tunnel. There are no obvious anomalies in the apparent resistivity images corresponding to several other transverse sections.

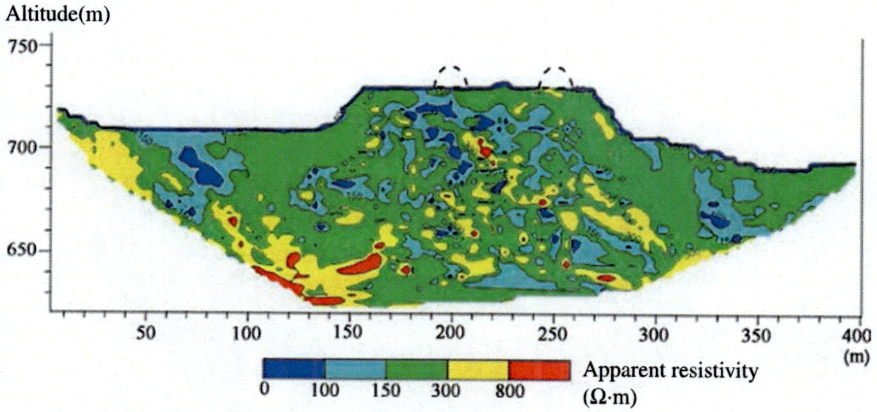

Fig. 5.12 Distribution image of apparent resistivity of section EH1 (scale 1:1000)

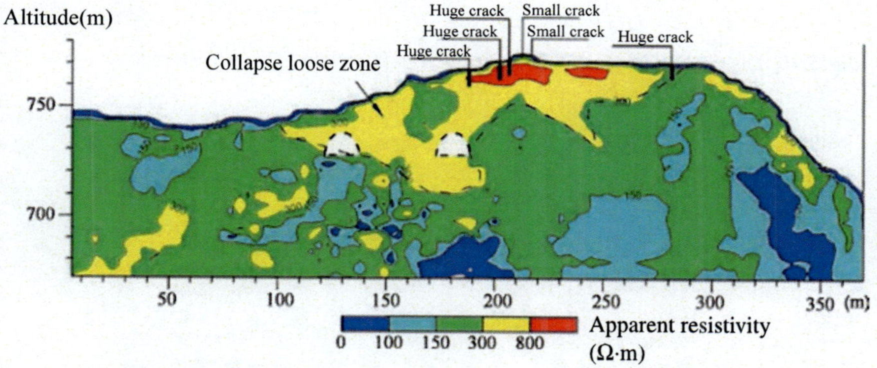

Fig. 5.13 Distribution image of apparent resistivity of section EH2 (scale 1:1000)

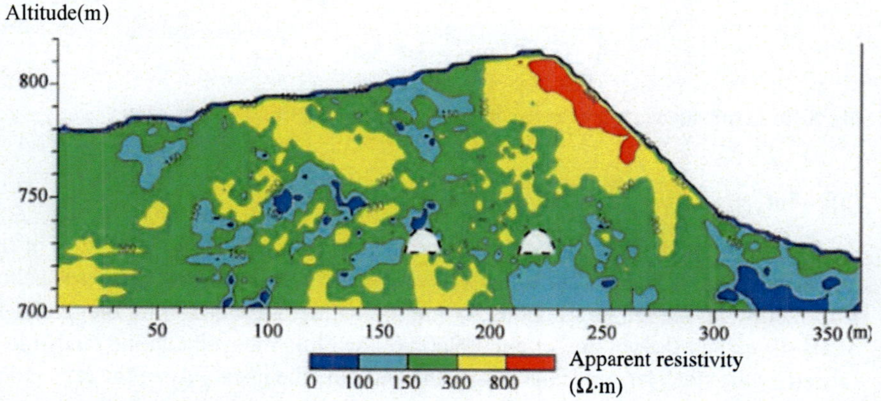

Fig. 5.14 Distribution image of apparent resistivity of section EH3 (scale 1:1000)

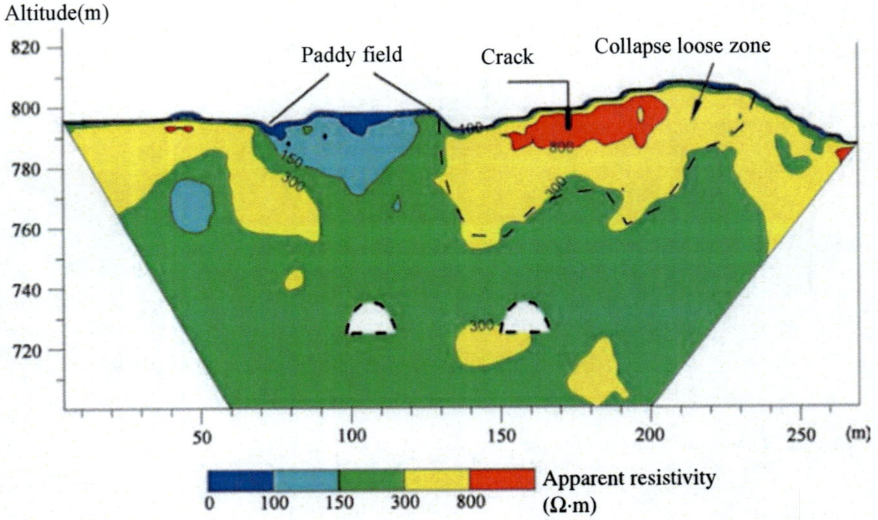

Fig. 5.15 Distribution image of apparent resistivity of section EH4 (scale 1:1000)

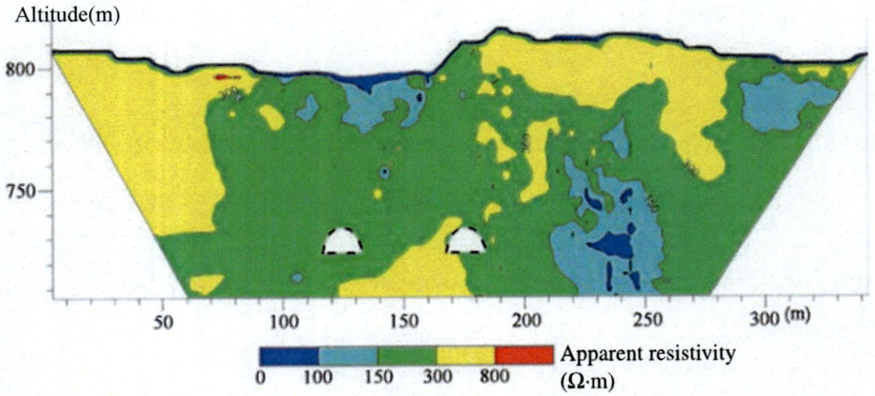

Fig. 5.16 Distribution image of apparent resistivity of section EH5 (scale 1:1000)

(5) Interpretation of the result

According to the comprehensive analysis, there is a landslide in the section of the tunnel entrance. The range of the landslide is from the section K45+000 to the section K45+250, with the length of about 200 m, the width of about 150 m, the thickness of 10–40 m, and the volume of about 900,000 m^3. It is inferred that the landslide is caused by the steep terrain at the tunnel entrance and the instability of the excavation of the toe of the tunnel entrance. The horizontal displacement of the landslide is not large, and the landslide is mainly shallow dumping and collapse. In addition to the slight sliding of the slope, the landslide extends laterally and deeply affected by the

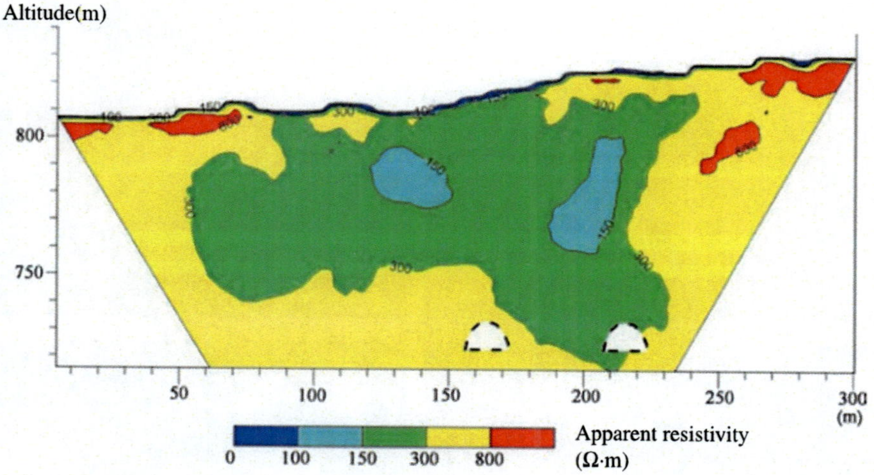

Fig. 5.17 Distribution image of apparent resistivity of section EH6 (scale 1:1000)

two sets of structural joints. The right tunnel is greatly affected, while the left tunnel is only affected at the entrance of tens of meters.

5.2.2 *Application of High-Density Resistivity Method in a Tunnel*

(1) **Project overview**

The tunnel is a multi-arch tunnel on an expressway with a total length of 185 m and a maximum buried depth of 24.03 m. It is a typical shallow-buried tunnel. The tunnel is located on the slope of the mountain. The surrounding rock of the tunnel is sericite schist, which is severely metamorphic. The rock mass is broken, weathered severely, and the joint fissures develop. There is fissure water overflowing in the surrounding rock at the entrance of the tunnel.

Several collapses occurred during the excavation of the middle guide tunnel. Large collapses occurred at K565+383 to +385 and K565+373 to 375. The roof collapsed near K565+383, and the surface collapsed caves were jar shaped. A ground fissure appeared at a horizontal distance of about 30 m above the roof fall (Fig. 5.18). It is suspected that the tunnel is located on the landslide and the excavation has been stopped.

In order to ascertain the engineering geological and hydrogeological conditions of the area where the tunnel line is located, especially whether there are unfavorable geological phenomena such as fault structures, water-rich sections and concealed landslides, provide reliable information for the next design and construction, the

Fig. 5.18 Tunnel disease situation

tunnel construction party entrusts to investigate the geological conditions of the tunnel area.

(2) **Engineering geological conditions**

① Lithology. The surface of the tunnel area is composed of brown-gray, gray-yellow, and brown-yellow cultivated soil containing gravels, and below the surface is mostly composed of brown-gray, gray-yellow, and dark-gray gravel, which is slightly wet with the thickness of 5.0–8.0 m. The bedrock is grayish-green, grayish-yellow, dark-gray sericite schist, which is strong weathered, with layers of thin to medium thick. It is soft rock. After strong weathering, it becomes into pieces (stone).

② Geological structure. Due to the strong influence of the regional Nujiang fault, secondary small faults and joint fissures develop. The secondary Huchonghe fault passes through the tunnel exit, making the surrounding rock of the tunnel broken and loose. According to the on-site geological survey, the schistosity occurrence at the tunnel entrance K565+273 is 70°–80°∠60°. There are two main types of joints. The one is $J_1 = 310°\angle 48°$. The frequency of occurrence is 6 strips/m. The other is $J_2 = 200°\angle 57°$, and the frequency of occurrence is 6 strips/m. The schistosity occurrence of the rock at the ground collapse is 115°∠54°. There are two main types of joints. The one is $J_1 = 338°\angle 34°$. The frequency of occurrence is 6 strips/m. The other is $J_2 = 245°\angle 80°$, the frequency of occurrence is 6 strips/m. The schistosity occurrence at the exit of the tunnel K565+445 is 123°∠37°. There are two main types of joints. The one is $J_1 = 330°∠69°$. The frequency of occurrence is 8 strips/m. The other is $J_2 = 259°∠66°$, and the frequency of occurrence is 8 strips/m. The surrounding rock becomes abnormally broken and unstable under the cutting of multiple sets of joints and cracks, so it is easy to fall off and collapse after excavation.

③ Hydrogeological conditions. The surface layer of the tunnel area is composed of pore water. The lower part is composed of bedrock fissure water, and the burial is deep. Both tows are affected by atmospheric precipitation. The water inflow in the rainy season is large. Groundwater in this area is not corrosive to concrete.

④ Earthquake. The seismic intensity of the area where the tunneling area is located is 8°, and the peak acceleration coefficient of ground motion is 0.30 g.

(3) **Program layout**

According to the purpose and task of the survey, a total of six sections of high-density resistivity method were placed. The layout of the survey lines is shown in Figs. 5.19 and 5.20. The electrode spacing is 3 m. Two sections were placed along the central axis of the tunnel: EH1 and EH2. EH1 was placed directly above the axis of the middle guide tunnel, with the length of 192 m. 64 electrodes were placed and 20 layers were collected. The detection depth is 60 m. EH2 was placed directly above the axis of the upline of the tunnel, with the length of 192 m. 64 electrodes were placed, and 20 layers were collected. The detection depth is 60 m. Four sections were

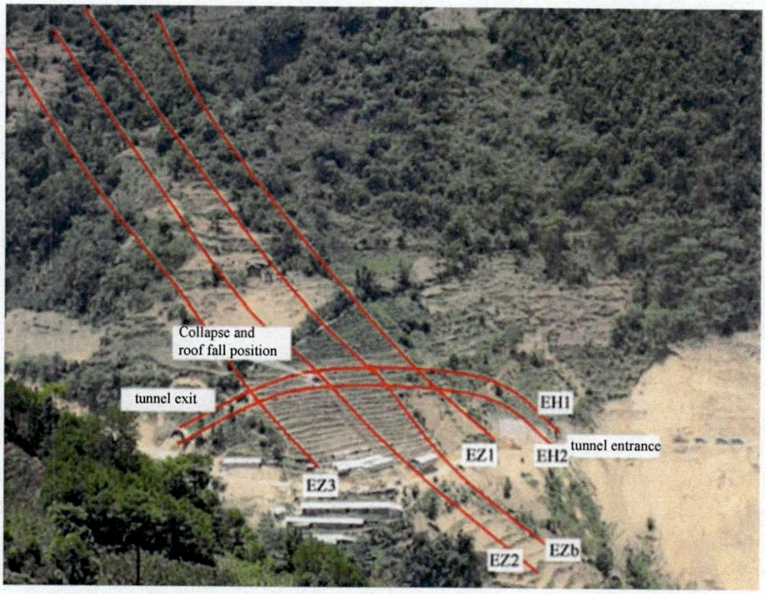

Fig. 5.19 Schematic diagram of the layout of the survey lines of the Huchong River No. 3 tunnel

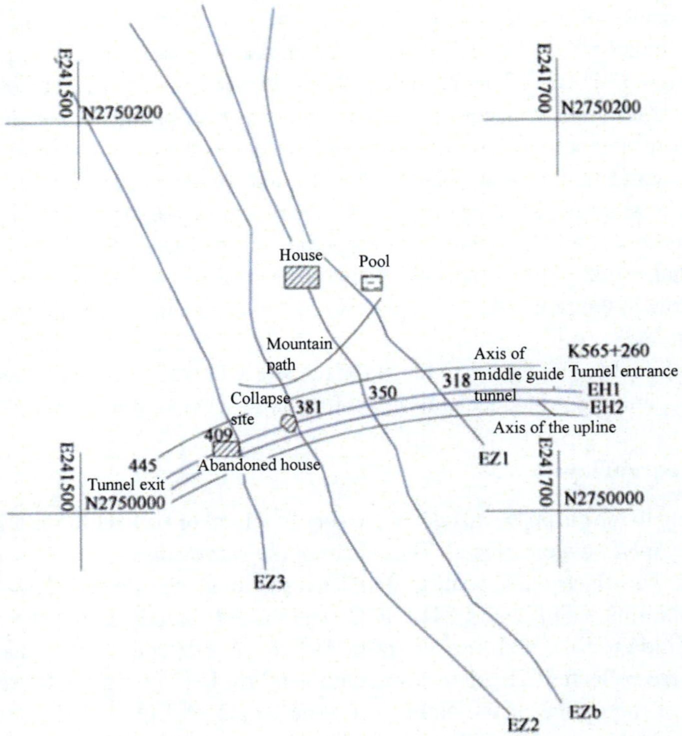

Fig. 5.20 Layout of survey lines of high-density resistivity method (scale 1:2000)

placed perpendicular to the direction of the central axis of the tunnel: EZ1, EZ2, EZ3, and EZb. EZ1 passed through K565+318, with the length of 336 m. 112 electrodes were placed, and 30 layers were collected. The detection depth is 90 m. EZ2 passed through K565+381, with the length of 480 m. 160 electrodes were placed and 30 layers were collected. The detection depth is 90 m. EZ3 passed through K565+409, with the length of 336 m. 112 electrodes were placed, 30 layers were collected. The detection depth is 90 m. EZb is a longitudinal section added later, passing through K565+350, with the length of 480 m. 160 electrodes were placed and 30 layers were collected. The detection depth is 90 m.

(4) **Detection result**

Figures 5.21, 5.22, 5.23, 5.24, 5.25, and 5.26 show the apparent resistivity images of the six sections of the high-density resistivity method. They visually reflect the distribution of apparent resistivity in the subsurface. The red area in the figure is a high-resistance area, which represents the rock mass with good structural integrity, low weathering strength and poor water content. The dark blue area and the light blue area are low-resistance areas, representing the water-bearing, loose, and all weathered to strong weathered sections. From red, yellow to green, the apparent

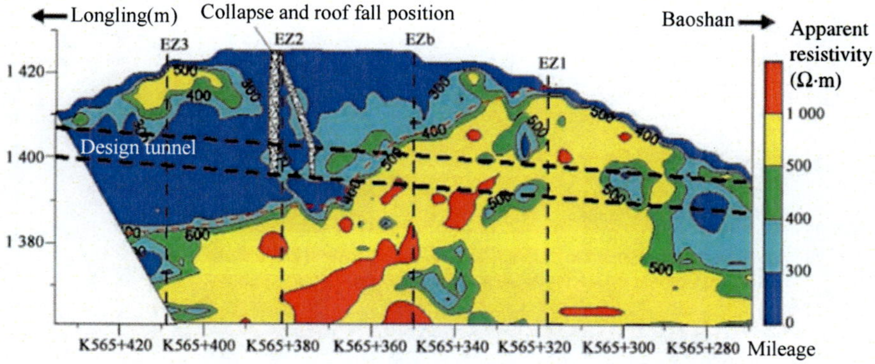

Fig. 5.21 Distribution image of apparent resistivity of section EH1 (along the positive direction of the axis of the middle guide tunnel)

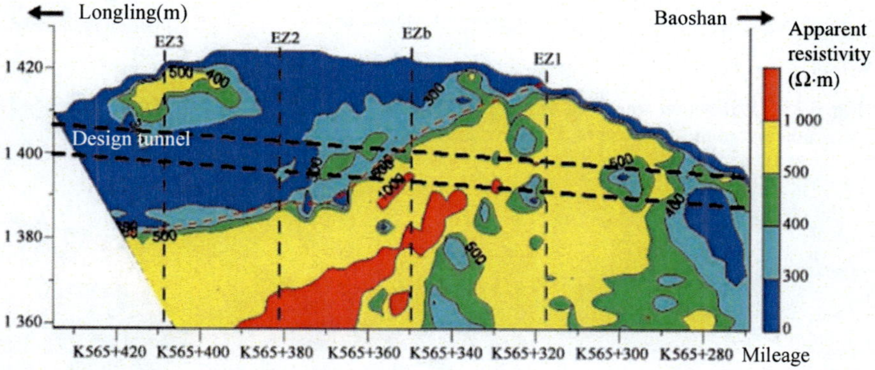

Fig. 5.22 Distribution image of apparent resistivity of section EH2 (along the positive direction of the axis of the upline)

resistivity changes gradually from high to low, and the degree of weathering of the rock is gradually strengthened.

Since the sections EH1 and EH2 are very close (only about 5 m), the apparent resistivity images of the two transverse sections are very similar (Figs. 5.21 and 5.22). It can be seen from the two figures that both ends of the tunnel are in a low-resistance area, and the surrounding rock is fully weathered to strong weathered. The apparent resistivity values of the tunnel entrance to K565+300 are low, which are less than 400 Ω·m. The surrounding rocks are broken and watery, and water seepage is found during excavation. It is estimated that these sections are in the fracture zone. The apparent resistivity values of the tunnel exit to K565+360 are less than 300 Ω·m. These sections are the zones affected by the fault zone, called fracture zones. The surrounding rocks are loose, broken, and watery, and they are easily collapsed. Collapses have occurred many times during the process of excavation. The apparent resistivity values of K565+300 to K565+360 are mostly higher than 500 Ω·m, which

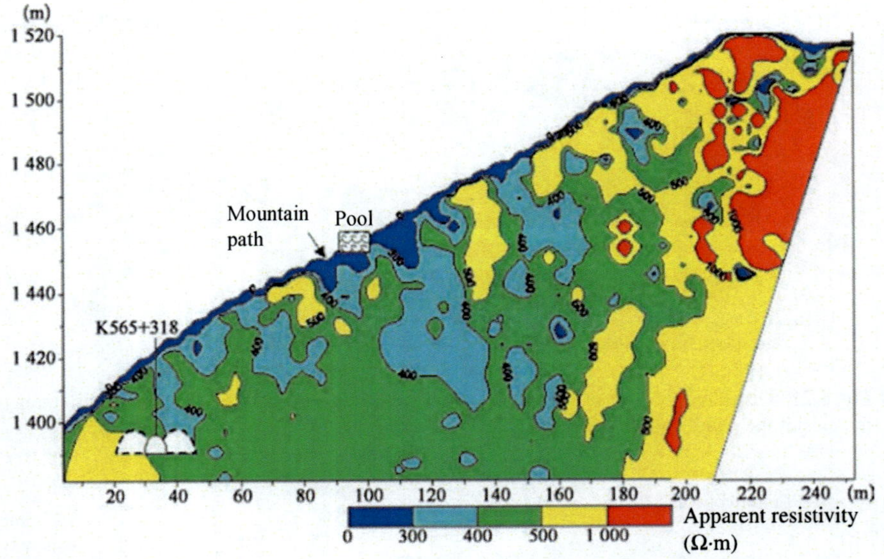

Fig. 5.23 Distribution image of apparent resistivity of section EZ3 (pass through K565+318 perpendicular to the direction of the tunnel axis)

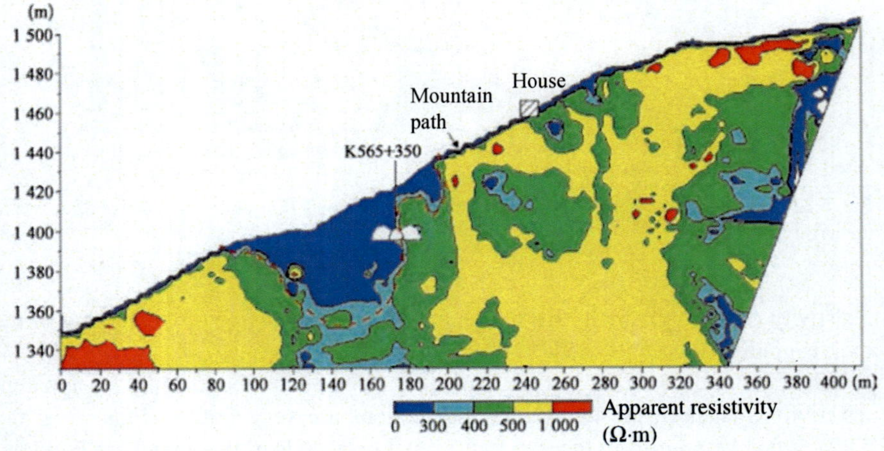

Fig. 5.24 Distribution image of apparent resistivity of section EZb (pass through K565+350 perpendicular to the direction of the tunnel axis)

indicates that the rock mass is slightly intact and the weathering is weak, which is a favorable construction area.

In Fig. 5.23 (EZ1 section), the low-resistance zone extends discontinuously from the surface to the deep, which indicates that the section of the section is unevenly fractured and the degree of weathering is different.

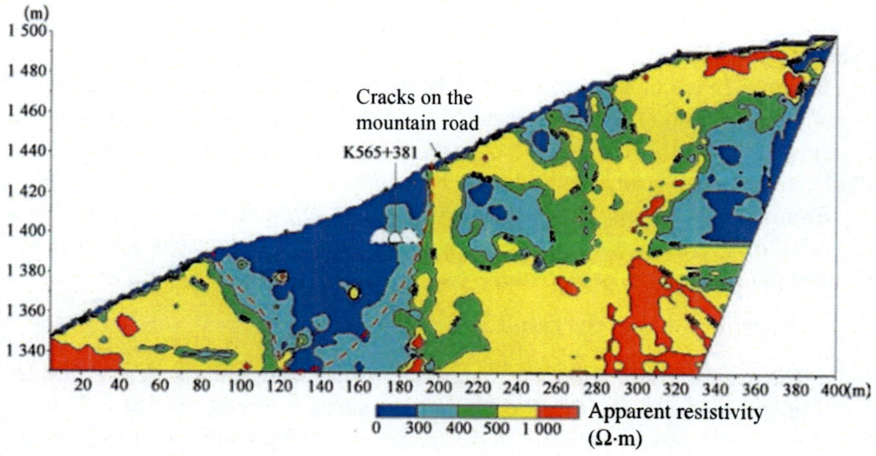

Fig. 5.25 Distribution image of apparent resistivity of section EZ2 (pass through K565+381 perpendicular to the direction of the tunnel axis)

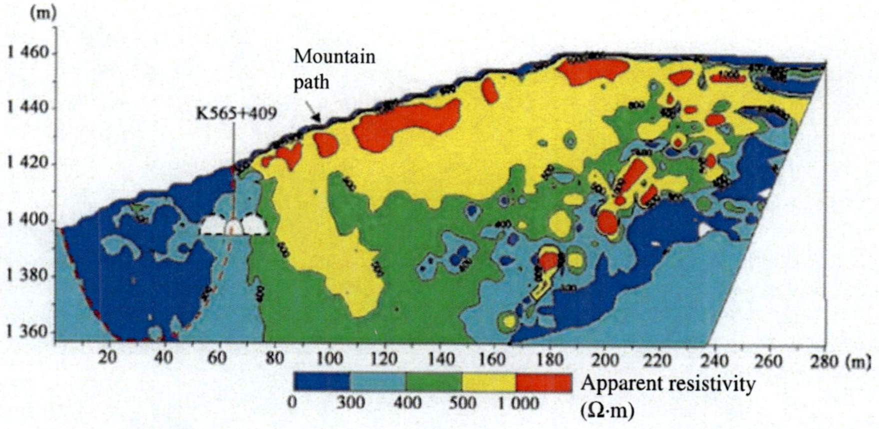

Fig. 5.26 Distribution image of apparent resistivity of section EZ3 (pass through K565+409 perpendicular to the direction of the tunnel axis)

There is funnel-shaped low-resistance zone both in Figs. 5.24, 5.25 and 5.26, which is circled by a red dotted line. This zone is the fracture zone. The tunnel is located at the edge of the fracture zone, and the downline is in the fracture zone. The surrounding rock conditions are poor. The upline is outside the fracture zone, and the surrounding rock conditions are good. The location of the fracture zone is consistent with the location of the fracture zone in the images of EH1 and EH2, and the mileage is K565+360 to K565+430. The Huchonghe fault passes through the exit end of the tunnel. This low-resistance zone is a branch fault of the Huchonghe fault zone. Affected by this fracture, the surrounding rock of this section is loose and broken, and it is easily collapsed.

Figure 5.27 is a three-dimensional schematic diagram of apparent resistivity. The spatial distribution of apparent resistivity can be visually seen from the figure. The section of exit in the figure is the zone affected by the fault zone. The rock mass is fully weathered to strongly weathered, loose, and watery.

From the apparent resistivity images of the four longitudinal sections, no obvious signs of low-resistance sliding surfaces can be seen. It is inferred that there is no hidden landslide in the exploration area.

(5) **Interpretation of the result**

① The overlying stratum of the tunnel is a two-layer structure with a surface layer of soil with gravels of 5–8 m. Below the surface is composed of full-strongly weathered sericite schist, and the tunnel is in a fully strongly weathered rock mass. Affected by the activity of the Nujiang River large fault, the secondary faults and joint fissures in the surrounding rock of the tunnel are extremely

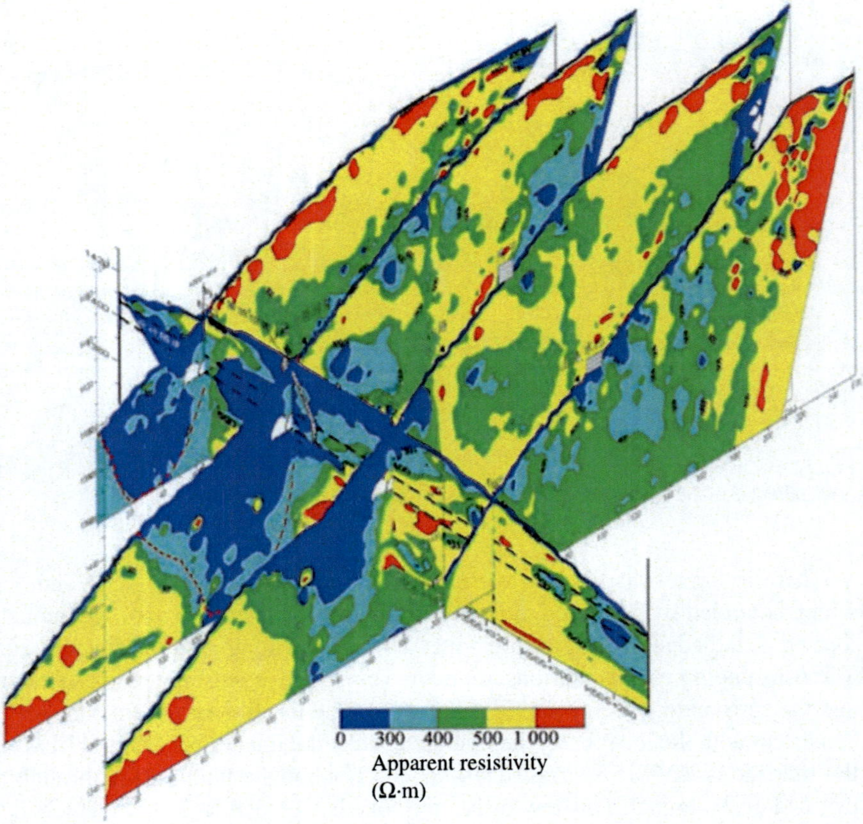

Fig. 5.27 Three-dimensional schematic diagram of apparent resistivity

developed. Many sets of joint fissures are interlaced and cut, which makes the rock mass very broken and gravel-like.

② The tunnel passes through two sections of fracture zone. The first section is the section K565+260 to K565+300 at the entrance. The length is 40 m, and the scale is small. The second section is the section K565+350 to K565+430 at the exit. The length is 80 m, and the scale is larger. There is groundwater overflow along the fault. The rock mass is severely weathered and fractured. The rock mass is weak, and the stability is poor, which is extremely unfavorable for construction.

③ The loose layer of surrounding rock of the tunnel is thin. No weak sliding surface exists in the rock layer. The mountain is stable and the landslide is unlikely.

Chapter 6
Three-Dimensional Detection Technology of TBM in Tunnel

With the rapid development of infrastructure technology in China, China has entered the peak period of urban subway construction, and more than 40 cities are currently constructing subways. Beijing has been building subways since 1968 and has a mileage of about 550 km. Subway construction is scheduled to continue for another 10–20 years. It is preliminarily speculated that in the next 10–20 years, China's subway construction will still maintain a relatively large scale. The investment in subway construction is huge, about 2 billion yuan/km, which has a major impact on the development of the national economy.

6.1 Shield Method on Urban Subways

The traditional tunnel excavation construction method is mainly the drilling and blasting method. It has an absolute advantage in the tunnel excavation market in China because of its economic, practical, simple, and flexible advantages. At present, this method is used in the construction of most highway and railway tunnels in China. However, in many years of engineering applications, the drilling and blasting method has also exposed many disadvantages. The main disadvantages are its poor controllability, prone to over-under-excavation, large consumption of explosives required, ineffective control of costs, and slow construction speed. Therefore, during the construction of urban subways, the drilling and blasting method has great limitations on subway construction, which may cause large safety hazards to nearby buildings. Meanwhile, the generated noise will affect the normal life of urban residents. As an emerging tunnel construction technology, the shield method has become the main means for urban subway construction.

The shield method is one of the underground tunnel excavation methods. It is to use the TBM to excavate underground and use the shield shell to prevent the excavation surface from collapsing and keep the excavation surface stable. Tunnel excavation

G. Cao and H. Liu, *Three-Dimensional Exploration Technology of Tunnel Geology*, Key Technologies for Tunnel Construction under Complex Geological and Environmental Conditions, https://doi.org/10.1007/978-981-16-9225-3_6

and lining operations are carried out in the machine to construct a tunnel. Its main control objective is not to disturb the surrounding rock as much as possible, thereby minimizing the impact on the ground structures and buried objects in the strata. At present, the slurry balance shield and the earth pressure balance shield are most commonly used in subway construction. Because of the combination of excavation and stable excavation surface, these two types of machines can adapt to strata with large changes in geological conditions without other auxiliary construction measures.

The main advantages are as follows:

First, most of the construction operations are completed underground, which has little impact on the environment, and the construction is generally not affected by the factor of depth and thickness of the overburden. Second, the construction operations are carried out underground (except for shaft construction), so ground transportation and facilities are not greatly affected by it. Third, the main procedures of shield advancing, excavating, and lining assembling are carried out circularly. The construction is easy to manage, with less construction personnel and fast construction speed. Fourth, the amount of earthwork is small, and the noise and disturbance during construction are relatively small. Fifth, the navigation is not affected when crossing the river course. The construction is not affected by the climate. Finally, the TBM has a higher economic superiority under specific conditions such as crossing waters and swamps in tunnel construction.

6.2 The Basic Principle of Seismic Reflection and Scattering Combined Imaging Technology

The construction of the subway provides a rare opportunity for the development and application of engineering geophysical exploration technology. There are requirements for engineering geophysical exploration technology in engineering geophysical exploration and survey, construction quality inspection, geological environmental impact monitoring and detection, and engineering treatment effect detection and evaluation of subway lines. However, unlike the construction environment of mountain tunnels, the urban environment is complex, with densely populated buildings, crisscrossing roads, heavy traffic, and strong interference, making it difficult for conventional geophysical prospecting methods such as seismic method and electrical method to obtain good results.

The depth of subway construction is generally shallow, and the geological problems it faces are also different from those in deep geology. Generally, near-surface materials have the characteristics of extremely strong heterogeneity, obvious anisotropy, large velocity gradient, high Poisson's ratio, and complex terrain. Therefore, it is particularly important to develop shallow exploration methods with high resolution and high signal-to-noise ratio for the characteristics of near-surface strata. Shallow seismic exploration is mainly used in engineering surveys to clarify the stratum interface, bedrock undulations, and to find out the location, width, dip, and inclination of faults. Compared with traditional electrical survey methods, its advantage lies in the reduction of human interpretation errors in data processing. In addition, ground penetrating radar technology is widely used in engineering construction, underground buried object monitoring, reinforced concrete structure monitoring, and other fields. It is currently the most developed geophysical method, which can supervise the activity of underground rock layers in real time. The ground penetrating radar usually performs non-destructive testing along the tunnel surface in a scanning manner. By cutting a section perpendicular to the lining surface along the survey line of the tunnel, the accurate data volume of the construction section is obtained. In general, due to the large difference in dielectric constant between air and concrete and tunnel surrounding rocks, the detection imaging of radar has a clearer reflection on cavities and larger gaps.

In order to meet the needs of urban subway construction, the technology of combined imaging of shallow seismic reflection and scattering and the ground penetrating radar three-dimensional imaging technology are combined with the urban environment to conduct in-depth research, making the technology of combined imaging of shallow seismic reflection and scattering and the ground penetrating radar three-dimensional imaging technology have been made innovative progress in urban environment, anti-interference, and geotechnical wave velocity imaging and important progress in subway basic survey, cavity detection, grouting effect evaluation, and lone stone detection.

Shallow seismic reflection is a method of using the elastic difference of the stratum medium to detect the stratum. The conventional method is to artificially excite seismic waves on the ground. When the seismic waves encounter the stratum interface with different elasticity or density, the reflection of the wave will be produced. People analyze the characteristics of the time domain and frequency domain of the reflected wave to ascertain the structural characteristics of the stratum. When the strata is complex, when seismic waves encounter irregular anomalies or the edge of the stratum lens, reflected waves will be generated at the same time as scattered waves. When processing data, the scattered wave is usually suppressed and eliminated as an interference signal, so that some useful information is lost when interpreting the data, and the interpretation error and misjudgment are increased. The method of the combined imaging of shallow seismic reflection and scattering increases the mathematical model for processing the scattered signal when processing the reflected signal, which can better reflect the stratum structure.

(1) **Longitudinal wave reflection**

The basic principle of shallow seismic imaging is as follows:

Firstly, the seismic wave is excited on the surface by artificial method. Then, according to the laws that the wave reflection and scattering will occur when the elastic wave encounters the interface with different wave impedance in the process of propagation in the medium, through the analysis and profile imaging of the received reflected waves and scattered waves, the purpose of obtaining the distribution state and physical property parameters of the subsurface medium can be achieved. According to Snell's law, like light, seismic waves will reflect and transmit

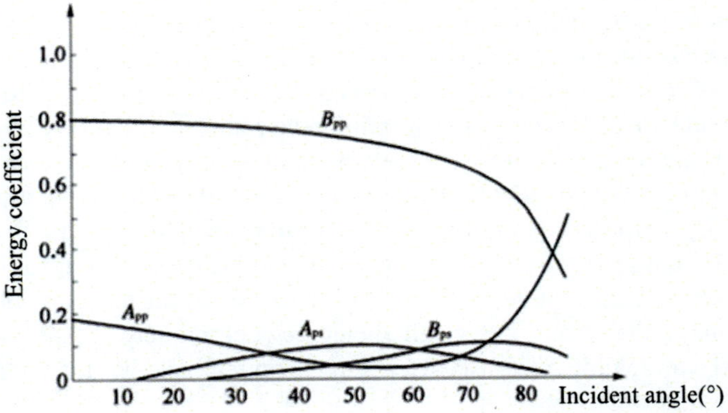

Fig. 6.1 Relationship curve between the incident angle of the longitudinal wave and the reflection coefficient

when they encounter an elastic interface. Its dynamic characteristics will further change, and the energy of the wave will be redistributed. The seismic reflection wave method usually uses the reflected signal on the interface.

Plane waves are an abstraction of mathematics, and this assumption does not lose its generality when it is far from the source area. Assuming that the plane simple harmonic longitudinal wave is incident on the interface at an angle of α, four secondary waves of reflected longitudinal wave A_{pp}, transverse wave A_{ps}, transmitted longitudinal wave B_{pp}, and transverse wave B_{ps} will be generated. According to Zoeppritz equation, the energy distribution of various waves can be solved. Figure 6.1 shows the relationship between reflection coefficient and transmission coefficient and the angle of incidence of longitudinal waves. The elastic parameters are $V_{p2}/V_{p1} = 0.5$, $\rho_2/\rho_1 = 0.8$. Since it is incident from a medium having a large wave impedance to a medium having a small wave impedance, there is no critical angle, and the curve is continuous.

It can be seen from the figure that the energy is mainly distributed on the transmitted wave, and the change of the incident angle has little effect on the transmission coefficient B_{pp}. The curve changes smoothly. The reflection coefficient A_{pp} of the longitudinal wave decreases with the increase of the incident angle. The reflection coefficient is large in the range of $\alpha < 15°$, and there is no interference from other converted waves. Therefore, this is the most favorable part of the seismic longitudinal wave reflection method, that is, the small offset measurement method near the shot point is adopted. When the incident angle $\alpha = 0°$, the Zoeppritz equation can be expressed as

$$\left.\begin{aligned} A_{\text{ps}} &= B_{\text{ps}} = 0 \\ A_{\text{pp}} &= \frac{\rho_2 V_{\text{p2}} - \rho_1 V_{\text{p1}}}{\rho_2 V_{\text{p2}} + \rho_1 V_{\text{p1}}} \\ B_{\text{pp}} &= \frac{2\rho_1 V_{\text{p1}}}{\rho_2 V_{\text{p2}} + \rho_1 V_{\text{p1}}} \end{aligned}\right\} \tag{6.1}$$

Analyzing formula (6.1), it can be concluded that there is no converted wave when it is incident vertically, but the same kind of reflected wave and transmitted wave. This is very advantageous for prospecting by seismic vertical reflection method. At the same time, only when the wave impedance of the upper and lower media ($Z = \rho V$) is not equal, can there be reflected wave, and the larger the difference of wave impedance, the stronger the energy of reflected wave. Therefore, only when the difference of wave impedance between abnormal body and surrounding rock is large, the stronger vertical reflection signal can be obtained.

(2) **Longitudinal wave scattering**

With the understanding of the importance of scattered waves, the research on the theory and application of scattered waves are constantly deepening. On the one hand, it is recognized that the application of scattered waves can better invert and interpret stratum abrupt points and smaller anomalies, thereby achieving super resolution seismic exploration. On the other hand, to make better use of scattered waves, it is necessary to study the formation and distribution law of scattered waves.

In seismic wave scattering theory, geological bodies are assumed to be heterogeneous, and the wave impedance in the geological body varies in space. In heterogeneous geological bodies, the propagation of seismic waves becomes very complicated. Rushan Wu et al. used the perturbation method to study in detail the scattering equations of seismic longitudinal wave and transverse wave propagation in heterogeneous media. The expression form is quite complicated, and it is difficult to quote here. N. Blastain et al. used a relatively simple wave equation in the study of seismic scattering without losing the correctness of the physical meaning, so the equation is often quoted. He expressed the wave field u as a combination of the incident wave μ_{I} and the scattered wave μ_{s}:

$$\mu = \mu_{\text{I}} + \mu_{\text{s}} \tag{6.2}$$

The equation that satisfies the scattered wave μ_{s} has the following form:

$$\nabla^2 \mu_{\text{s}} - \frac{\partial^2 \mu_{\text{s}}}{v_0^2 \partial t^2} = -\frac{\alpha(r)}{v_0^2} \frac{\partial^2 \mu_{\text{I}}}{\partial t^2} \tag{6.3}$$

$$\alpha(r) = \frac{v^2 - v_0^2}{v^2} \tag{6.4}$$

where

v the wave velocity of rock mass, which is a spatial variable;
v_0 the average wave velocity in the background of the area;
$\alpha(r)$ the percentage anomaly of the square difference of wave velocity.

It characterizes the magnitude of the change of wave velocity in space. It takes the extreme values at the location of the geological interface. It is a physical quantity that uses the difference in physical properties to characterize the existence of the geological interface.

Equation (6.3) is the inhomogeneous wave equation. The right term is the source intensity of the scattered wave, also called the scattering intensity, which is actually a passive source. The physical meaning of the equation is that under the excitation of the incident wave, the wave velocity anomaly α emits scattered waves to the surroundings. The intensity of the passive source is proportional to the product of the acceleration of the incident wave and the anomalous amplitude α. In seismic wave scattering theory, the interface replaces the continuous conditional constraints in the reflection theory by the action of the scattering source. Seismic scattering exploration technology studies the inverse problem. Scattering occurs in the seismic wave propagation excited by the surface, and some scattered waves return to the surface. Using the received scattered wave, the wave velocity distribution of the geological structure and the spatial distribution of the wave velocity anomaly α are reconstructed. The wave velocity anomaly α can reveal the morphology of the formation interface, including the position and the magnitude of the difference in wave impedance. Wave velocity is a physical quantity that characterizes the mechanical modulus of geotechnical media. It can show the mechanical properties of the media more clearly. It is an important basis for the classification of geotechnical engineering and has more practical significance for geotechnical engineering applications.

(3) **The combined imaging of shallow seismic reflection and scattering**

Seismic imaging is the ultimate goal of processing seismic reflection data. Conventional shallow seismic imaging methods are CMP dynamic correction superposition and diffraction scanning offset superimposition imaging. However, these two methods have obvious shortcomings and limitations of application conditions. The basic theory of CMP signal dynamic correction superposition is carried out under the assumption that the surface is a horizontal plane, and the near-surface medium is uniform. When the ground conditions are inconsistent with the hypothesis, CMP signal superposition cannot achieve ideal results. Although diffraction scanning offset superimposition imaging solves the problem of dynamic and static correction, the method used in the process of extracting common scattering points is relatively simple, which also affects the imaging accuracy.

The combined imaging of shallow seismic reflection and scattering is proposed after careful analysis of the above two methods. Combined imaging fully absorbs and draws on the advantages of each method. The basic principle of combined imaging of reflection and scattering is shown in Fig. 6.2. The formation profile is first divided into grids and then scanned along the grid points. For any scanning point P, the records of common reflection point and common scattering point are extracted by

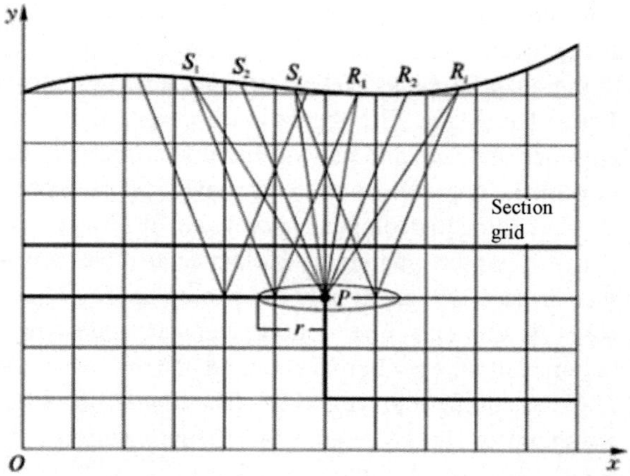

Fig. 6.2 Schematic diagram of the principle of seismic imaging

calculation, and the reflected signal and scattering signal on the records are focused to their spatial position at the same time. Then, the average amplitude (or correlation coefficient of instantaneous value, etc.) is calculated according to the focusing signal, and its relative size is expressed by different colors (e.g., the point with high signal reflection energy is represented by dark blue. On the contrary, the points with low reflection energy are represented by light green), so that the profile image reflecting geological structure can be obtained. This imaging method has the advantages of high speed, high accuracy, and strong anti-interference ability.

6.3 Typical Cases

6.3.1 Application of the Combined Imaging of Shallow Seismic Reflection and Scattering in a Shallow Tunnel

(1) **Project overview**

A municipal engineering tunnel in Kunming is located on the right side of a tunnel existing in the Chengkun line. The entrance and exit of the tunnel are adjacent to the existing line, and the distance from an existing tunnel is gradually changed from about 17 m at the entrance to 140 m at the tunnel body, and then to about 32 m at the exit. The entrance and exit mileage of the newly built tunnel is D2K1082+575 to D2K1084+890, with the length of 2315 m, the maximum buried depth of 100 m of the tunnel, and the minimum buried depth of 6 m. The tunnel area belongs to the plateau low-middle mountain stripped landform. The tunnel passes through Xishan, a famous scenic spot in Kunming. The ground elevation is 1900–2090 m, and the maximum relative height difference is 190 m. The terrain is undulating. The natural crossslope is generally 10°–45°, and the local terrain is steep. Most of the slopes are covered by the Quaternary soil layer, and the thickness is about 2–6 m. The slopes at the entrance and exit are mostly reclaimed as paddy fields or dry fields. The bedrocks in the steep slopes and valleys are sporadically exposed. The caves are mainly Xishan Forest Park Area. The vegetation is dense, mainly pine trees and cypress trees. The entrance and exit and part of the tunnel body are densely populated, with roads or pavements communicating and convenient transportation.

According to geological surveys and drilling reveals, the surface of the tunnel area is overlying the Quaternary Holocene artificial refills (Q_4^{ml}), the alluvial-proluvial-lacustrine (Q_4^{dl+l}), the alluvial-proluvial (Q_4^{dl+p}), and the residual slope product (Q_4^{dl+el}) silty clay, fine round gravel soil, underlying Jurassic Middle Lufeng Group (J_2) mudstone, ore intercalated marl and the selected Xiaoxijieshan Basalt Formation ($P_2\beta$) basalt, fault breccia (Fbr). The tunnel site is located in the southeast margin of the Qinghai-Tibet Plateau, which is the junction of Sichuan-Yunnan rhombic block, South China block and Lincang Simao block. With the further uplift and the extrusion of the Qinghai-Tibet Plateau to the southeast, at the end of the early Pleistocene, the Sichuan-Yunnan rhombic block slid as a whole to the southeast with a dramatic rise in the crust. The Pudu River-Xishan Mountain fault zone is spread across the eastern edge of the Sichuan-Yunnan rhombic block. It is hidden under the Dianchi Lake and extends from north to south along Jingma Street, the foot of Xishan Mountain, and Jinning. Xishan Mountain and Dianchi Lake Basin are formed by the activities of the Pudu River-Dianchi Fault since Quaternary. The height difference between the mountain and the lake basin is 800 m, and a sedimentary layer of more than 100 m in the Dianchi Basin has been formed.

(2) **Program layout**

In the field detection, four geophones (numbered 1–4) are arranged at a time, with a distance of 2 m between the geophones, and two excitations each at 0.5 m beside the geophones 1 and 4. After the measurement, the geophones are moved forward 4 m each time, and the above work is repeated. The detection diagram is shown in Fig. 6.3. In this work, a seismic imaging survey section was arranged on the surface directly above the tunnel axis along the route. Its starting and ending mileage was D2K1os3+065 to D2K1083+180, and the length of the survey line was 115 m.

(3) **Detection result**

In the seismic imaging intensity map (Fig. 6.4), the green and blue regions represent strongly weathered rock formations or broken loose coatings. The yellow and brown

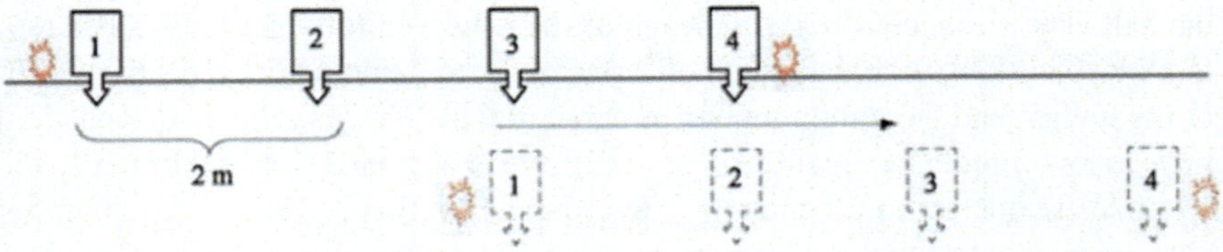

Fig. 6.3 Seismograph measurement scheme

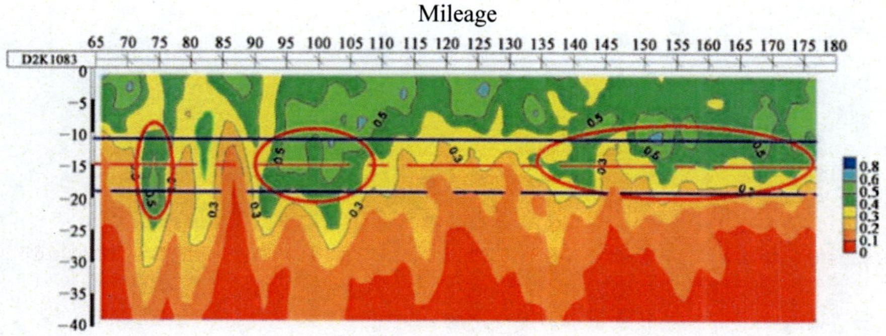

Fig. 6.4 Vertical seismic imaging intensity map of section D2K1083+065 to D2Kt083+180 in a tunnel

regions represent broken zones or weakly weathered rock formations. The red and orange regions represent unweathered rock formations or complete dense bedrocks. According to the engineering geological section diagram (Fig. 6.5) combined with the on-site geological survey, the entire mileage section of D2K1083+065 to D2K1083+180 is a shallow tunnel with an average buried depth of 15 m, of which D2K108+072 to D2K1083+078, D2K1083+092 to D2K108+110, D2K108+138 to D2K1083+1180 are loose and broken zones. When excavating, attention should be paid to strengthen the support to prevent collapse or even roof fall.

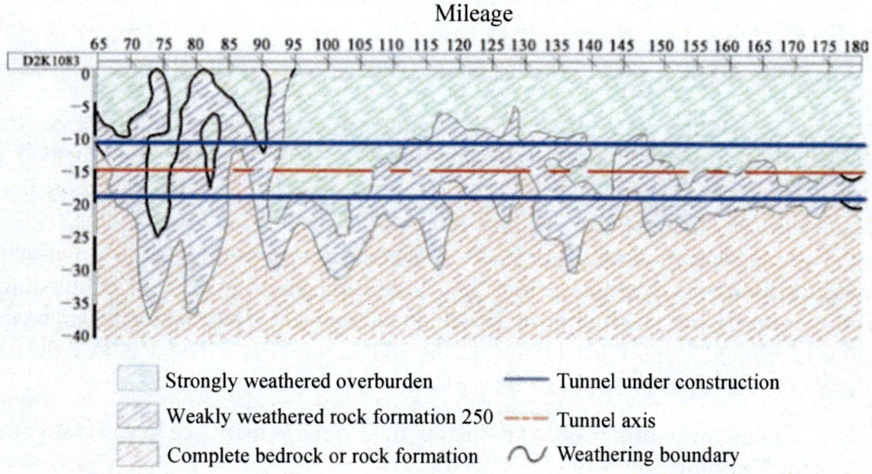

Fig. 6.5 Engineering geological image of section D2K1083+065 to D2K1083+180 in a tunnel

Fig. 6.6 Non-destructive three-dimensional detection method of radar on a certain subway line in Kunming (on-site photo)

6.3.2 Application of Ground Penetrating Radar 3D Detection Technology in Municipal Engineering

(1) **Three-dimensional radar detection of a subway line in Kunming**

In April 2016, the non-destructive three-dimensional radar detection method was tested on a subway line in Kunming (Fig. 6.6). The measurement method is designed to measure concentric circles along the tunnel wall. Multiple survey lines of concentric circles realize the three-dimensional acquisition of radar data.

From the detection results (Fig. 6.7), the green area represents the distribution of water behind the segment. The gray and white areas represent the distribution of the cavities or the uneven grouting areas behind the segment. The position at 0 m on the blue horizontal axis is the centerline of the 25th segment, and the position at the last 64.8 m is the centerline of the 77th segment.

(2) **Radar three-dimensional detection of surface subsidence in a certain area of Kunming**

The reasons for the collapse of urban roads mainly include the following aspects:

① The pipelines or well chambers under the road are damaged. The damage of pipelines or wells often causes the surrounding soil to be transported along the pipelines or wells, forming holes or even cavities under the roads, especially rain and sewage pipelines and rain and sewage wells and other pipelines and wells

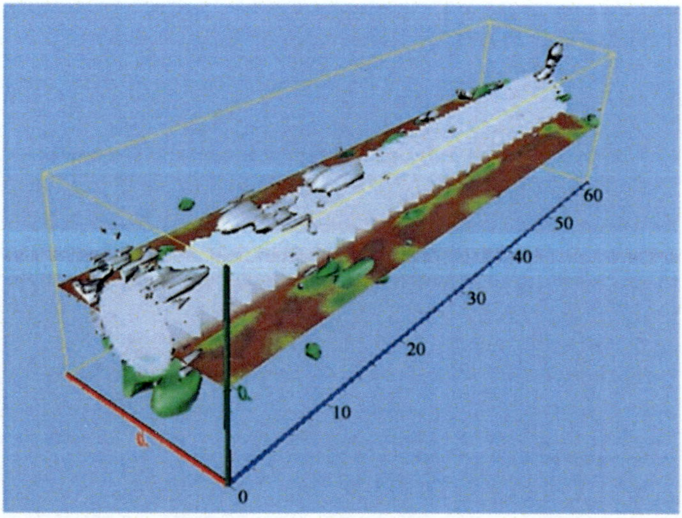

Fig. 6.7 Effect of ground penetrating radar three-dimensional detection of hidden defects in surrounding rocks behind subway shield segments

with water flowing inside. Under the action of water flow, the transporting effect of the soil is obviously intensified, and the transporting speed is accelerated, causing more serious diseases and leading to road collapse. The inefficiency of the urban drainage system causes the accumulation of water to seep, causing the soil under the road to loosen and even form cavities.

② The road is built in the karst subsidence area. The road collapses due to the destruction of the karst balance due to the establishment of the road and the load of road vehicles.

③ In cities and mining areas, large amounts of groundwater must be pumped or discharged due to water supply or deposit mining, forming a large area of goafs. In the rainy season, the increase in precipitation causes the overburden to be eroded so that the road surface collapses.

④ The construction of urban subways and the construction of large projects such as power tunnels, thermal tunnels, and underground buildings (including pipe jacking, shallow buried excavation and underground tunnel shield) greatly disturb the surrounding soil under the road, forming loose circles, which are easy to cause diseases such as looseness, voids, or even cavities under the road.

As shown in Fig. 6.8, the roadbed in a certain area of Kunming was hollowed out due to the burst of the drainage pipe in the rainy season, which caused the road collapse. In order to accurately delineate the size of the collapsed cavity and the scope of water leakage, a high-precision ground penetrating radar is used to detect the collapsed area in three dimensions. The final interpretation result is shown in Fig. 6.9.

Fig. 6.8 Site photo of road collapse

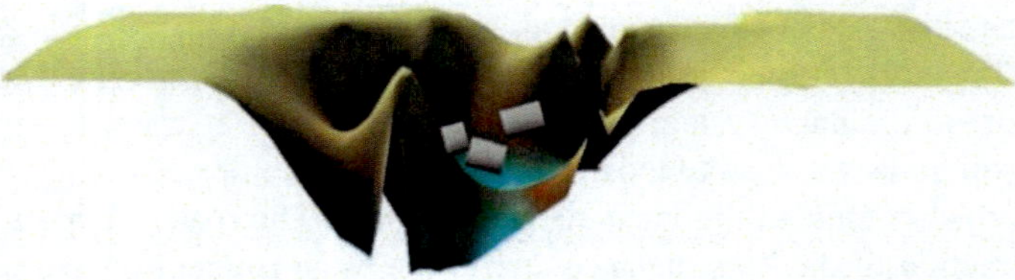

Fig. 6.9 Result map of the three-dimensional detection of the ground penetrating radar

Chapter 7
Three-Dimensional Detection Technology for Tunnel After Sudden Geological Disasters

The unfavorable geological disasters in tunnel construction mainly include karst, collapse, water gushing, mud outburst, etc. It can be seen from past engineering accidents that unfavorable geology is one of the main causes of geological disasters during tunnel construction. Only by strengthening the prediction and prevention of the unfavorable geological disasters and preventing them before they happen can the occurrence of engineering accidents be effectively reduced.

7.1 Three-Dimensional Detection Technology for Tunnel After Sudden Geological Disasters

After sudden disasters such as mud outburst and roof fall in the tunnel, special surveys are required to provide a basis for treatment. Due to the complex geological conditions, comprehensive geophysical methods are often used to achieve three-dimensional detection. Commonly used methods include high-density resistivity method, seismic CT detection, ground penetrating radar three-dimensional detection, etc.

7.1.1 Seismic CT Detection Technology

(1) **Principle of seismic CT technology**

Seismic CT is a new exploration method for engineering geophysical exploration, also known as tomography, which is an example of the application of CT technology in geophysics. It can accurately and reliably invert the visual image of the internal

G. Cao and H. Liu, *Three-Dimensional Exploration Technology of Tunnel Geology*, Key Technologies for Tunnel Construction under Complex Geological and Environmental Conditions, https://doi.org/10.1007/978-981-16-9225-3_7

structure of the geological body according to the external measurement data of the geological body (structure). It has the characteristics of high resolution, good reliability, and intuitive image, which is a very effective survey method in engineering exploration.

The working principle of seismic CT is like that of medical CT. Medical CT uses X-rays to penetrate the human body and image human tissues by observing the attenuation of the ray intensity. Seismic CT is the use of seismic wave rays to see through the geological body (structure) and to image the geological body (structure) through the observation of the travel time and energy attenuation of the seismic wave. When the seismic wave penetrates the rock and soil medium, its velocity is related to the elastic modulus, shear modulus, and density of the rock and soil medium. Rocks with high density and high strength have large modulus, high wave velocity, and short travel time, and vice versa. The complete hard rock mass has a high wave velocity, and the broken rock mass and the loose soil have a low wave velocity. The compact rock mass with small porosity has small energy attenuation, and the broken rock mass and the loose soil have large energy attenuation. Therefore, the seismic CT image can reliably reflect the distribution boundary of various types of rock and soil and the degree of fragmentation and distribution of the fractured rock mass. Seismic CT is especially suitable for studying the distribution boundaries and mechanical properties of various strata, structures, rock masses, and soils. It is often used to explore the position, shape, and mechanical strength of the geological bodies (structures), such as fault zones, dense joint zones, water-bearing zones, caves, weathering zones, etc.

In the exploration of engineering geological structure of tunnel surrounding rock, seismic CT is used as the preferred method to find out the geological cause and scope of collapse of tunnel surrounding rock, and the development, extension, and mechanical properties of fault zone in rock mass.

With the in-depth study and the increasingly widespread application of seismic CT methods, seismic CT has now developed into a series of methods. The imaging physical quantities include wave velocity, energy attenuation, Poisson's ratio, etc. The imaging method can use a combination of direct waves, reflected waves, refracted waves, surface waves, etc., and can use various observation conditions such as boreholes, tunnels, slopes, mountains, and ground to perform two-dimensional and three-dimensional geological imaging. In the railway, highway and hydropower construction in the mountainous areas of western China, seismic CT can be widely used in engineering geological exploration and disaster remediation of lines, sites, tunnels, slopes, and other projects to solve complex geological conditions.

(2) The working method of seismic CT

Depending on the purpose of the survey and the needs of the research, seismic CT surveys typically require the placement of two perspective sections, as shown in Fig. 7.1. In order to find out the engineering geological structure of the tunnel and the overlying rock mass, to find out the scope and geometric form of the tunnel collapse and roof fall and the mechanical strength of the collapsed body, to understand in detail the development and extension of the fracture and the relationship with the collapse of the surrounding rock, it is generally necessary to arrange two vertical seismic CT

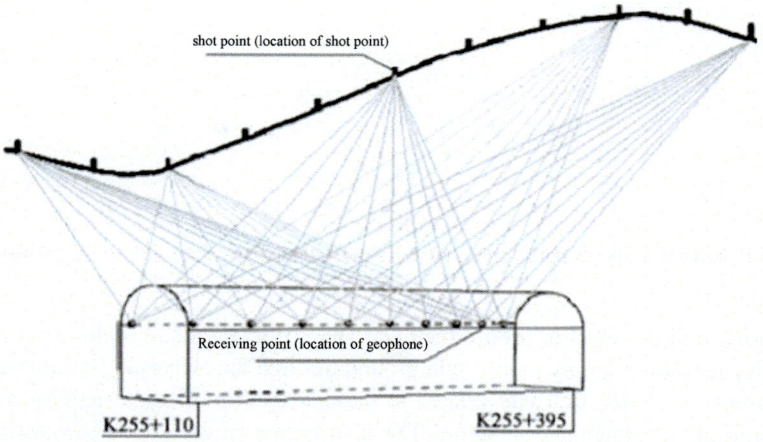

Fig. 7.1 Design of observation system for seismic CT

sections along the tunnel axis along the straight line section of the tunnel's upline and downline.

7.1.2 Detection Technology of High-Density Resistivity Method

As shown in Fig. 7.2, the high-density resistivity method is an array exploration method, also called the automatic resistivity system. It is a development of the direct current method, and its function is equivalent to the combination of the triode

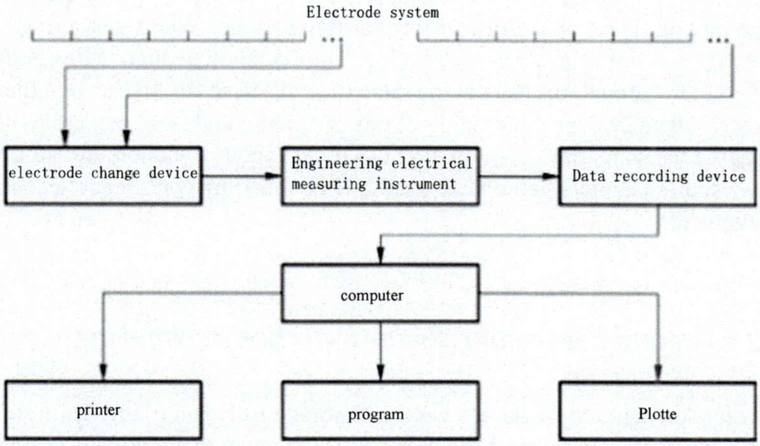

Fig. 7.2 Schematic diagram of working system of high-density resistivity method

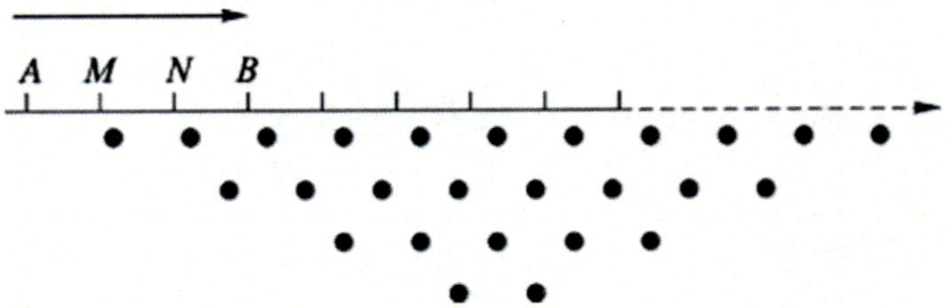

Fig. 7.3 Schematic diagram of the scanning of the Wenner device

sounding and the electric profile method. The artificial electric field is formed by the electric power supply to the underground through the electrode. The distribution of the electric field is closely related to the distribution of the resistivity ρ of the underground geotechnical medium. The distribution of the apparent resistivity ρ_s of the underground medium is understood through the measurement of the artificial electric field in different parts of the surface. The distribution of the apparent resistivity of the medium is inferred to explain the underground geological structure. The method is particularly sensitive to the water content of the surrounding rock. If the surrounding rock is broken and water-bearing, its apparent resistivity is significantly reduced. The apparent resistivity of intact and hard rock is significantly higher than that of surrounding rock in fault zone, broken zone, and water-rich zone. This method is clear in principle and intuitive in image. It is a geophysical method with high resolution. In recent years, with the improvement of computer data acquisition technology, the exploration efficiency has been greatly improved. The coverage area and detection depth of the profile have been increased. Reliable data can be obtained in a strong interference environment, which greatly improves the signal-to-noise ratio and can accurately detect geological bodies. This method has been widely and successfully applied in engineering and hydrogeological exploration and mineral and water resources exploration.

The most commonly used Wenner device (as shown in Fig. 7.3) is often selected for the measurement of high-density resistivity method. When measuring, AM = MN = NB = AB/3 is an electrode spacing. The detection depth is AB/3. *A*, *B*, *M*, *N* move point by point to the right at the same time to obtain the first section line. Then AM, MN, NB increases the spacing of one electrode, and *A*, *B*, *M*, and *N* move to the right at the same time point by point to obtain another section line. In this way, the inverted trapezoidal geological section is obtained by continuous scanning and measurement.

7.1.3 Ground Penetrating Radar Detection Technology

The ground penetrating radar detection technology is shown in Fig. 7.4.

Ground penetrating radar three-dimensional imaging technology is currently the main development direction of geophysical exploration. Using the directionality

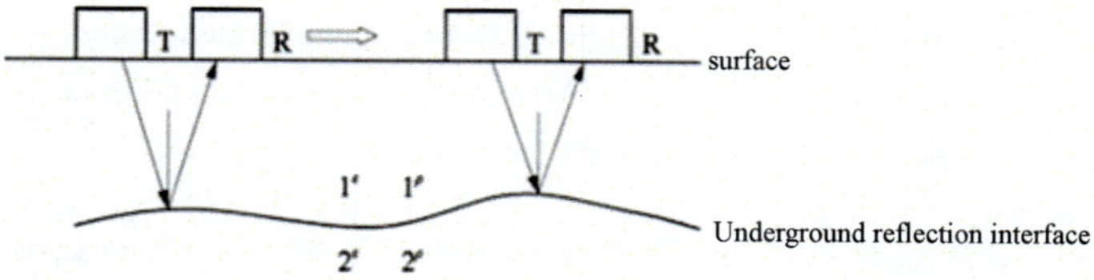

Fig. 7.4 Working principle diagram of ground penetrating radar detection

of the electromagnetic wave emitted by the ground penetrating radar antenna, the directional radiation method is adopted to collect the radar echo signals in front of the tunnel in a limited space, and qualitative analysis is performed on the target medium at different positions in the spatial region covered by the ray to invert the spatial position and scale shape of the target medium to achieve the purpose of visual imaging survey.

7.2 Typical Cases

7.2.1 Application of Seismic CT Technology and High-Density Resistivity Method in a Highway Tunnel in Yunnan

(1) **Project overview**

The tunnel is located on a certain expressway in Yunnan. It is a monolithic double-span double-arch single-hole single-way double-lane tunnel with separated left and right sections. The tunnel has a clear width of 24.1 m. The tunnel is a new tunnel that was rerouted to avoid landslides. The tunnel extends in a reverse *S* shape with a total length of 320 m.

Due to the serious influence of regional geological structure, the terrain in the section through which the tunnel crosses are steep. The section is steep slope with partial hanging slope, and the lithology is soft and hard alternately, which poses obvious potential threat to the tunnel construction and operation after completion. The area where the tunnel is located is affected by the deep fault of the Nujiang River and the structure is very complicated. The mountain where the tunnel is located is steep, with the unloading bandwidth of the air surface of 30–50 m, severely broken rock mass, large bias, and the unstable lateral. The entrance and exit of the left tunnel are in the accumulation layer, which is prone to collapse. After the middle guide tunnel has been excavated in the tunnel, the tunnel has been collapsed many times, as shown in Table 7.1.

Table 7.1 List of collapse positions during the excavation of middle guide tunnel in a tunnel

Mileages	Problems	Mileages	Problems
K562+858 to +938	Three major collapses	K563+033	A cave
K562+986 to +992	A major collapse	K563+090 to +093	Water seepage
K562+980 to K563+030	Shallow buried section, falling blocks, small collapse	K563+124 to +132	A major collapse
K563+004 to +010	Drop of water	K563+124 to exports	The surrounding rock is loose and broken

(2) **Engineering geological conditions**

Topography: The tunnel area is located at the southwestern end of the Hengduan Mountain and the southern extension of the Gaoligong Mountain. The terrain is high in the north and west and low in the south and east. The valleys are vertically and horizontally intersected. The highest elevation is 1342 m, and the lowest elevation is 1020 m. It belongs to the middle and deep cutting terrain and belongs to the tectonic and denuded alpine landforms. The tunnel has a steep terrain, with a slope of 40°–50°, and a part of it is a suspended slope. The horizontal distance from the river bed to the cave body is 150–200 m. The elevation of the riverbed is 1020 m, and the elevation of the tunnel is 1287.29 m, with a difference of 267.29 m. The slope zone is mainly composed of gravel subclay, gravel soil, block stone and other slope residual layers, and there is artificial accumulation on the upper part, showing the loose structure of block and gravel. The vegetation is not developed, mainly shrubs and grasses.

Lithology: The exposed strata in the tunnel area mainly include the Quaternary slope residual layer (Q_4^{dl+el}) and the Lower Paleozoic Cambrian Upper Lushui Formation ($\in_3 l$).

① Quaternary slope residual layer (Q_4^{dl+el}). It is mainly distributed in slopes with relatively slow slope and concave areas. The lithology is brown-yellow, brown-gray, brown-black clay, loam, and gravel, slightly to medium dense, slightly wet, and loose structure. According to the data revealed by detailed drilling holes, the maximum thickness is 10 m.

② The Cambrian Upper Lushui Formation ($\in_3 l$). It is composed of gray-white middle layer intercalated thick stratified marble and lime gray-to light-gray slight ~ thin stratified marble intercalated brown gray to gray green sericite slate and schist. The three tend to be in a transitional relationship, and the transitional segments tend to be interlayered. Gray-white middle layer intercalated thick stratified marble and lime gray to light-gray slight ~ thin stratified marble are relatively hard to hard rocks. The surface is moderately to weakly weathering. The dissolution pores with a diameter of 5–20 cm can be seen locally. Brown gray to gray green sericite slate and schist are relatively soft to soft rocks. The surface is strong to full weathering. The strike direction of the stratum is obliquely at a large angle with the central axis of the tunnel. The overall occurrence is N10° ~ 75°W·NE∠50°–84°. The entrance section of the tunnel is affected by faults, forming synclinal and anticlinal folds, and the reverse inclined occurrence is N10° ~ 75°W·SW Angle 50°–65°.

③ Geological structure. The tunnel area is located at the southwestern end of the Sichuan–Yunnan meridional tectonic system, the western part of the Nanling zonal tectonic system, and the complex parts of the three giant tectonic systems such as the Qinghai-Tibet-Burma-India Indonesian "歹"-shaped tectonic system. The geological structure is extremely complicated. The tectonic action of different forms and different stages forms three secondary faults directly related to the deep fault of the Nujiang River in the tunnel area, which is obliquely intersected with the central axis of the tunnel at an angle of 60°–85°. The rock mass is broken, which seriously affects the stability of surrounding rock and slope of the tunnel.

④ Hydrogeological conditions. The southeast side of the tunnel area is a tributary of the Nujiang River. It is a seasonal river in the mountains. It flows in the east–west direction and flows into the Nujiang River from west to east. There are two main types of groundwater in the tunnel area that are pore water and fissure water.

Pore water: It is mainly distributed in the subclay and gravel soil pores of the Quaternary slope. The source of recharge is mainly atmospheric precipitation. The pore water flows from the high to the low and is discharged into the valleys, rivers and underlying bedrock fissures. The chemical types of pore water are mainly HCO_3-Ca, HCO_3-Ca·Mg, and the degree of mineralization is 0.1–0.5 g/L.

Fissure water: It is mainly distributed in weathered fissures, tectonic fissures and fault zones of bedrock. The source of recharge is mainly atmospheric precipitation and pore water in the overlying loose layer. The chemical types of fissure water are mainly HCO_3-Ca and HCO_3-Ca·Na, with low salinity and good water quality, which is in line with industrial and agricultural production and living and drinking standards for human and livestock.

⑤ Climatic conditions. The tunnel area has a tropical and subtropical climate with abundant rainfall and distinct wet and dry seasons. The rainy season is mainly concentrated in May–October, with an average rainfall of 1200–1500 mm and an average annual evaporation of 1500–1600 mm. The hottest is from June to August, and the coldest is from November to February of the following year. There is no frost all year round. The climate in this area is clearly divided vertically, and there are significant differences in climate between different altitudes.

⑥ Regional earthquakes. The epicenter distribution shows that the Longling, Ruili, and Baoshan areas are high-risk areas, and the Nujiang and Lushui Ruili faults are active structures and are now in active activities.

According to the China Earthquake Parameter Zoning Map (GB18306-2001), the peak acceleration of ground motion in this area is 0.20 g.

(3) Program layout

In order to meet the survey and exploration requirements, a total of two seismic CT profiles and seven high-density resistivity detection profiles were arranged. Seismic CT focuses on the analysis of mechanical properties of the surrounding rock, including lithology distribution, fracture degree, fault structure, and stability of surrounding rock. The focus of resistivity imaging is to find out the water content, loose layer distribution, karst distribution, and fault structure of the surrounding rock. The combination of the two provides a scientific basis for formulating the treatment plan.

(4) The layout of the survey lines of seismic CT

① D1 section (horizontal section). The observation system is arranged along the middle guide tunnel in the tunnel and the left side access road outside the tunnel, as shown in Fig. 7.5.

In order to ensure the completeness of the observation system, the geophones for receiving in the observation are arranged inside and outside the tunnel at the same time, and the excitation source is arranged on the left side of the tunnel. A total of 120 geophones are arranged along the left wall in the middle guide hole in the tunnel, and the geophones are arranged on the same horizontal line. The spacing of the geophones is 2–3 m. A total of 48 geophones are arranged in the section K563+950 to K563+100 on the left side of the tunnel. The spacing of the geophones is 2–3 m.

The seismic excitation point is arranged along the left side of the tunnel, with a total of 140 excitation points and a shot interval of 3 m. The observation system obtained 23,520 seismic rays.

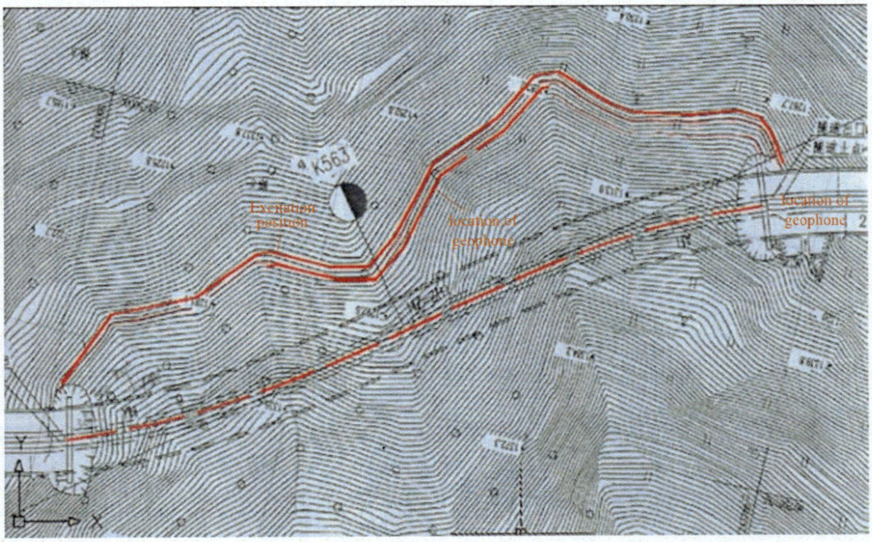

Fig. 7.5 Schematic diagram of the observation system of the D1 section of seismic CT

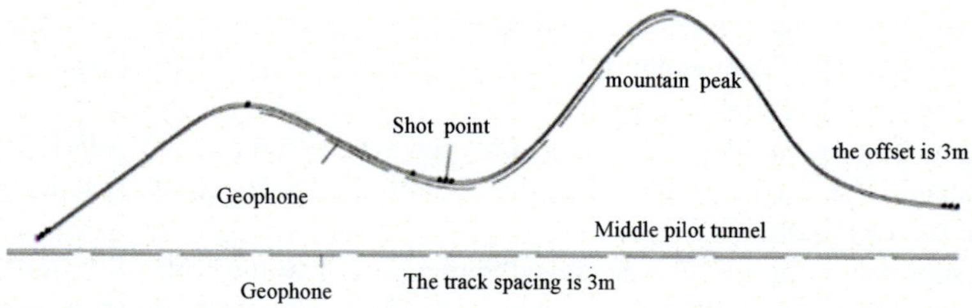

Fig. 7.6 Schematic diagram of the observation system of the D2 section of seismic CT

② D2 section (vertical section). The observation system is placed along the vertical section of the middle guide hole, as shown in Fig. 7.6. In order to ensure the completeness of the observation system, the receiving system is arranged in the tunnel and on the hillside at the same time, and the excitation source is arranged on the hillside. 120 geophones are arranged along the left wall in the tunnel, which are located on the same horizontal line with a track spacing of 2–3 m. On the hillside directly above the middle guide hole, 48 geophones are arranged between the two peaks of the section K562+930 to K563+070. The distance between the tracks is 2–3 m.

The earthquake excitation is arranged on the hillside directly above the middle guide hole. There are 150 excitation points with the shot interval of 3 m. The observation system obtains 25,200 seismic rays.

(5) **The layout of the survey lines of high density resistivity method**

A total of seven survey sections are arranged for high-density resistivity imaging, including four longitudinal sections EZ1–EZ4 parallel to the tunnel axis, and three transverse sections EH1–EH3 perpendicular to the tunnel direction. Among the four sections parallel to the tunnel axis, two are vertical sections EZ3 and EZ4 along the left and right tunnel axes. The other one is the section EZ2 along the horizontal plane of the tunnel and the access road, and the last one is the vertical section EZ1 near the bottom of the access road. The three sections perpendicular to the tunnel direction are located at the tunnel entrance EH1, exit EH3 and the middle part EH2, and are arranged along the hillside.

Longitudinal section EZ1 is arranged along the left side of the construction road. The length of the survey line is 384 m. The electrode spacing is 3 m. There are 128 sounding points.

Horizontal section EZ2 is arranged along the right side of the construction road. The electrodes are vertically arranged on the mountain wall. The length of the survey line is 384 m. The electrode spacing is 3 m. There are 128 sounding points.

Longitudinal section EZ3 is arranged along the mountain directly above the axis of the left tunnel. The length of the survey line is 456 m. The electrode spacing is 3 m. There are 152 sounding points.

Longitudinal section EZ4 is arranged along the mountain directly above the axis of the right tunnel. The length of the survey line is 456 m. The electrode spacing is 3 m. There are 152 sounding points.

Transverse section EH1 is arranged perpendicular to the K562+895 mileage of the tunnel. The length of the survey line is 408 m. The electrode spacing is 3 m. There are 136 sounding points.

Transverse section EH2 is arranged perpendicular to the K562+995 mileage of the tunnel. The length of the survey line is 312 m. The electrode spacing is 3 m. There are 104 sounding points.

Transverse section EH3 is arranged perpendicular to the K562+135 mileage of the tunnel. The length of the survey line is 288 m. The electrode spacing is 3 m. There are 96 sounding points.

Each high-density resistivity profile collects 30 layers with a sounding depth of 90 m. The high-density resistivity profiles have a cumulative length of 2688 m and 896 depth sounding points.

(6) Detection result

After analysis and processing, the wave velocity distribution image of seismic CT is shown in Fig. 7.7. The ray density distribution image of seismic CT is shown in Fig. 7.8. The ray orthogonality image of seismic CT is shown in Fig. 7.9.

(7) Exploration result of seismic CT

From the wave velocity distribution image of the horizontal section D1 of seismic CT, it can be clearly seen that the rock on the left side of the tunnel is extremely

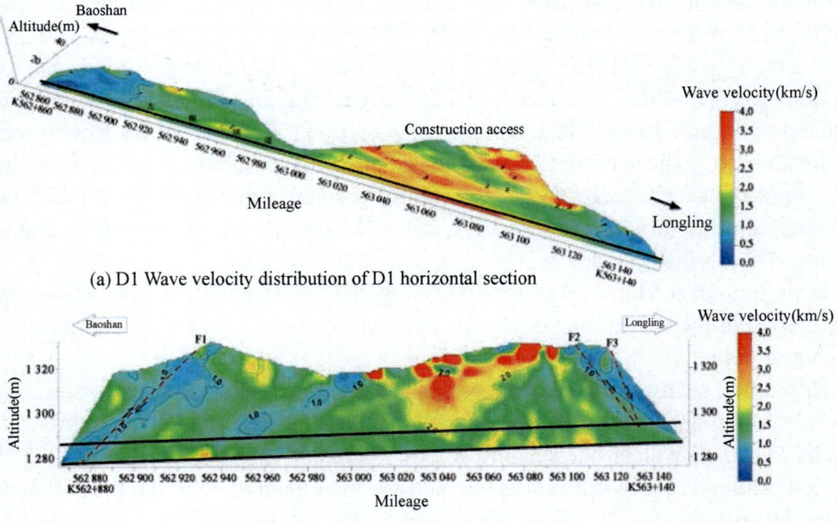

(b) D2 Wave velocity distribution of D2 vertical section

Fig. 7.7 Wave velocity distribution image of seismic CT

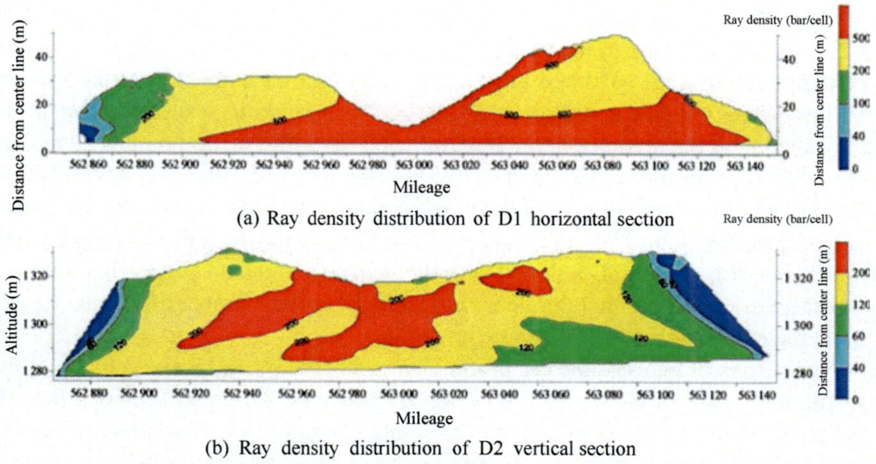

(a) Ray density distribution of D1 horizontal section

(b) Ray density distribution of D2 vertical section

Fig. 7.8 Ray density distribution image of seismic CT

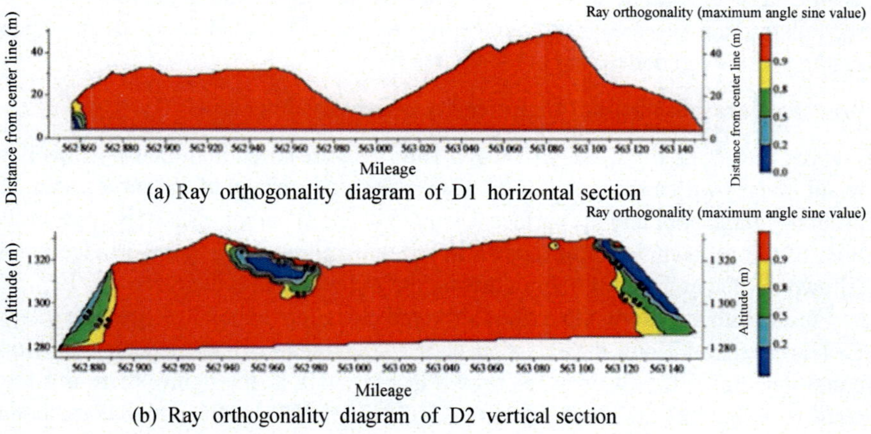

(a) Ray orthogonality diagram of D1 horizontal section

(b) Ray orthogonality diagram of D2 vertical section

Fig. 7.9 Ray orthogonality image of seismic CT

loose and broken. The seismic wave velocity of the two sections of 175 m long from k562+845 to k63+020 at the entrance and 55 m long from k563+110 to k563+165 at the exit is lower than 1.5 km/s. In particular, the 65 m-long section at the entrance and the 40 m-long section at the exit have wave speeds less than 1.0 km/s, which are characteristics of loose accumulation and fracture zone, and may be broken through. In the 175 m-long section of the entrance and the 55 m-long section of the exit, the mountain body is loose and broken, with the crack develops, and it is unstable under the bias pressure. After the tunnel excavation, due to the biasing effect, the mountain body is easily cracked and the lining is destroyed. It is a hidden area of geological disasters, which must be treated.

From the wave velocity distribution image of the vertical section D2 of seismic CT, it can be seen that in the section within the length of 165 m from the entrance of the tunnel to k563+010 and the section of 35 m after k563+130 at the exit are all low-velocity regions, with the velocity less than 1.5 km/s, especially the 45 m-long section at the entrance K562+845 to K562+890, the section within 10 m near the K562+950 in the middle, and the 35 m-long section at the exit K563+130 to K563+165. The wave velocity of the three sections is less than 1.0 km/s. These low-velocity areas extend from the surface of the hillside down to the direction of the tunnel in a strip pattern. Combined with the surface geological survey, it is inferred that there are faults passing through, with faults F1 at the entrance and faults F2 and F3 at the exit.

The results of two seismic CT profiles show that the overall wave velocity of the surrounding rock of the tunnel is low, mostly less than 2.0 km/s. The rock mass is broken. The fissure is developed, and the loose accumulation is dominant. Most of the surrounding rock of the tunnel belongs to class II, and the individual sections are less than class II. At 175 m at the entrance and 55 m at the exit, the rock mass is weaker and has fracture development. There are three most serious areas: 65 m at the entrance, 55 m at the exit, and 10 m near the K562+950 in the middle. These sections are the key areas for engineering treatment. Each anomaly area has been marked with a red dotted line in the figure.

(8) **Exploration result of high-density resistivity detection**

Figures 7.10, 7.11, 7.12, 7.13, 7.14, 7.15, and 7.16 are seven high-density resistivity result images with a scale of 1:1000. Due to the steep terrain of the survey area, the groundwater is poor and the surface is very dry. The apparent resistivity is generally high of the geotechnical soil (shown in red and yellow), and the resistance is low (shown in blue and green) only in the middle of the tunnel (near K563+000).

For the four longitudinal sections arranged along the tunnel direction, the section EZ1 is arranged along the left side of the construction road to detect the slope conditions under the roadway, as shown in Fig. 7.10. In the figure, there is a low-resistance anomaly area in the section K562+975 to K563+005, which corresponds

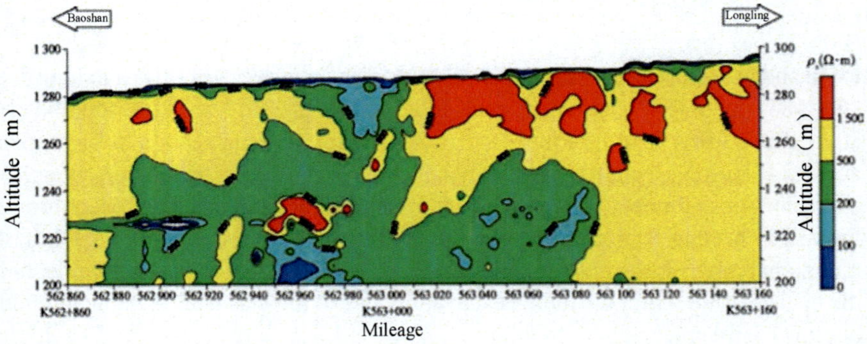

Fig. 7.10 Apparent resistivity result image of section EZ1 (K562+860 to K563+160)

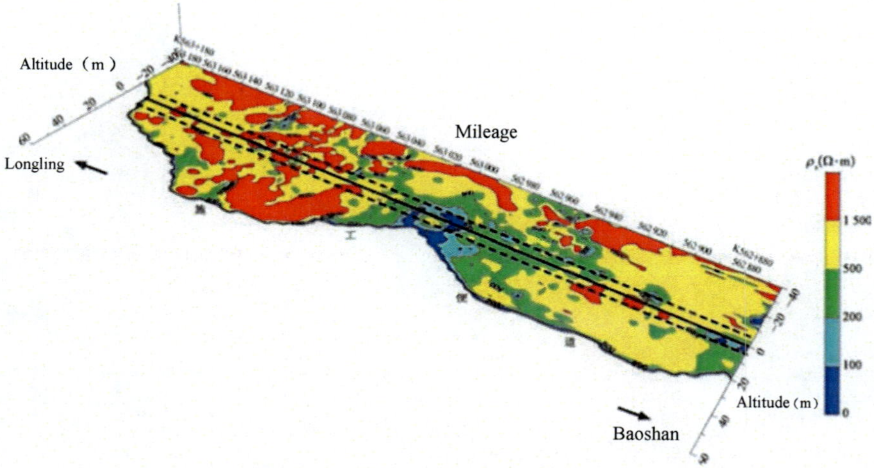

Fig. 7.11 Apparent resistivity result image of horizontal section EZ2 (K562+865 to K563+180)

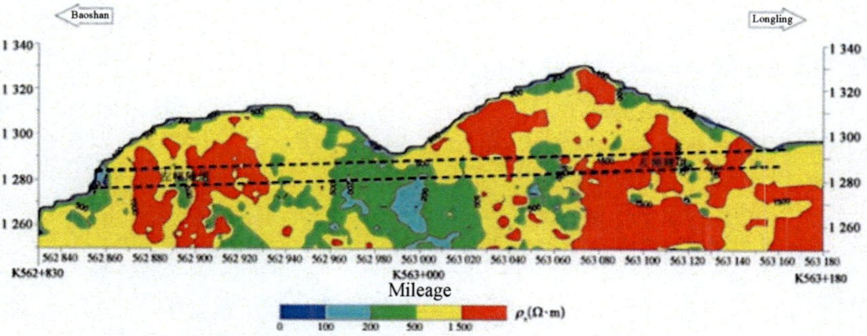

Fig. 7.12 Apparent resistivity result image of section EZ3 (K562+830 to K563+180)

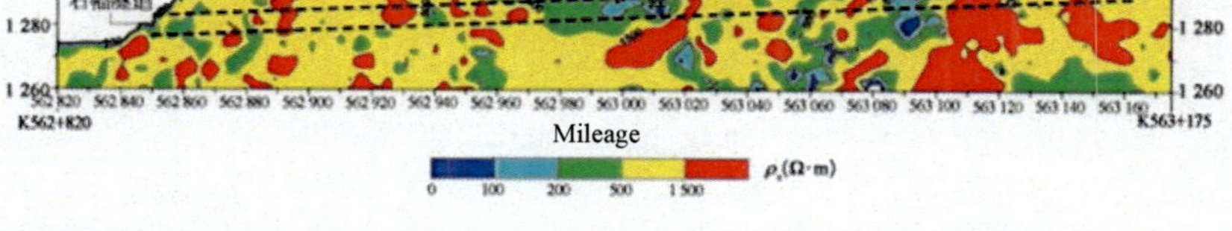

Fig. 7.13 Apparent resistivity result image of section EZ4 (K562+820 to K563+175)

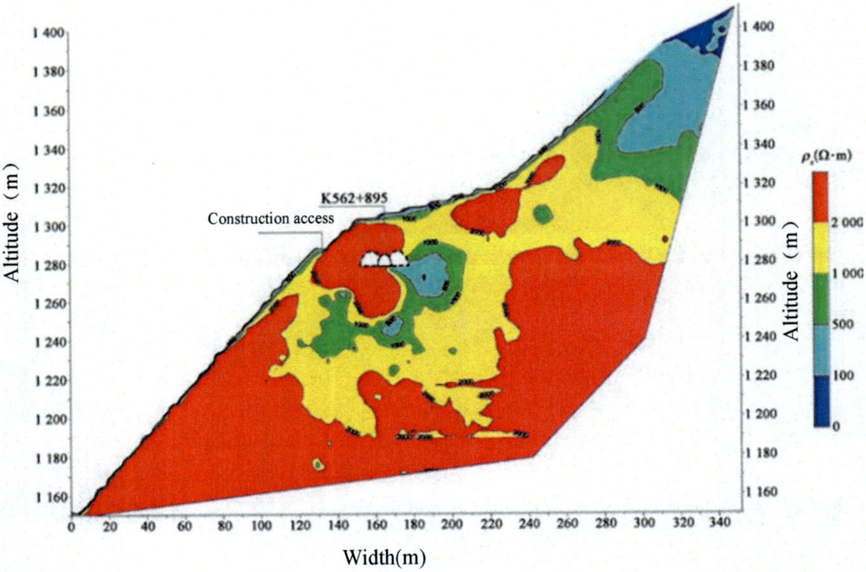

Fig. 7.14 Apparent resistivity result image of section EH1 (K562+895)

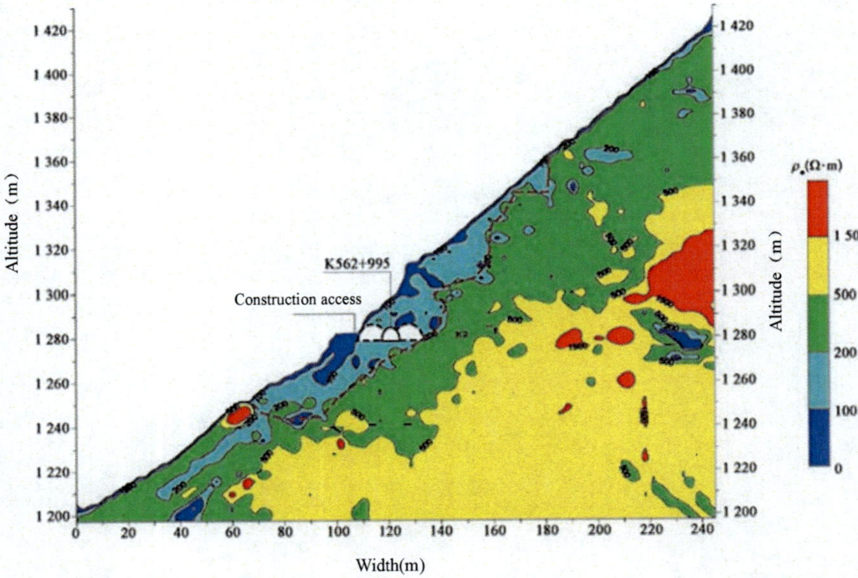

Fig. 7.15 Apparent resistivity result image of section EH2 (K562+995)

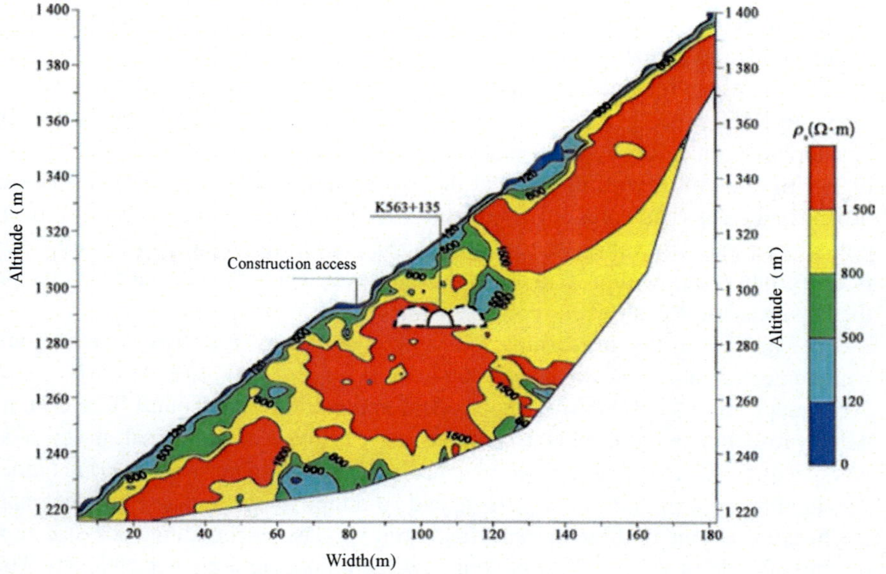

Fig. 7.16 Apparent resistivity result image of section EH3 (K563+135)

to position of the valley. Geological survey shows that the lithology of this section is sericite schist with low resistivity, which is inferred to contain water. The section EZ2 is arranged along the stone wall 1 m above the right side of the construction road, and the electrode is vertically driven into the stone wall to obtain a horizontal section map (Fig. 7.11). It can be seen from the figure that there is a low-resistance zone in the corresponding valley location of section K562+980 to K563+010, which is presumed to be water-bearing sericite schist. The sections EZ3 and EZ4 are arranged directly above the left and right tunnels respectively, corresponding to Figs. 7.12 and 7.13. In both figures, there are low-resistance zones in the corresponding valley locations of sections K562+960 to K563+020. It is speculated that the surrounding rocks are hydrated sericite schist. The tunnel is shallowly buried in this section, which is unfavorable for construction.

The three transverse sections EH1–EH3 arranged perpendicular to the tunnel direction pass through K562+895, K562+995, and K563+135, respectively. The corresponding apparent resistivity images are shown in Figs. 7.14, 7.15, and 7.16. From the comparison of the three figures, the apparent resistivity values of the two sections EH1 and EH3 at the entrance and exit of the tunnel are relatively high near the tunnel. Combined with the analysis of the corresponding low-velocity zone in the seismic CT profile, this is an area where the fissure zone is concentrated, with a large proportion of voids in the rock mass, and the tunnel is in the fissure zone. The middle section EH2 is located near the valley and has a low apparent resistivity value. The shallow layer contains water within 30 m and the tunnel is in the loose aquifer. It can be seen from the three figures that the mountain where the tunnel is located is very steep, and there is a low-resistance zone on the surface, which is 20–30 m deep. It is assumed that the tunnel is located in the unloading zone of the free-face.

Overall resistivity imaging survey results show that from the entrance of the tunnel to K562+895, there are 20 m of loose accumulation, followed by the fissure development zone. The rock mass is extremely broken, and the void ratio is very high. From the mileage K563+130 at the exit section to the exit, the rock mass is broken and cracks and voids are developed, which are the characteristics of a typical unloading zone. The valley location near k562+990 in the middle of the tunnel is rich in groundwater within 30 m of the shallow layer. These three sections should be the key areas to be dealt with.

The survey results of seismic CT and high-density resistivity method are complementary and basically consistent.

Through the comprehensive geophysical exploration of seismic CT and high-density resistivity method and the comprehensive analysis of geological survey data, it is speculated that the main unfavorable geological phenomena in the area are tunnel surrounding rock fragmentation, fissure and void unloading zone development, and fracture development at the entrance and the exit. The surrounding rock structure on the left side of the tunnel is broken, and the unloading zone is developed. The bias pressure in the tunnel excavation will lead to the cracking of the mountain and the damage of the lining. The surrounding rock at the entrance, exit and central part must be prereinforced.

(9) **Conclusion**

① The tunnel is located in a composite part of three giant structural systems with complex structures and broken rock mass. The two regional faults of Nujiang deep fault and the Lushui–Ruili fault are active fault zones and are earthquake-prone areas. There are faults F1, F2, F3 passing through the entrance and exit of the tunnel.

② The terrain of the mountain that the tunnel traverses is steep. The unloading zone is developed. The cracks and voids are dense. The unfavorable structural plane combination is obvious. The rock mass is broken. It is unfavorable to the stability of the tunnel surrounding rock and the slope. The tunnel slope is an unstable slope.

③ The results of seismic CT survey show that the overall wave velocity of tunnel surrounding rock is low, most of which are less than 2.0 km/s. The structure of rock mass is broken, and the cracks and voids are developed. Most of the surrounding rocks of the tunnel are of Type II, and individual sections are not of Type II. The rock mass is weaker and has fracture development at 175 m in the entrance section of the tunnel and 55 m in the exit section. Among them, there are three sections with the most severe rock fragmentation, 65 m at the entrance section, 55 m at the exit section, and 10 m near K562+950 in the middle.

④ The survey results of resistivity imaging show that from the K562+845 at the entrance of the tunnel to K562+895, there are 20 m of loose accumulation, followed by the fissure development zone. The rock mass is extremely broken, and the void ratio is very high. From the mileage K563+130 to K563+165 at the exit section, the rock mass is broken and cracks and voids are developed, which are the characteristics of a typical unloading zone. The valley location near k562+990 in the middle of the tunnel is rich in groundwater within 30 m of the shallow layer.

⑤ The surrounding rock of the mountain on the left side of the tunnel is severely broken at 65 m of the entrance section of the tunnel K562+845 to K562+910, the 50 m of the K562+970 to K563+020 in the middle, and the 55 m of the K563+100 to K563+165 in the exit section. The crack is developed, and the thickness is thin. The slope is extremely unstable. Bias pressure during tunnel excavation may cause mountain cracks and lining damage, and engineering reinforcement must be carried out.

7.2.2 Application of Seismic CT Technology in a Highway Tunnel in Yunnan

(1) **Project overview**

As shown in Fig. 7.17, the expressway tunnel is a multi-arch tunnel with separate upper and lower lines. The tunnel mileage is from K235+745 to K235+980, with the total length of 235 m. The tunnel is located on the circular curve of $R = 150$ m, $L_s = 120$ m, $I = 10–2.37\%$. The clear width of the tunnel is 25.24 m, the thick of the middle wall 2 m, the maximum buried depth 63.96 m.

Fig. 7.17 Location of the tunnel

During the construction of the tunnel, the section K235+879 to K235+889 of the upline occurred collapse and roof fall. In the exit section of the tunnel (K235+920 to K235+970), the side walls, vaults, and partition walls were displaced, deformed, and damaged. The quality and construction safety of the tunnel were seriously affected.

There are three main types of tunnel disasters. The first type is partial collapse and roof fall. The second type is tunnel lateral displacement, deformation, and cracking. The third type is collapse of slope top. The collapse and roof fall of the tunnel occurred in the section K235+879 to K235+889 of the upline. The lithology of the collapse section is mainly sand and clay, with a small amount of rock mixed. The surface subsidence has a diameter of 14.6 m and a depth of 12.3 m. Affected by this, the initial support of the section K235+889 to K235+899 also occurred displacement and cracking. The lateral displacement, deformation, and cracking of the tunnel mainly occurred in the exit section K235+920 to K235+970, and the more serious part is the section K235+940 to K235+965. Cracks and offsets occurred in the side walls, vaults, and partition walls. The middle wall of the section K235+940 to K235+965 was offset to the left side of the route by 5 cm from the end of February to May 20th. In the ten days from July 20th to the beginning of August, the section K235+957 to K235+965 was offset to the left of the route by 12.5 cm. The maximum misalignment of the side wall crack was 15 cm, and two large cracks appeared in the wall of the hole. The cracking of the retaining wall of the open hole is 6.5 cm, and the crack distribution is shown in Fig. 7.18. The damage of the tunnel side wall, the vault and the middle partition wall has three characteristics. First, the degree of cracking is heavier near the hole K235+965 and is relieved in the tunnel. The second is that the

Fig. 7.18 Photos of on-site disasters

upper part of the tunnel is heavily damaged and the lower part is lightly damaged. The third is that the cracking is left-shifting, and the upper side of the crack and the side of the exit both move to the left. The slide and collapse of slope top mainly occurred on the slope of the tunnel entrance section, where the slope is steep and the loose layer is thick, and there are multiple collapses. Landslides and mudslides may occur in every rainstorm.

(2) **Engineering geological conditions**

① Lithology. The strata of the tunnel area are mainly:

The Quaternary artificial fill layer (Q^{ml}): It is dominated by broken stones with gneiss composition, with a thickness of less than 5 m, and a diameter of 5–10 cm. It contains sandy silt and is loose and has a loose structure.

The Quaternary alluvial gravel layer (Q^{al+pl}): It is composed of gray, medium-dense pebbles and gravels. The composition of the gravel is gneiss. The size of the gravel is uneven. The filling is fine sand, silty sand, with the thickness of less than 4 m, and it is mainly distributed on the floodplain at the bottom of the slope.

The Quaternary Collapse Slope (Q^{col+dl}): It is mainly composed of grayish-yellow and grayish-white stone, containing sandy silt, slightly dense, softened with water. The composition of the stone is gneiss, and the diameter of the soil skeleton stone varies between 10 and 30 cm. It has an overhead structure and is prone to loosening and sliding.

The Upper Jurassic bedrock (Ptx^b) of the Xiaoyangjie Formation in the Proterozoic: It is mainly composed of medium to thick layered gray-white granitic gneiss, with an overall tendency to the northwest, contrary to the tendency of the slope. The rock mass has a piece of hemp structure, and the joint fissure is relatively developed. The strongly weathered rock mass is fragmented to a loose structure, with low strength, loose structure, and localized weathered sand, which is prone to collapse. The medium weathered rock mass has a block-to-fragment structure with certain mechanical strength and slightly better stability. Due to the difference in weathering, there are locally strong weathered inclusions and there are overhead phenomena.

② Geological structure. The surface layer in the tunnel area is the Quaternary sloped loose layer, and the underlying bedrock is the granitic gneiss (Ptx^b) of the upper subgroup of the Xiaoyangjie Formation in the Proterozoic. The overall tendency is to the northwest, which is opposite to the slope tendency. The occurrence of rock strata is generally 220°–235°∠28°–39°. The interlaminar fold structure and joints and fissures are developed. The rock mass in the tunnel area is strongly weathered and fractured. The weathered layer is thick, and the surrounding rock stability is poor.

③ Hydrogeological conditions. The groundwater in the tunnel area is mainly bedrock fissure water, and the water-bearing medium is weakly water-rich granitic gneiss. The vertical infiltration of atmospheric precipitation is the main recharge route of groundwater, and the water level is buried deep and unstable, and the hydraulic connection is weak. Excretion is controlled by topographical

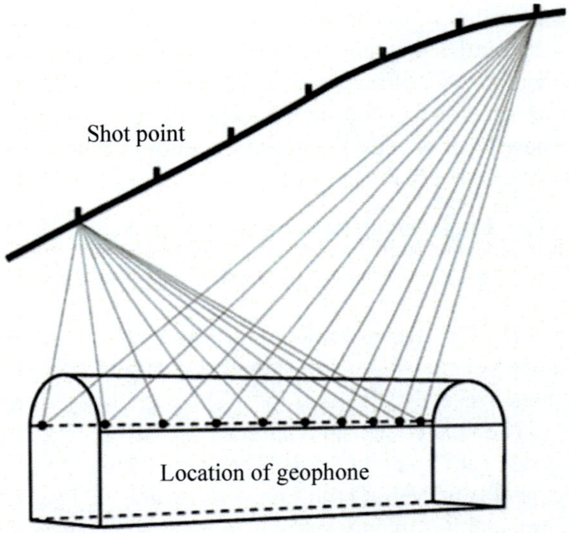

Fig. 7.19 Observation layout of seismic CT

conditions. The bottom of the river or the partial cliff foot is a common excretion site, and the regional drainage level is a river channel.

(3) **Program layout**

According to the purpose of the exploration and the research needs, six exploration sections were arranged for the seismic CT survey, including three transverse sections, two longitudinal sections, and a lateral section of the tunnel and the river channel. The specific position of the section is shown in Fig. 7.19, and the schematic diagram of the observation system is shown in Fig. 7.20.

① K235+950 transverse section. At the K235+950 position of the exit section of the tunnel, perpendicular to the central line in the upline and the downline, 24 detector points are arranged along the side wall and the arch ring, and a total of 48–60 Hz geophones are used to receive. The excitation points are arranged on the mountain, and the excitation points and the receiving points in the tunnel are kept in a vertical plane. There are 101 shot points with the spacing of 2 m, and the length of the ground survey line is 200 m. The explosion source is used. In order to accurately ascertain the thickness and distribution of the loose and weathered layers along the slope, 48 geophones were arranged on the side of the mountain below the ground excitation survey line with a distance of 2 m. A total of 8 seismic sources were excited on the side of the mountain.

② K235+910 transverse section. At the K235+910 position of the exit section of the tunnel, perpendicular to the central line in the upline and the downline, 24 detector points are arranged along the side wall and the arch ring, and a total of 48–60 Hz geophones are used to receive. The excitation points are arranged on the mountain, and the excitation points and the receiving points in the tunnel are kept in a vertical plane. There are 101 shot points with the spacing of 2 m,

Fig. 7.20 Layout diagram of survey lines

and the length of the ground survey line is 200 m. The explosion source is used. Same as the previous section, in order to accurately ascertain the thickness and distribution of the loose and weathered layers along the slope, 48 geophones were arranged on the side of the mountain below the ground excitation survey line with a distance of 2 m. A total of 8 seismic sources were excited on the side of the mountain.

③ K235+799 transverse section. The K235+799 section is arranged perpendicular to the tunnel axis along the K235+799 mileage. Vertically down from the ground, it is located outside the boulder at the entrance of the tunnel and diagonally crossing the hillside. The section observation consists of three arrangements. The first arrangement is to arrange 48 geophones in the uplink tunnel, and to shoot from the bottom to the top of the hill. The distance between the shots is 2–4 m, and there are 100 shot points. The second arrangement is to move the receiving point to the guide hole in the side wall of the downline, which is received by 24 geophones. The seismic source is excited on the hillside. The shot points were placed at the foot of the mountain and on the top of the mountain, and a total of 48 seismic sources were excited. The received points of the third arrangement were placed at the foot of the mountain, which is received by 24 geophones. The shot points are on the hillside, and there are 20 shot points and 6192 effective rays.

④ Longitudinal section at the middle wall and middle road pit of the upline. At the section K235+897.5 to K235+947 of the upline, 24 geophones are arranged in the corner of the middle wall. A total of 48 geophones are arranged in the right corner of the section K235+779 to K235+873 of the middle road pit. A total of 72 channels are received. The track spacing is 2 m, and the length of the survey line is 168 m. In the K235+759 to K235+965 section, the excitation point is arranged along the receiving direction of the hole. The excitation point is kept in a vertical plane with the receiving point in the tunnel. The distance between the shot points is 2 m. The length of the survey line is 206 m. The explosion source is used. A total of 104 seismic sources were excited.

⑤ Longitudinal section at the corner on the right side of the downline. 48–60 Hz geophones are arranged in the right corner of the section K235+894 to K235+959.2 of the downline, with a track spacing of 2 m, recorded by the SE2432 comprehensive engineering surveyor produced by GEOPEN. In the K235+759 to K235+965 section, the excitation point is arranged along the receiving direction of the hole. The excitation point is kept in a vertical plane with the receiving point in the tunnel. The distance between the shot points is 2 m. The length of the survey line is 206 m. The explosion source is used. A total of 104 seismic sources were excited.

⑥ Lateral section of the upper middle wall and the middle road pit and the river channel of the upline. At the section K235+897.5 to K235+947 of the upline, 24 geophones are arranged in the corner of the middle wall. A total of 48 geophones are arranged in the right corner of the section K235+779 to K235+873 of the middle road pit. A total of 72 channels are received. The track spacing is 2 m, and the length of the survey line is 168 m. The explosion source is used. A total of 195 seismic sources were excited.

(4) **Detection result**

The observation results of the six seismic CT sections of the tunnel clearly show the characteristics of engineering geological structure around the tunnel, as well as the distribution of mechanical strength of rock mass. The results of seismic CT and geological analysis show that the joints and fissures of the granite gneiss around the tunnel are well developed and seriously weathered, and they belong to the type of jointed fractured rock mass. The joint weathering zone has a bandwidth of several meters to ten meters, and it is filled with sand and clay. The landslides at K235+879 to K235+889 are related to the joint zone. The type of surrounding rock at the exit section of tunnel is slightly different, which is loosely packed. The accumulation body is composed of rock blocks, sand and clay, and the content of sand and clay is more. The wave velocity of the surrounding rock of the tunnel is low, and the overall integrity is poor, especially at the top of the ascending tunnel.

The wave velocity distribution of the surrounding rock of the tunnel is roughly as follows. The wave velocity of the rock mass with good integrity such as microweathering and unweathering is 2.0–4.0 km/s, and the wave velocity of the medium weathering rock is 1.3–2.0 km/s. The wave velocity of strongly weathered rock masses is 1.0–1.3 km/s. Compared with weathering layers of the same class of granite, they are all about 0.5 km/s lower. The wave velocity of the loose soil layer is less than 1.0 km/s. The detailed results of the six sections are described below:

① Geological conditions at the exit section of the tunnel and detection result of seismic CT section 1 (K235+950). The location of the seismic CT section 1 is at the exit section K235+950, and the section is upright orthogonal to the tunnel axis. The wave velocity image and geological interpretation are shown in Figs. 7.21 and 7.22. It can be clearly seen from the figure that the wave velocity of the surrounding rock at the location of the tunnel at the exit section is very low, which is less than 1.0 km/s. It is the loose accumulation in front of the slope.

The accumulation body is composed of massive granite gneiss, sand and clay, and the content of sand and clay is more. It has low mechanical strength and poor stability, especially when it is saturated with water, it is prone to displacement and deformation. There are also similar accumulations on the left side of the upper line of the tunnel with low strength and poor stability. The right side of the downward tunnel is

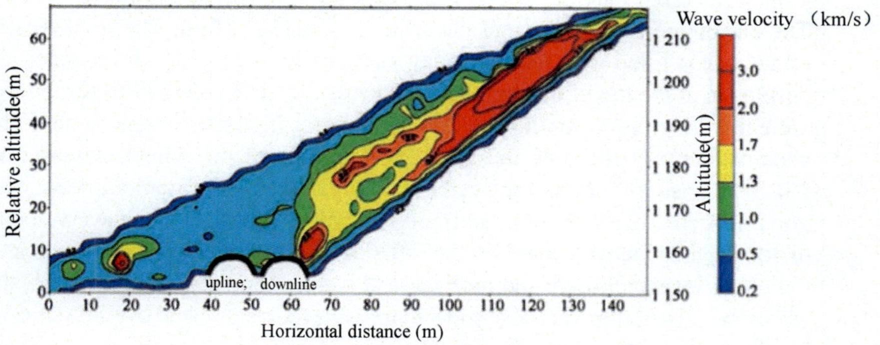

Fig. 7.21 Velocity image of the transverse section of the tunnel K235+950

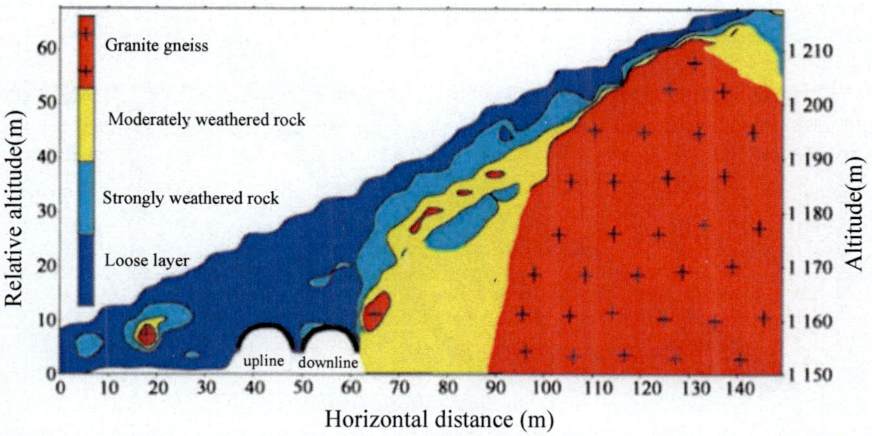

Fig. 7.22 Geological interpretation of the transverse section of the tunnel K235+950

moderately weathered rock with good rock strength and stability, but the top hillside has 5–10 m loose layer with poor stability and easy sliding. Beyond 30 m on the right side of the tunnel, the wave velocity of the rock mass reaches 2.0–3.0 km/s. It is the microweathered and unweathered rock with good integrity and stability. The instability of the surrounding rock at the exit of the tunnel is mainly caused by the gravity slip of the loose deposits around the tunnel under saturated conditions.

② Geological conditions in the middle of the tunnel and detection result of seismic CT section 2 (K235+910). Seismic CT section 2 is located at K235+910, and the section is upright orthogonal to the tunnel axis. The wave velocity image and geological interpretation are shown in Figs. 7.23 and 7.24. The distribution of wave velocity and characteristics of geological structure are slightly different from those in section 1. It can be clearly seen from the figure that the wave velocity of the tunnel is greatly changed, and the low-velocity zone is linearly distributed, indicating that the joints are developed and belong to the type of jointed fractured rock mass. The width of joint weathering zone varies greatly. The narrow part is 3–5 m, and the wide part reaches 20 m. The joint weathering zone is filled with sand and clay with low wave velocity. Joint zones are developed at the top of the arch, and the landslides at K235+879 to K235+889 are caused by the instability of loose fillings in high-angle joint zones. The wave velocity of the rock mass on the right side of the tunnel increased to 2.0–3.0 km/s. It is the microweathered and unweathered granite gneiss with good integrity, high strength, and good stability. The high-low wave velocity distribution of the rock mass on the left side of the ascending tunnel indicates the uneven weathering of the rock mass, the development of joint fissures, the embedded distribution of loose soil and massive rock mass, and overall stability, but the loose part of the vault is locally unstable.

③ Geological conditions at the entrance of the tunnel and detection result of seismic CT section 3 (K235+779). Seismic CT section 3 is located at K235+799, and

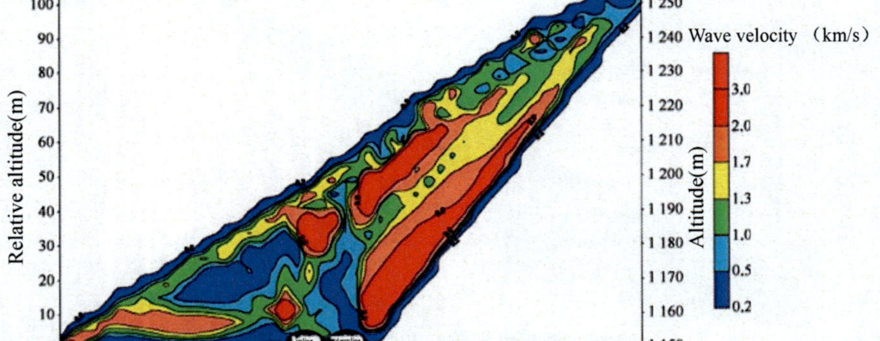

Fig. 7.23 Velocity image of the transverse section of the tunnel K235+910

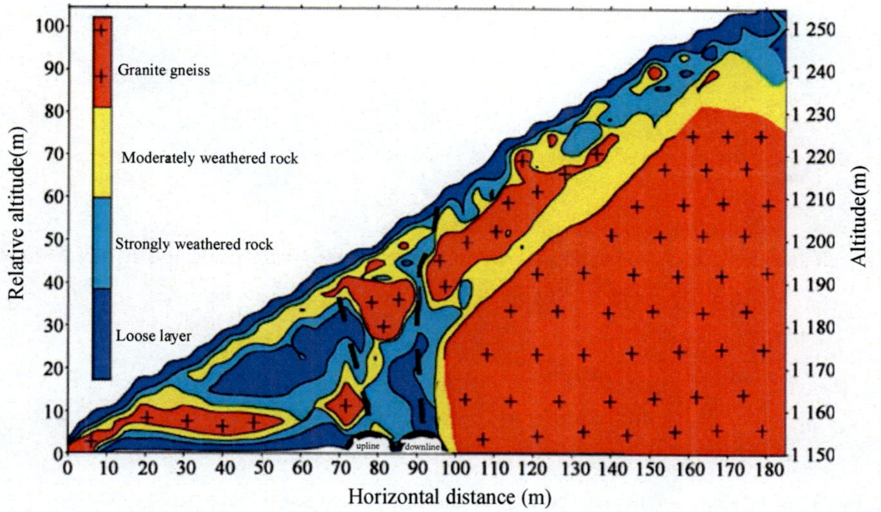

Fig. 7.24 Geological interpretation of the transverse section of the tunnel K235+910

the section is upright orthogonal to the tunnel axis. The wave velocity image and geological interpretation are shown in Figs. 7.25 and 7.26. This section is located outside the big boulder at the entrance, cutting the hillside obliquely. The lithology of the shallow part of the section cut is mainly loose deposits. The wave velocity is mostly below 1 km/s. Underneath the loose deposits are strongly weathered granite gneisses. The rock mass is broken, and the wave velocity is about 1.3 km/s. There are no moderately weathered rocks. Compared with the K235+910 section, the surrounding rock condition of this section is worse, and most of the rocks in the area are type I–II rock masses. The mountain at the top of the tunnel has a V-shaped low wave velocity fracture zone filled with fully weathered loose material. The V-shaped spine is a high wave velocity

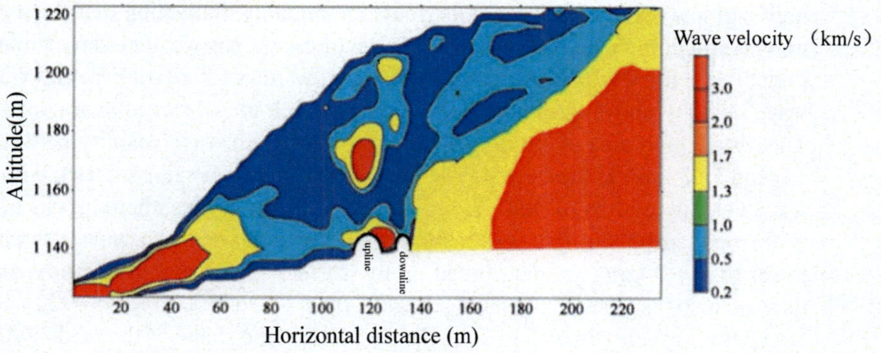

Fig. 7.25 Velocity image of the transverse section of the tunnel K235+799

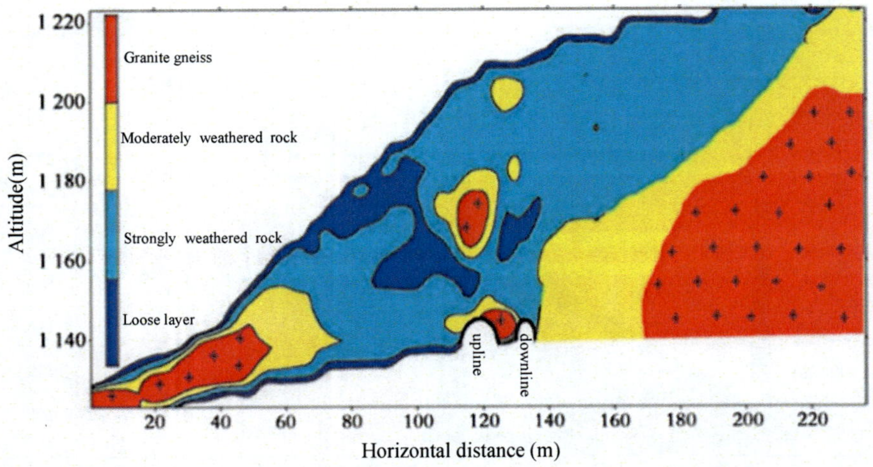

Fig. 7.26 Geological interpretation of the transverse section of the tunnel K235+799

rock mass, which is a moderately weathered granite gneiss mass connected to a large solitary rock at the entrance. Big boulder is a granite gneiss body with a thickness of 40–50 m. The joints are developed. The joint surface is almost vertical and slightly inclined to the tunnel, which is stable in the natural state. After tunnel excavation, the lower support of the bunker is affected and needs to be strengthened.

④ The characteristics of the longitudinal surrounding rock at the top of the tunnel and detection result of seismic CT section 4 and 5. Seismic CT section 4 is arranged along the centerline of the tunnel and cuts vertically up to the surface. The seismic CT section 5 is arranged along the right wall of the downline and cuts vertically up to the surface. The horizontal distance between the two sections is about 12 m. The wave velocity image and geological interpretation are shown in Figs. 7.27, 7.28, 7.29, and 7.30. The wave velocity distribution of the rock mass in the seismic CT section 4 at the centerline of the tunnel is very uneven, and the high and low wave velocity are distributed alternately, indicating that the rock mass is still dominated by joint fracture structures, and the weathered rock mass joint zone is filled with loose matter. There are few intact, hard rock masses with wave velocity above 2.0 km/s in the profile, which are scattered sporadically. The distribution area of strongly weathered rocks with wave velocity between 1.0 and 1.13 km/s is the largest. The area of moderately weathered rock with a wave velocity of 1.3–2.0 km/s is second only to strongly weathered rocks and is surrounded by strongly weathered rocks. Loose layers with wave velocity lower than 1.0 km/s are distributed on the surface of the mountain body, and they are also distributed in the tunnel vaults from K235+800 to K235+920 with thicknesses ranging from 5 to 20 m. In the section, the shape of the landslide at K235+879 to K235+889 mileage is clearly visible. This is the intersection of a

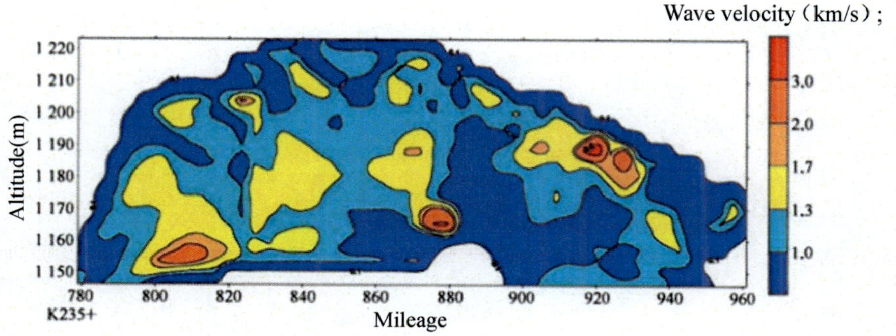

Fig. 7.27 Wave velocity image of the longitudinal section of the tunnel's upward line and the middle guide pit

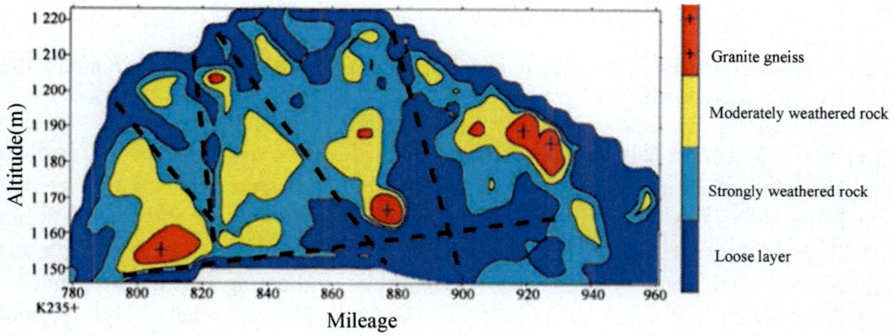

Fig. 7.28 Geological interpretation of the longitudinal section of the tunnel's upward line and the middle guide pit

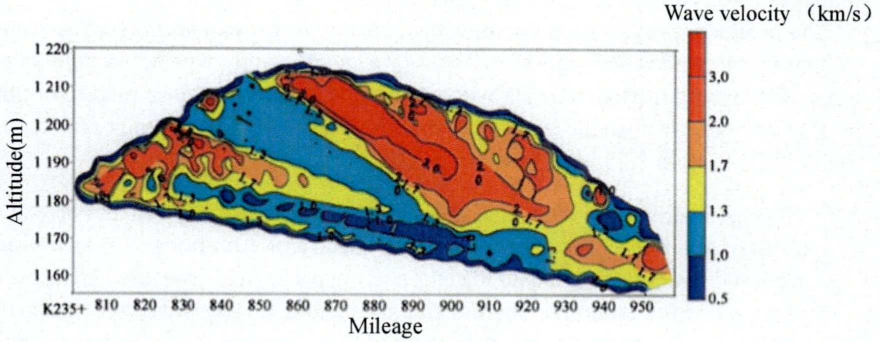

Fig. 7.29 Wave velocity image of the longitudinal section of the tunnel's downward line

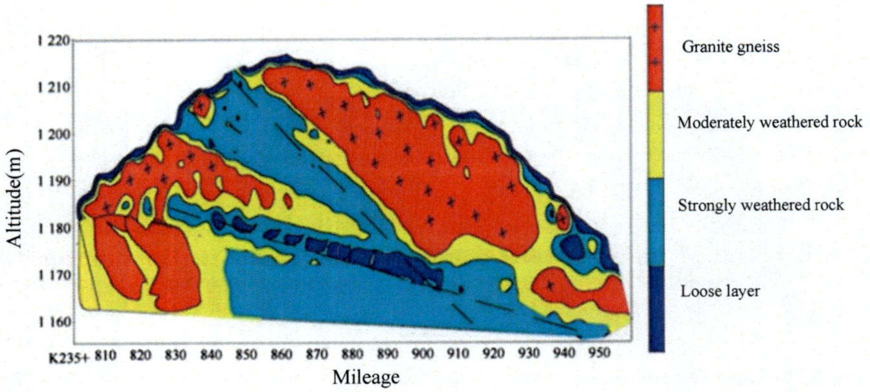

Fig. 7.30 Geological interpretation of the longitudinal section of the tunnel's downward line

high-angle joint zone and a nearly horizontal joint zone. The joint zone is filled with loose objects.

Seismic CT section 5 is located on the right wall of the downline, which is closer to the uphill of the mountain than section 4. The distribution area of high-velocity rock masses in the section increased significantly, and the integrity, strength, and stability of the rock masses become better. There are two weathered joint zones in the section with wave velocities of 1.0–1.3 km/s, indicating that the rock mass in the joint zone is strongly weathered rock. The upper joint zone has a slightly higher angle, starting from the slope with a mileage of K235+840 and extending obliquely downward to the K235+930 vault. The upper part is 20 m wide and the lower part is about 10 m wide. The lower one is almost horizontal and is distributed between mileages K235+850 and K235+920. This is similar to section 4.

The results of the two sections show that, except for the exit section of the tunnel (K235+920 to K235+965), the surrounding rocks are mainly weathered joint structures. The mountain is stable as a whole. The part filled with loose material in the joint fissure zone is unstable, especially the surrounding rock at the top of the tunnel in the K235+840 to K235+920 section is loose and the stability is poor.

⑤ The characteristics of surrounding rock of the tunnel foundation and detection result of seismic CT section 6. Seismic CT section 7.1 is arranged as an oblique section in order to understand the engineering geological structure and stability of the slope foundation on the left side of the tunnel. The wave velocity image and geological interpretation are shown in Figs. 7.31 and 7.32. It can be clearly seen from the image that the wave velocity of the rock mass in the lower part of the left slope is very high. The wave velocity in most regions is higher than 2.0 km/s, 50% of which are greater than 3.0 km/s, which indicates that the rock mass has low weathering degree, good integrity, high strength, and good stability. No large-scale low-velocity zones, no fault structures, and no weak zones are found in the area. Under the background of high wave velocity, it

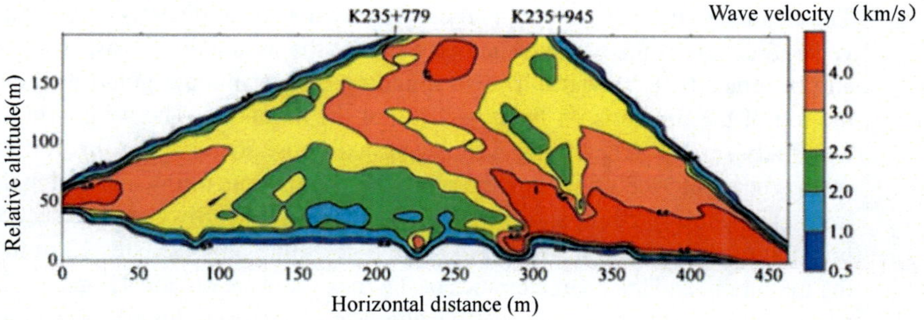

Fig. 7.31 Wave velocity image of the section of the tunnel's upward line, middle guide pit, and river channel

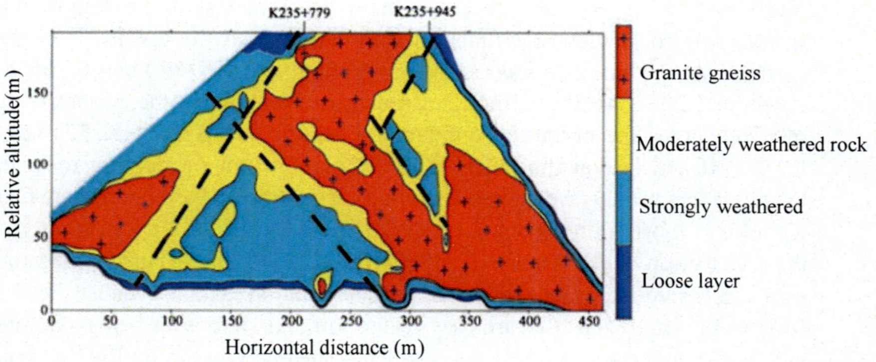

Fig. 7.32 Geological interpretation of the section of the tunnel's upward line, middle guide pit, and river channel

can be clearly found that there are two sets of conjugate shear joints developed in the direction of NE and NW, but the weathering of the joint zone is slight. The result shows that the slope foundation on the left side of the tunnel is very stable, and there is no landslide condition on the tunnel foundation.

(5) Conclusion

Summarizing the survey results of seismic CT, we can have a clear understanding of the engineering geological environment in which the tunnel is located. Its basic engineering geological structural characteristics can be summarized as follows:

① The engineering geological structure around the tunnel belongs to seriously weathered jointed fractured rock mass on the whole. Two sets of conjugate joints develop in the surrounding rocks (granite gneiss), and the directions are from north to east and from north to west. The weathering along the joint zone is severe. The width of the weathering zone varies greatly. The narrow part is 3–5 m, and the wide part is 15–20 m. It is filled with sand and clay.

② The characteristics of surrounding rocks have changed significantly in the transverse direction of the tunnel. The outer side of the mountain is loose, broken, and becomes better inward. The surrounding rock at middle guide wall and the top of the upline is on the outer side of the mountain. The weathering is extremely serious. The strong weathering and loose materials in the vertical section account for 80% of the total area. The width of the filling material in the joint zone is up to 20 m, and the stability is poor after full water. The section position of the wall on the right side of the downline only deviates 12 m into the mountain, and the characteristics of the rock mass are obviously better. The moderately weathered and slightly weathered rocks account for 60–70% of the total area of the section, and there are few loose materials. The strongly weathered rocks only develop along two joint zones, and their stability is better. To the right, the lithology of the mountain is more complete.

③ There are also great differences in the characteristics of the longitudinal surrounding rocks along the tunnel, which can be basically divided into three sections. At the entrance section, before the upline K235+820 and before the downline K235+840, the surrounding rock is moderately weathered rock with a relatively complete structure. At the middle section of the upline K235+820 to K235+940 and the downline K235+840 to K235+940, the surrounding rock near the tunnel is seriously weathered, especially inside the upward line, where there are a lot of loose fillings, which are easy to deform and have poor stability. At the exit section of the mileage K235+940 to K235+970, the surrounding rock near the tunnel is dominated by loose accumulation. The low-density blocks stick to the sand, and they are easy to deform and slide after being saturated with water, and the stability is poor, which is the main reason for the excursion and destruction of the tunnel.

④ The vertical changes of the surrounding rocks have certain regularity. The foundation rock structure of the tunnel is complete. The surrounding rock at the top of the tunnel has a large loose zone with a thickness ranging from 5 to 20 m. It mainly develops between K235+820 and K235+940. It may be the combined effect of weathering joint zone and loosening of excavation. The integrity of the overlying rock mass in the loose zone has improved slightly, especially near the inside of the mountain on the right.

7.2.3 Application of High-Density Resistivity Method and Seismic CT in a Tunnel

(1) **Project overview**

As shown in Fig. 7.33, the tunnel is located between the K88+670 and K88+885 of the 4th contract section of Pingyuan Street and Suolongsi Expressway, with a total length of 215 m. The whole tunnel is located on the left-turn curve of $R = 341$ m, $L_s = 100$ m and $I = 5$. The longitudinal slope is -0.7%. The maximum buried depth

Fig. 7.33 Overview of the tunnel site

of the tunnel is 71 m, and it is designed as a rectified double-arch structure with a cast-in curved wall. Its effective net width is 9.75 m, and effective net height is 5 m.

The upline of the tunnel has been constructed first, and the second lining has been completed. During the construction of the downline, cracks appeared at the top of the second lining vault of the upward line near the retaining wall, which ranged from K88+680 to K88+840, especially the upper part of the middle wall of the sections K88+800 to K88+810. The plaster strip is not broken, indicating that the deformation of the tunnel has tended to stop. The uplift and cracking of the inverted arch are mainly in the sections of K88+780 to K88+860, and the maximum deformation reaches 2 cm (Fig. 7.34). According to the monitoring data, the inverted arch continues to uplift, but the velocity of uplift is much slower than before.

(2) **Engineering geological conditions**

The stratum lithology of the tunnel is the Middle Triassic Niaoge Formation (T3n) yellow-gray, dark-gray argillaceous siltstone with a thin layer of dark-gray, gray-black fine sandstone, and gravel soil. Due to the structural influence, the rock mass is broken shows the shape of gravel. The rocks are strongly weathered. The occurrence of rock strata is $20° < 49°$.

The tunnel is located in the area affected by the Kaiyuan mountain-shaped structure. Under the influence of the structure, the rock mass is broken. The rocks are strongly weathered, and the cracks are developed.

The lithology of the tunnel is single, and the hydrogeological conditions are relatively simple. The type of the groundwater of tunnel surrounding rock is bedrock

Fig. 7.34 Tunnel disasters

fissure water. The elevation of the tunnel is about 1377 m, and the water level of the Lushuitang River in the valley outside the tunnel is 132.5 m. Drilling revealed that the lower bedrock is sandstone and belongs to a permeable layer. The terrain at the tunnel is relatively steep, and the drainage of groundwater is unobstructed. The tunnel is above the groundwater level.

The slope at the entrance (K88+675) and exit (K88+880) of the tunnel is steep, and the upper overburden is gravel soil, which is slightly dense, moist to saturated, and the overburden is thick. The high slope is very unstable.

(3) **Program layout**

According to the needs of the research purpose, three high-density resistivity sections and one seismic CT transmission section were arranged in this work (Fig. 7.35).

The high-density resistivity method EZ1 survey line is arranged at the upper left of the tunnel upline in the direction of the central axis of the tunnel, and it is 2 m away from the retaining wall. The distance between the electrodes is 3 m. There are 120 electrodes. The section length is 310 m, and the sounding depth is about 90 m (the maximum buried depth of the tunnel is about 70 m). The EH1 survey line passes through K88+800, and the section is basically vertical to the center axis of the tunnel. The electrode distance is 3 m. There are 104 electrodes. The section length is 280 m, and the sounding depth is about 90 m. the EH2 survey line passes through K88+760. The section is basically vertical to the central axis of the tunnel.

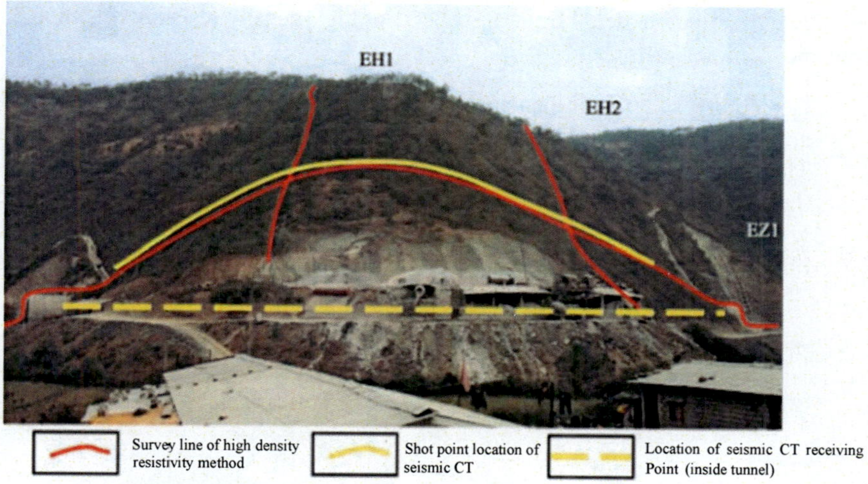

Fig. 7.35 Schematic diagram of survey line layout

The electrode distance is 3 m. There are 104 electrodes. The section length is 270 m, and the sounding depth is about 90 m.

A section (two arrays) was arranged in the seismic CT, and the geophones were arranged along the direction of the upline vault (2 m away from the retaining wall) with a track spacing of 3 m. There were 48 channels in each array and the two arrays overlapped 34 arrays to ensure the orthogonality of rays and the requirement of ray density. The shot points were arranged on the high-density resistivity EZ1 survey line with an interval of 3 m. There are 97 shots in total (Fig. 7.36).

(4) **Detection result**

After on-site data collection, indoor data processing, and geological interpretation, the engineering geological environment in which the tunnel is located has been identified, and the cause of the tunnel disaster has been deduced. In the following,

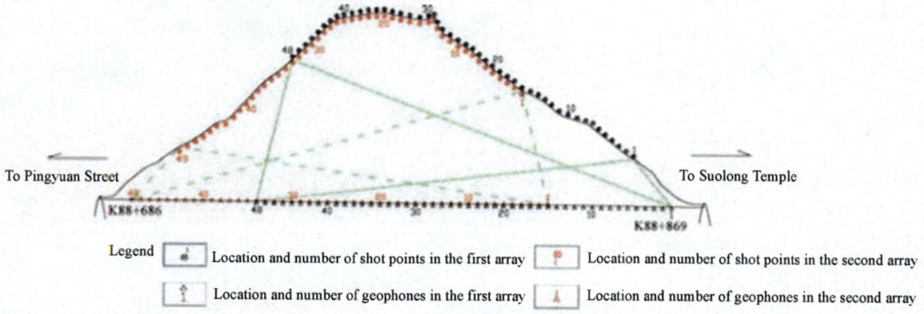

Fig. 7.36 Schematic diagram of the layout of geophones and shot points

it will be described in detail from the aspects of high-density resistivity method and seismic CT.

① Detection result of high-density resistivity method. Figures 7.37, 7.38, and 7.39 are the apparent resistivity images of the three high-density resistivity sections. They intuitively reflect the distribution of underground apparent resistivity. In the figure, the red areas are high-resistance areas, which represent rock masses with good structural integrity and poor water content. The dark blue and light blue areas are low-resistance areas, which represent water-containing and loose

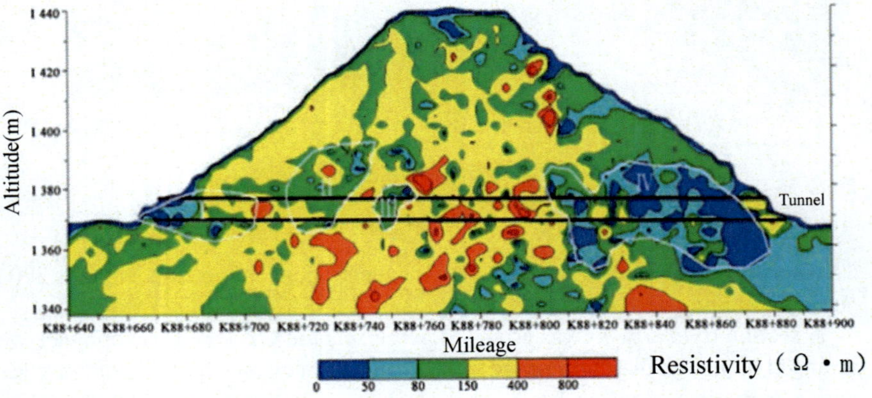

Fig. 7.37 Apparent resistivity image of EZ1 profile (Scale 1:1000)

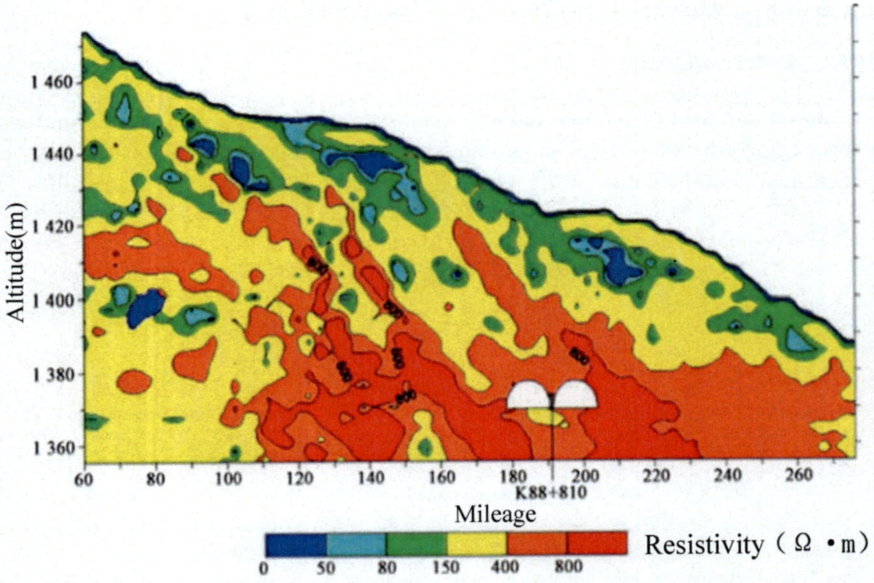

Fig. 7.38 Apparent resistivity image of EH1 profile (Scale 1:1000)

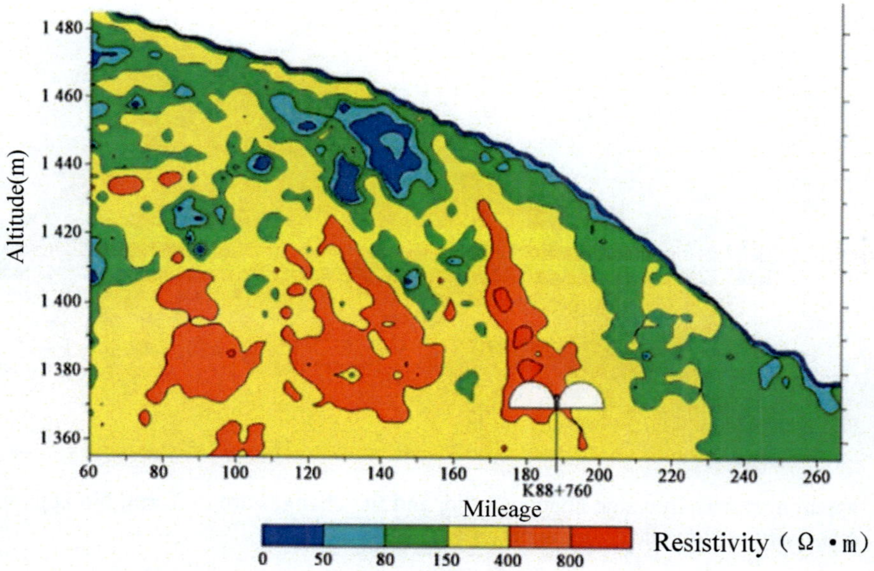

Fig. 7.39 Apparent resistivity image of EH2 profile (Scale 1:1000)

areas. The changes from red and yellow to green indicate the gradual change of resistivity from high to low, and also the gradual change of water content from weak to strong, and the structural integrity of rock mass from good to poor. The red areas appearing above the tunnel in the three figures are local dryout areas.

It can be seen from the three maps that the overall apparent resistivity value is low, most of which are less than 800 Ω·m. Based on geological data, it is known that the lithology here is mainly low-resistance fine sandstone and siltstone.

The apparent resistivity of the EZ1 profile (Fig. 7.37) varies little, and the resistance value is mainly 200 Ω m. There are four resistivity anomalies along the tunnel axis. The anomaly No. 1 is located at K88+665 to K88+705, and the apparent resistivity value is mostly less than 150 Ω m, locally less than 50 Ω m, which is presumed to be a loose fracture zone. The anomaly No. 2 is located at K88+715 to K88+735, and the apparent resistivity value is less than 150 Ω m, which is presumed to be loose fracture zone. The anomaly No. III is located at K88+740 to K88+755, and the apparent resistivity value is less than 150 Ω m, which is presumed to be a loose fracture zone. The anomaly No. IV is located at K88+805 to K88+880, and the apparent resistivity value is mostly less than 80 Ω·m, locally less than 50 Ω·m, which is presumed to be a water-containing loose zone. There is a low-resistance zone in the area of K88+810 to K88+820, which corresponds to the location of the crack in the tunnel.

The transverse EH1 section (Fig. 7.38) and EH2 section (Fig. 7.39) pass through K88+800 and K88+760, respectively. From the two apparent resistivity images, the

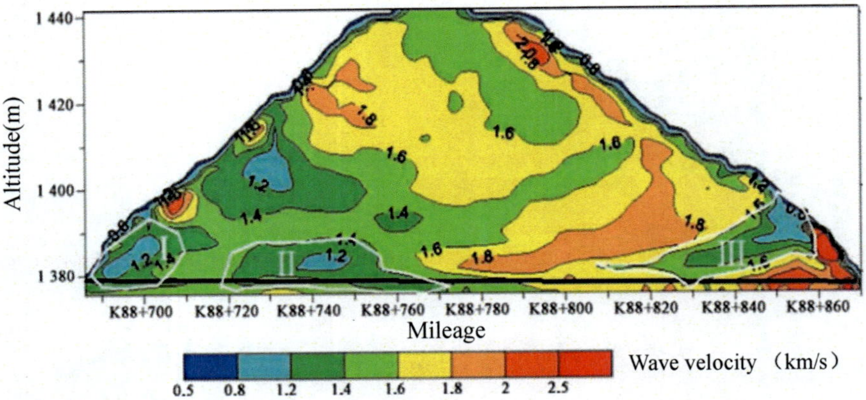

Fig. 7.40 Seismic CT wave velocity image

apparent resistivity value changes little, and the changes are uniform. No apparent resistivity is found near the tunnel.

② Detection result of seismic CT. Figure 7.40 is the wave velocity image of the seismic CT. It can be seen from the figure that the wave velocity of the surrounding rocks does not change much and is generally low, ranging from 1.4 to 1.8 km/s, which is a typical characteristic of surrounding rock of II type. It shows that the surrounding rock joints are developed and the weathering is severe.

The wave velocity distribution image of the seismic CT image shows that, with the boundary of K88+770, the wave velocity at the entrance of the small mileage is lower and the integrity of the rock mass is poor. The integrity of the rock at the exit of large mileage is slightly better than the former.

There are three low wave velocity anomalies along the tunnel axis, corresponding to three rock mass fracture zones. The anomaly No. I is located at K88+690 to K88+710, and the wave velocity is less than 1.4 km/s, locally less than 1.2 km/s. It corresponds to the location of the low apparent resistivity anomaly No. I. The width of the rock mass fracture zone is 20 m. The anomaly No. II is located at K88+715 to K88+770, and the wave velocity is less than 1.4 km/s, locally less than 1.2 km/s, it corresponds to the location of the low apparent resistivity anomalies of II and III. The width of the rock mass fracture zone is 55 m. The anomaly No. III is located at K88+810 to K88+860, and the range of the wave velocity is from 1.2 to 1.6 km/s. It corresponds to the location of the low-resistance anomaly No. IV. The width of the rock mass fracture zone is 50 m.

(5) **Conclusion**

Combining seismic CT and apparent resistivity images and geological surveys, a comprehensive interpretation result map of geophysical prospecting is finally obtained (Fig. 7.41). The result shows that no obvious fault zones and voids are found in the area. The wave velocity of the surrounding rocks is between 1.4 and

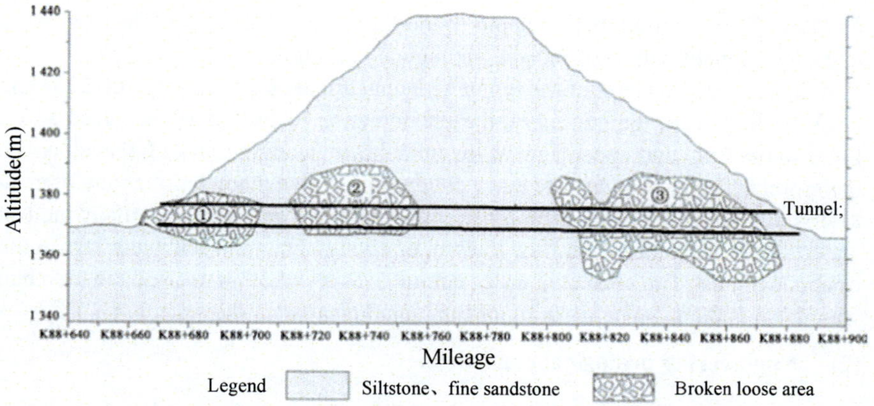

Fig. 7.41 Comprehensive result map of geophysical prospecting

1.8 km/s, which is a typical characteristic of type II surrounding rocks. The rock mass has developed joints, with dense cracks and severe weathering. From the perspective of sections, the integrity of the rock mass is unevenly distributed, and the integrity of the surrounding rock in the small mileage section is even worse. There are three sections with relatively broken rock masses distributed along the axial direction of the tunnel.

Combining with the comprehensive result map of geophysical prospecting (Fig. 7.41), the following conclusions are reached.

① The engineering category of the surrounding rock of the tunnel belongs to category II, and no obvious fault zone or cavity is found.

② There are three loose fracture zones along the tunnel axis (①, ②, ③ in Fig. 7.41), and the corresponding mileage is K88+665 to K88+705, K88+715 to K88+755, K88+805 to K88+880.

③ The damage of tunnel vault lining is mainly caused by the large bias pressure of the mountain and the unreasonable construction procedure of the upline and the downline of the tunnel. At the same time, the existence of sections with relatively weak geological conditions is also one of the important factors causing tunnel cracks.

7.2.4 Application of High-Density Resistivity Method in a Tunnel

(1) **Project overview**

The tunnel is a left and right separated tunnel. The starting and ending mileage on the left is ZK281+528 to ZK289+108, with a total length of 7580 m. The starting and ending mileage on the right is K281+525 to K289+100, with a total length of

7575 m. The maximum buried depth of the tunnel is about 730 m. The tunnel is an extra-long tunnel with complicated geological conditions.

When the left entrance of the tunnel was constructed to the mileage of K281+821 (293 m away from the entrance), the phenomenon of mud burst occurred several times in the front and upper right of the tunnel face, leading to roof fall accidents on the ground. The ground surface had a collapse pit with a diameter of about 10 m and a depth of about 6 m. During the construction period, there have been many mud and water inrush accidents, each time interval of about 12 h, and each water inrush time of about 10 min. The estimated water volume was about 5000 m^3, and the mud burst was about 1000 m^3. The situation of the tunnel disaster is shown in Fig. 7.42.

(2) **Engineering geological conditions**

① The entrance of the tunnel has an elevation of 2258–2664 m and a relative height difference of 406 m. It belongs to the structure denudation and erosion of the mountainous landforms. The terrain has an undulating slope of 30°–50°. The local terrain is steep and the valleys crisscross. Most of the slope are covered by larch forest, and part of the land is reclaimed into farmland. Dirt roads on the hillside are connected to the tunnel roads. The topography of the tunnel area is shown in Fig. 7.43.

② The tunnel area has a subtropical plateau monsoon climate with abundant rainfall, with an average annual rainfall of 723 mm. The valley water system is

Fig. 7.42 Situation of tunnel disasters

Fig. 7.43 Topography of the tunnel area

developed. The catchment area is wide, and the hydraulic slope is relatively gentle, which is conducive to the replenishment of groundwater.

③ The tunnel body traverses a variety of lithologies, including granite porphyry, limestone, dolomite limestone, dolomite, gabbro, diabase, and sand shale, etc., and most of them are interbedded. The geological characteristics of various lithologies are different, and the degree of weathering and fragmentation is different. At the same time, the development of unfavorable geologies of structural faults, such as broken zone, water-rich zone, soft and hard interbed, karst, etc., has resulted in geological disasters such as mud outburst, water gushing, landslides, roof fall, and karst during tunnel excavation, seriously affecting the safety and progress of tunnel construction.

④ The tunnel is located between Chenghai fault and Erhai fault, and the entrance of the tunnel is close to Chenghai fault. Affected by regional structure, secondary faults developed in the tunnel area. The fault distribution of Jiudingshan Tunnel is shown in Fig. 7.44.

(1) Lithology. The overlying strata at the entrance of the tunnel are mainly Quaternary residual slope strata (Q_4^{el+d1}), and the lower bedrocks are Yanshanian granite porphyry (γ_{53}), Devonian Qingshan Formation (D_1q) limestone, and Kanglang Formation (D_1k) Dolomite limestone, and dolomite.

① Quaternary residual slope layer (Q_4^{el+d1}). The main lithology is brown-yellow, brown–red clay, silty clay with 5–10% crushed stone and breccia, with the diameter of 1–8 cm and angular shape. The main composition is parent rock granite porphyry, limestone weathering, and denudation products, which are slightly wet, with the hard, hard plastic, and plastic shape and the thickness of 1–10 m.

② Yanshanian granite porphyry (γ_{53}). It is grayish-white medium to acidic rock. Phenocrysts are potassium feldspar. Mineral composition is sodium plagioclase,

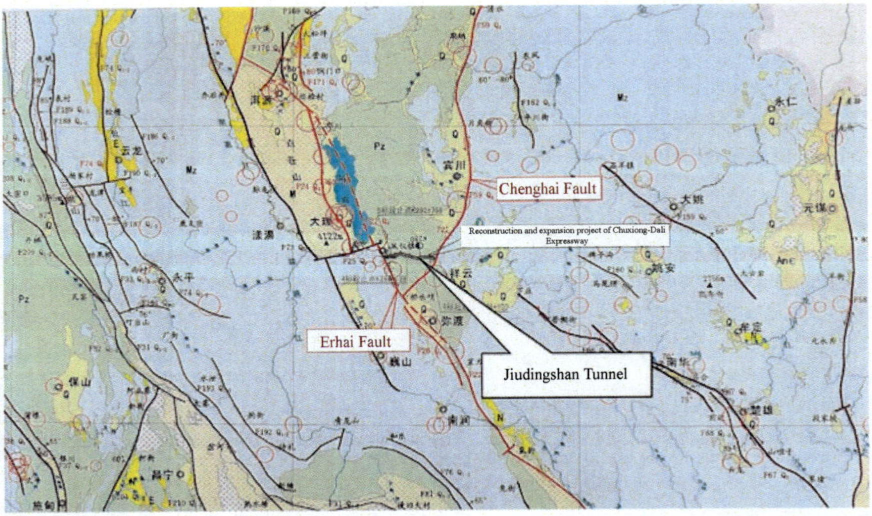

Fig. 7.44 Regional fault distribution map

potassium orthoclase, biotite, amphibole and a small amount of quartz. The rock mass is full to strong weathering, mainly full weathering, with joint fissures developed, especially invisible micro fissures. The rock mass is broken, showing a debris to chipped structure.

③ Devonian Qingshan Formation ($D1q$). The main lithology is limestone. It mainly distributes in the ZK281+920 to ZK282+350 section. The occurrence of the rock layer is 330°∠15° gray, dark gray to black, and thin to medium thick, with small siliceous intermittent bands, fragments, and breccia structures, and a small amount of mud. Strong to moderate weathering, joint fissures developed, the rock mass is broken and fragmented, and surface tension dissolution fissures developed, extending downward for nearly 100 m.

④ Devonian Kanglang Formation (D_1k). The main lithology is dolomite limestone and dolomite, which are mainly distributed in the ZK282+350 to ZK282+650 section. Light gray to gray white, thin to medium-thick layered structure, strong to moderate weathering, surface erosion cracks developed, with knife-shaped dissolution ditch.

(2) Geological structure. The Devonian strata in this area underwent magmatic activities and crustal fault structures in the Hualixi, Yanshan, and Himalayan stages, with frequent crustal activities and frequent earthquakes. According to the high-density resistivity survey combined with the on-site geological survey, a total of 6 multi-stage fault structural zones were found from the entrance of Jiudingshan Tunnel to K283+000 section, numbered F_2, xf_1, xf_2, xf_3, xf_4, xf_5. Three of them (F_2, xf_1 and xf_2) cut through the tunnel and caused serious damage to the tunnel. The fault structure distribution at the entrance of Jiudingshan Tunnel is shown in Fig. 7.45.

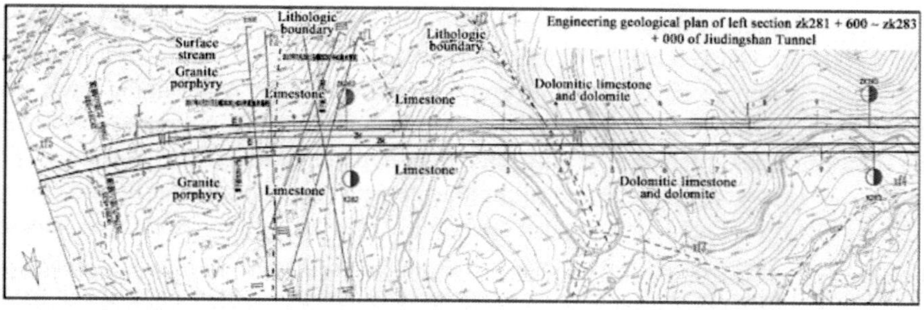

Fig. 7.45 Distribution map of fault structure at the entrance of Jiudingshan tunnel

① F_2 fault structure. The F_2 fault structure is located at the mileage ZK281+850 in the Jiudingshan Tunnel, and the occurrence state is NE15°∠80°. The main lithology of the upper wall of the fault structure is granite porphyry and the lower wall is limestone. The fault location is consistent with the lithological contact zone. Affected by it, a fracture zone is formed. The affected area is the left section ZK281+670 to +K281+920, and the axial width of the tunnel is about 250 m. It is inferred that the geological environment of the right tunnel is basically the same as the left tunnel, and both of them are unfavorable sections of engineering geology and hydrogeology.

② xf_1 fault structure. The xf_1 fault structure intersects with an axis mileage of ZK282+100 in Jiudingshan Tunnel, and the occurrence state is NE7° ~ 73°. Affected by the fault, the left side of the tunnel entered the fracture zone from mileage ZK282+090.

③ xf_2 fault structure. The xf_2 fault structure intersects with the axis of the Jiudingshan Tunnel at a distance of about ZK282+350, and the occurrence state is NW11°∠80°. The fault location is consistent with the lithological contact zone of the formation. Affected by the interaction of this fault and the xf_1 fault, a large-scale fault fragmentation zone appears on the left side of the tunnel, with a range of ZK282+090 to ZK282+410, with an axial width of about 320 m, which is a poor area of engineering and hydrogeology.

(3) Hydrogeological conditions

① Surface water. In the vicinity of the tunnel area, three streams and feathery tributaries are mainly distributed. The recharge area is large, and it accepts atmospheric precipitation. There is surface water infiltration in the rainy season to recharge groundwater. Among them, the seepage of stream water in ${}_{X}F_2$ fault is related to tunnel water inrush.

② Types of groundwater and recharge, runoff, and drainage conditions.

(a) Pore water. Pore water is deposited in the Quaternary residual slope deposits and loose piedmont deposits. Because the loose body is 1–10 m thick, it receives meteoric water supply, and the pore water is relatively developed. The spring water can be seen gushing out at the foot of the slope, and the general water inflow is 1–3 m^3/h. Because the pore water replenishment area is relatively wide and the water volume is abundant, it supplies karst water and granite porphyry fissure water in the lower part.

(b) Fissure water. It is mainly contained in diagenetic cracks, weathering cracks, tectonic cracks, and fractured rocks of granite porphyry. **The fissures** intersect and communicate with each other. Their fissure, uniformity, and openness are all good. They form a groundwater network water storage area system that accepts atmospheric precipitation and groundwater recharge around the tunnel. It has certain static and dynamic reserves. The water diversion system runs off and drains from high to low and flows into the area of the palm surface of the airspace, with different degrees of gushing and water inrush.

(c) Karst water. According to geological survey and the detection of the electrical method, in the limestone, dolomite limestone, and dolomite areas, karst holes, small karst caves, karst trenches, and karst tanks are developed, and they are connected to the siphon network system of karst water pipelines. Especially in the fracture zone of the fault structure, the karst water develops abnormally and the water richness is very strong. The invaded granite porphyry rock body fracture water supplies a water-rich system in the tunnel, which is very detrimental to the stability of the surrounding rocks.

(3) **Program layout**

According to the purpose of the survey, a total of six geophysical prospecting sections were arranged in this geophysical survey. The geophysical survey uses the high-density resistivity method, as shown in Table 7.2.

(4) **Detection result**

Table 7.2 Survey line layout table of high-density resistivity method

Serial number	Number of survey line	Location of survey line	Arrangement
1	I–I'	Left ZK281+600 to ZK282+770 axis surface	Vertical arrangement
2	II–II'	Left ZK281+825 intersects right YK281+820	Horizontal arrangement
3	III–III'	Left ZK281+926 intersects right YK281+899	Horizontal arrangement
4	IV–IV'	Left ZK281+950 intersects right YK281+950	Horizontal arrangement
5	V–V'	The left ZK281+975 intersects right YK281+947	Horizontal arrangement
6	VI–VI'	Center line K281+650 of the left and right tunnel—K282+438 surface	Vertical arrangement

The geophysical survey method used this time is the high-density resistivity method, and the result map obtained is the high-density resistivity imaging map (Figs. 7.46, 7.47 and 7.48). The red, yellow, and green areas in the figure are relatively high-resistance areas, which represent rock masses with relatively dense, good integrity, and low water content. The blue areas are relatively low-resistance areas, which represent rock masses with water content, loose fragmentation, or weathering. The uneven distribution of the apparent resistivity indicates that the physical properties of the rock masses are very different. It is presumed that the rock mass is structurally developed and loosely broken. The horizontal distribution of the apparent resistivity is uniform, indicating that the rock masses are uniform.

The apparent resistivity value is 0–150 Ω·m. It is a residual slope layer, a high weathering degree of broken rock mass, and an area affected by fracture structure. It

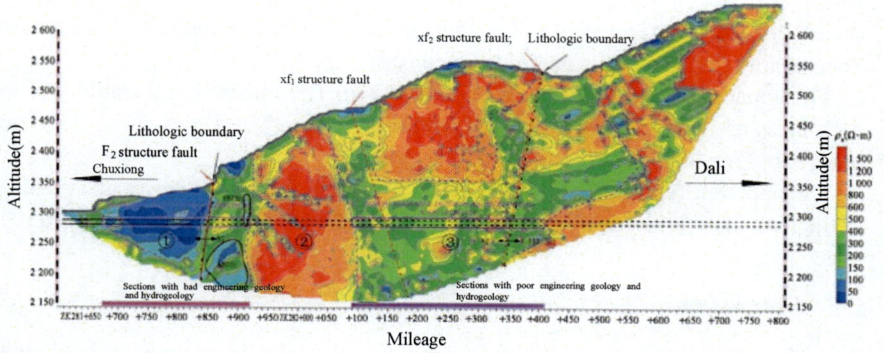

Fig. 7.46 Apparent resistivity imaging map of the longitudinal section of the left axis of the high-density resistivity method

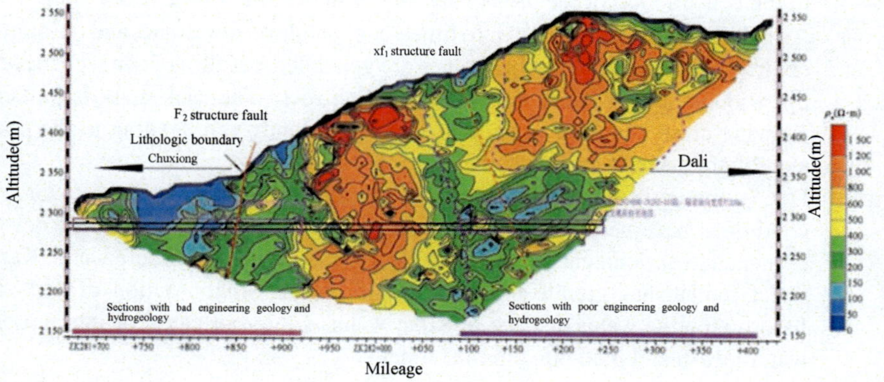

Fig. 7.47 Apparent resistivity imaging map of the longitudinal section of left and right midline of high-density resistivity method

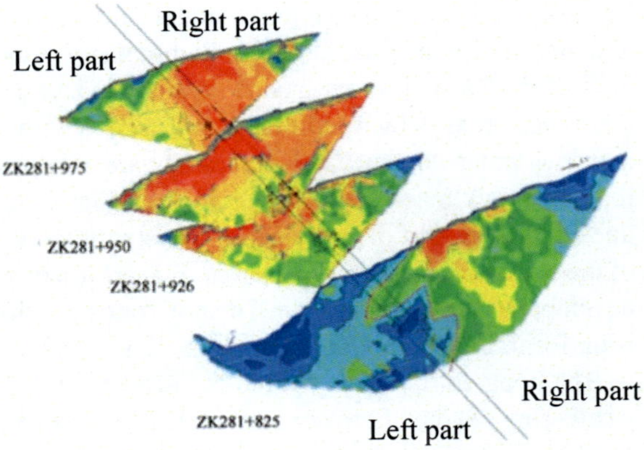

Fig. 7.48 Apparent resistivity imaging map of the transverse section of high-density resistivity method

is rich in fissure water and is a poor location of engineering geological conditions. Special attention should be paid during construction.

The apparent resistivity value is 150–400 Ω·m. The integrity and stability of the rock mass are average. Most of them are strongly weathered rock mass. The rock mass is broken, and joint fissures are developed.

The apparent resistivity is 400–1500 Ω·m and above. The integrity and stability of the rock mass is relatively good, and most of them are moderately weathered rock mass.

(5) **Conclusion**

① The strata in this area are the limestone, dolomite limestone, and dolomite deposited in the Paleozoic Devonian Lower Age, invaded by the Mesozoic Yanshanian granite porphyry. The skarn mineralization fracture zone with complex lithology has many secondary minerals, and the most unfavorable to the construction of the project are rock fragmentation zone and kaolin.

② As granite porphyry intrudes into limestone, dolomite limestone, and dolomite, six fractured structural fracture zones are generated, resulting in densely developed joints and fractures in the surrounding rock. The rock mass is broken, showing clastic and gravel structure, which is sandy soil, leading to the poor stability of surrounding rock.

③ Due to the development of fault structure and rock fragmentation, favorable conditions have been created for groundwater activities, leading to the development and enrichment of karst water and granite porphyry fissure water. Karst water replenishes granite porphyry fissure water. Siphon action of bedrock fissure pipeline water or karst pipeline water is the root cause of intermittent water inrush and mud burst in the tunnel.

7.2.5 *Application of Ground Penetrating Radar Method in a Tunnel in Hebei*

(1) **Project overview**

The total length of the left line of the tunnel is 5656 m, and the total length of the right line is 567.7 m. It is a six-lane expressway super-long tunnel separated from the up and down.

(2) **Program layout**

① The distribution of the collapse cavity of the tunnel, the three-dimensional spatial position, and the scope of influence;
② Distribution of groundwater and unfavorable geological bodies in surrounding rock;
③ Three-dimensional geological situation outside the tunnel outline within 80 m in front of the collapsed surface of the tunnel.

The principle of survey line layout is circular layout. The hoop section is arranged at the collapsed working surface with angles of 0°, 20°, 45°, 70°, and 90°, respectively, and the detection ray spacing on the section is 0.5 m, as shown in Fig. 7.49.

(3) **Detection result**

Fig. 7.49 Schematic diagram of the radar wave directional radiation distribution data acquisition system

The detection range of GPR 3D imaging is centered on the intersection of the section K103+607 and the axis of the tunnel. The interpretation range is 30 m on the left and right, 40 m in height, and within the region of K103+607 to K103+567 (40 m). The detection results are shown in Figs. 7.50 and 7.51.

It can be seen from Fig. 7.51 that the landslide area is in the fault fracture zone, and the surrounding rocks are loose and broken. The groundwater is located directly above the collapsed cavity, which is a rich water-bearing area. The relatively cavity area formed by the collapse is located in front of the K103+607 accumulation body above left and right sides.

It can be seen from Fig. 7.52 that the cavity is located on the left and right sides of

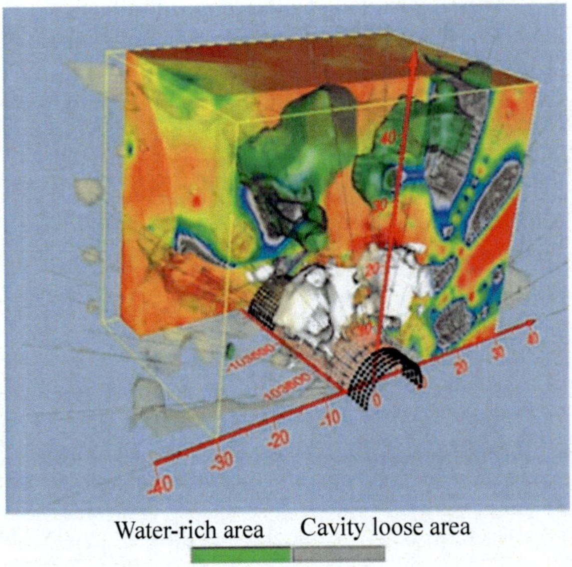

Fig. 7.50 Comprehensive result map of three-dimensional imaging detection of ground penetrating radar on the right of the tunnel

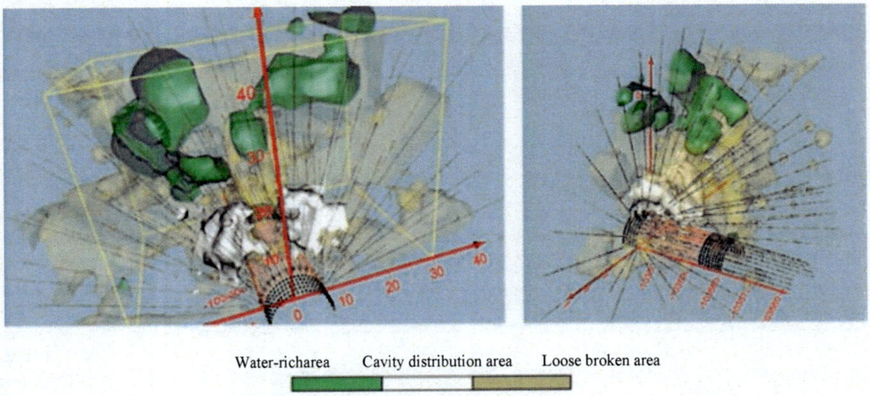

Fig. 7.51 Distribution of unfavorable geological bodies in the tunnel

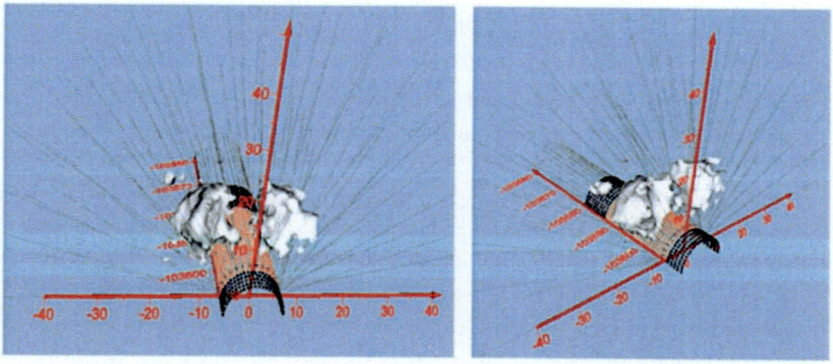

(a) Three-dimensional diagram of collapsed cavity

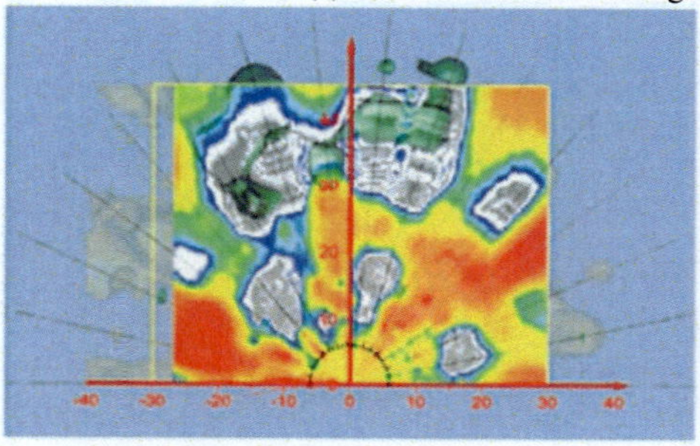

(b) Mileage K103+603 transverse section

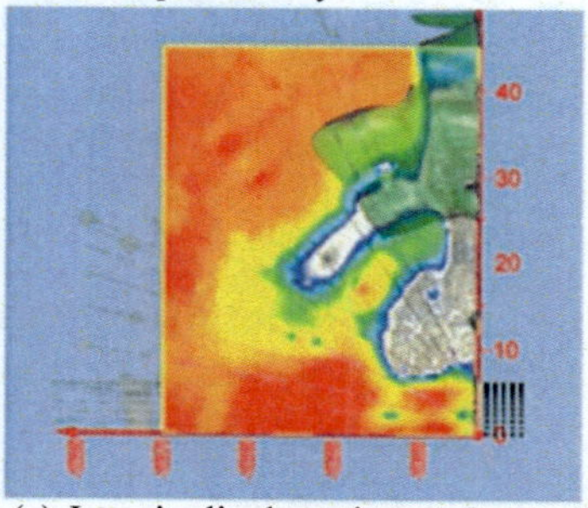

(c) Longitudinal section at 5m to the right of thecentral axis of the tunnel

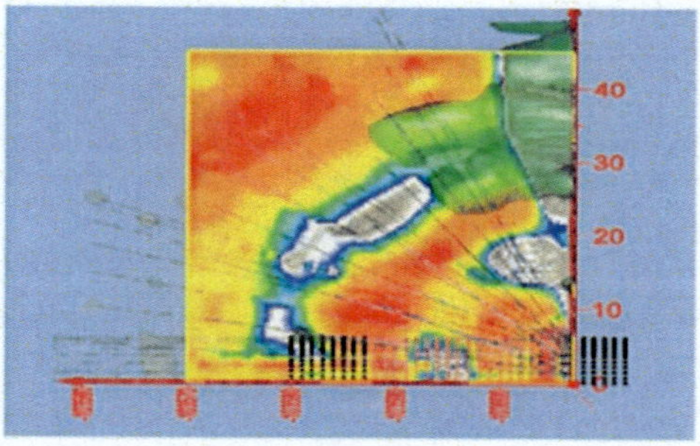

(d) Longitudinal section at 10m to the left of the central axis of the tunnel

Fig. 7.52 Scale distribution of collapsed cavity in surrounding rock of tunnel

the front and top of the K103+607 accumulation body. The volume of the left cavity is about 1000 m^3, and the volume of the right cavity is about 500 m^3. Cavity range: longitudinal K103+609 to K103+586 (about 23 m), horizontally from about 18 m to the left of the axis of the cave to about 15 m to the right of the axis (the middle is not completely continuous). It extends vertically to about 23 m above the vault.

(4) **Conclusion**

Through special detection of the collapsed section of the right exit of the tunnel, the conclusions are as follows:

① The collapse cavity of the tunnel is distributed on the left and right sides of the front and top of the K103+607 accumulation body. Cavity range: longitudinal K103+609 to K103+586 (about 23 m), horizontally from about 18 m to the left of the axis of the cave to about 15 m to the right of the axis (the middle is not completely continuous). It extends vertically to about 23 m above the vault.

② Distribution of surrounding groundwater and unfavorable geological bodies: The landslide area is in a fault fracture zone, and the surrounding rocks are loose and broken. The groundwater is located directly above the collapsed cavity, which is a water-rich area.

③ Three-dimensional geological situation outside the tunnel outline within 80 m in front of the collapsed surface of the tunnel: The surrounding rocks of K103+616 to +576 (40 m) tunnel are broken. The sections are prone to collapse. The surrounding rocks of K103+576 to K103+536 are relatively good.

7.2.6 *Application of Ground Penetrating Radar Method in a Tunnel*

(1) **Project overview**

During the tunnel excavation, a large-scale karst cave was found at K40+230. In order to further investigate the morphology, scale characteristics and controlling factors of the east of the cave, an underground karst cave survey was conducted to investigate its shape, scale and hole-forming factors.

(2) **Engineering geological conditions**

Lithology: The exposed strata in the tunnel crossing area are the Quaternary (Q_4) and the Triassic Yongningzhen Formation (T_1y). The Quaternary System (Q_4) is a slope residual layer, which is mainly gravel and subclay gravel soil and pebble soil, with a thickness of 0–17.6 m. The Triassic Yongningzhen Formation (T_1y) mainly has the following two lithologies.

① Gray off-white thin medium-thick layered dolomite rock. Weathered in the ground, weak weathered in the tunnel, it is hard rock. Surface dissolution pores are widely developed, especially in the joint dense zone, which are controlled by the double factors of rock layer and joint, and distributed along them. The occurrence state of the strata is N55° W·NEZ33°.

② Light gray-gray-green sericite slate. The surface is strongly weathered, and the tunnel is strongly weathered. It is extremely soft rock, and it is easy to soften, swell, loose, and extremely unstable in contact with water. The occurrence state of the strata is N55° ~ 60° W·NEZ33°.

Geological structure: The investigation area is a compound part of three giant tectonic systems: Qinghai–Tiang–Yunnan, Burma–Indonesia, Sichuan–Yunnan meridional tectonic system, and Nanling zonal tectonic system. The Honghe deep and large fault spreading in NW–SE direction and the newly emerging estuarine broom-like structure are particularly serious. The low-grade and low-order fault folds are developed, and the occurrence of the strata changes greatly in strike and dip. The joints are densely integrated in local areas, such as near the left exit of the left side of the tunnel and outside the karst cave at K40+230. The breaking of surrounding rock by tectonic action not only affects the stability of surrounding rock, but also affects the flow and drainage of groundwater, and at the same time, it also affects the occurrence of dissolution of surrounding rock by groundwater. In surface surveys, dissolution holes are commonly found along the layer of rock formations and joint exhibition.

Hydrogeology. The survey area mainly has two types of groundwater: pore water and karst water.

① Pore water. It is mainly present in the Quaternary slope remnants and alluvial floods. The source of recharge is atmospheric precipitation, and the pore water is runoff and discharged from high to low. The type of water quality is mainly HCO^{3-}, Ca^{2+}, and Mg^{2+}. The salinity is generally 0.1 g/l.

② Karst water. The karst water in the survey area exists in the dolomite limestone of the Lower Triassic Yongningzhen Formation. Because the area is in a strong crustal rising condition, the karst water is mainly characterized by vertical seepage and erosion. The type of water quality is mainly HCO^{3-}, Ca^{2+}, Mg^{2+}.

(3) Program layout

① Geological survey. Three times of intra-cave and out-cave geological survey were carried out, with a total area of 0.4 km^2.

② High-density resistivity method. A high-density resistivity method profile E1 is arranged on the ground directly above the left tunnel axis along the line direction.

③ Ground penetrating radar. Survey lines are placed on the road surface of the lower guide pit that is not excavated. A total of 7 survey lines are arranged along the axis of the tunnel with a distance of 2 m. A total of 11 survey lines are arranged perpendicular to the tunnel axis, with a survey line spacing of 2 m. A total of 272 m of survey lines have been completed, with a total of 2720 measuring points. The spatial distribution of survey lines is shown in Fig. 7.53.

(4) Detection result

Through geological surveys and measurements inside and outside the cave, the basic characteristics of the cave are as follows:

The horizontal section of the karst cave at K40+230 on the left side of the tunnel has a cone shape at an elevation of 880 m, wide from south to east, and sharp from north to west. Its long axis direction is S7°E ~ N7°W. The horizontal length is 33 m.

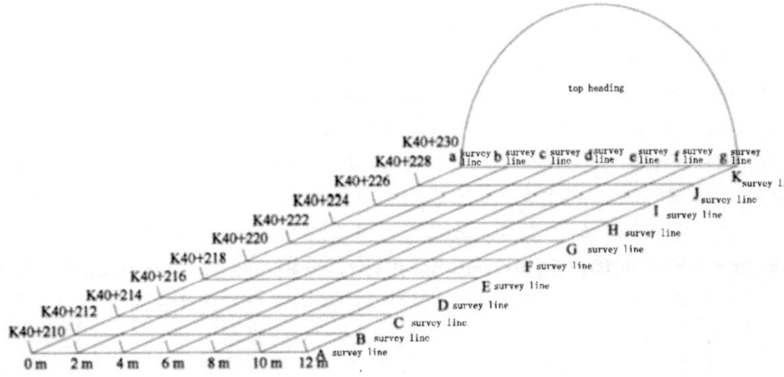

Fig. 7.53 Schematic diagram of tunnel karst cave detection radar line

It obliquely crosses the central axis of the tunnel (145°) at an angle of 28°, with a maximum width of 18 m. From the central axis of the tunnel K40+230 to the northeast side wall B–C of the cave, the horizontal distance is 17 m. The karst cave A–A' is perpendicular to the section morphology. The long axis of the cave (B–B') is perpendicular to the section morphology. Overall, the height difference between the top wall of the cave and the floor of the tunnel is 16.5 m. The height difference between the bottom wall of the cave and the floor of the tunnel is 10 m.

Geological characteristics of karst surrounding rocks

① The morphologic transition of the cavity side wall is basically consistent with the structural plane (fault joint plane) and the rock layer plane, indicating that the dissolution of the cave is controlled by the structural plane and the rock stratum level.

② At the top of the cavity and the steep suspension of the side wall, the stone bell is well developed, and the corresponding stalagmites at the bottom are developed, indicating that the underground seepage flow and dripping are rich, and the erosion is strong.

③ The surrounding rock at the top of the cavity and the steep suspension of the side wall has poor stability. It is easy to fall off, and the local unfavorable structural surface combination is easy to collapse.

④ The bottom of the cavity is mainly the accumulation of block stones, which are the accumulation of collapsed bodies during the formation of the cave.

⑤ There is running water under the stone accumulation body at the bottom of the cavity, and the sound of running water can be heard.

⑥ A small karst cave (affiliated cave) is also seen along the southeast extension of the main karst cave, with a height of about 3 m, a width of about 2.5 m, and an extension of more than 20 m.

(3) Surface geological characteristics corresponding to karst caves. The surface geological survey can confirm that the underlying strata of the dolomite limestone in K40+230 caves in the left panel of the tunnel are sericite slate. Slate is

a water-impermeable layer relative to dolomitic limestone, so groundwater in the lower part of dolomitic limestone is relatively abundant, and its dissolution may be stronger, and it is more likely to form underground caves.

Figure 7.54 is the result image of apparent resistivity. It can be seen from the figure that there is an abnormal zone of apparent resistivity, which basically coincides with the location of the cave.

In order to find out whether there are karst caves 15–20 m below the lower guide hole of section K40+210 to +230 of the tunnel, the geological radar work has been done. The survey results show that the sections K40+210 to K40+230 of the tunnel are thin to medium layered dolomite limestone, with broken rock masses, more mud, cracks, and a small amount of water seepage in the cave. It is estimated that the left side of K40+227 to +230 is 0–5 m, and the depth is 15–20 m. Karst caves are developed 8–11 m to the right of K40+222 to +229, with a depth of 15–20 m. The detection result map of ground penetrating radar is shown in Figs. 7.55, 7.56, and 7.57.

(5) **Conclusion**

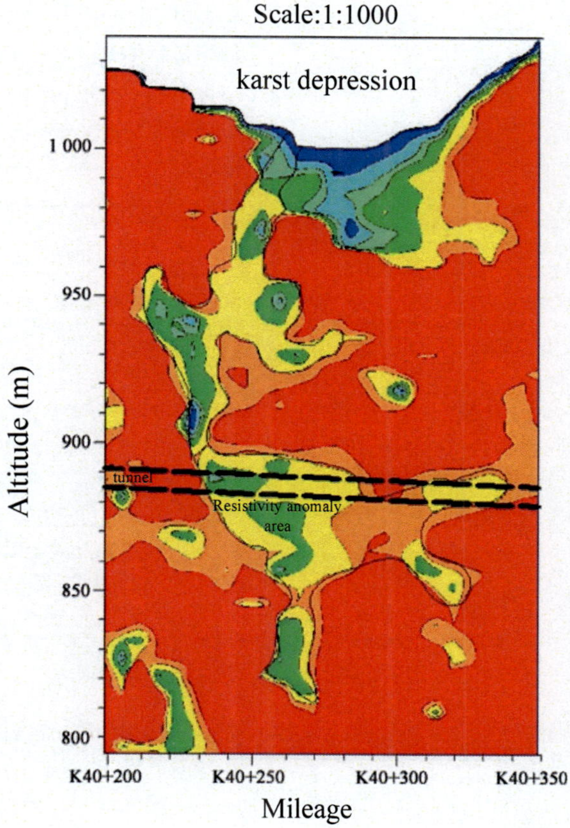

Fig. 7.54 Apparent resistivity distribution of K40+230 karst cave on the left of the tunnel

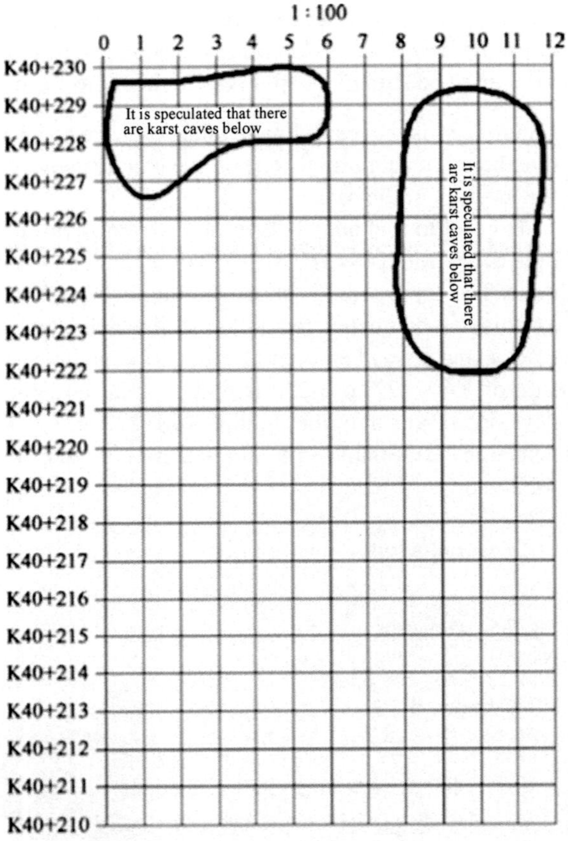

Fig. 7.55 Radar interpretation plan for tunnel cave detection (Unit: m)

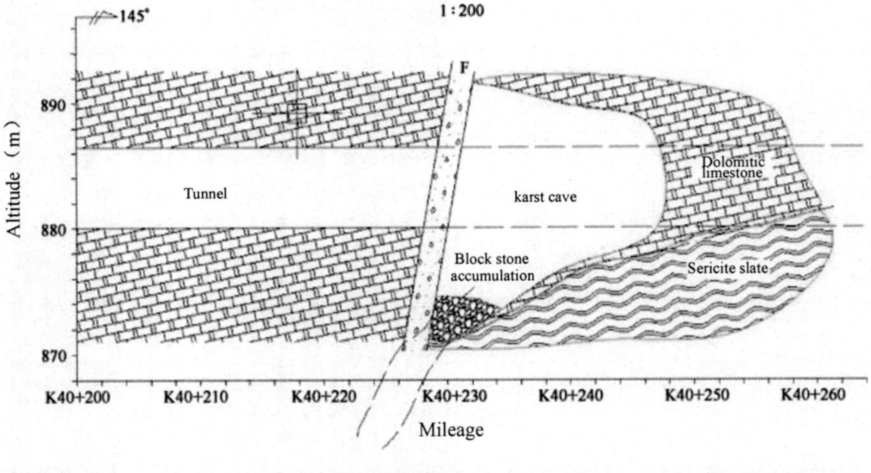

Fig. 7.56 Comprehensive section plan of K40+230 karst cave survey on the left of the tunnel

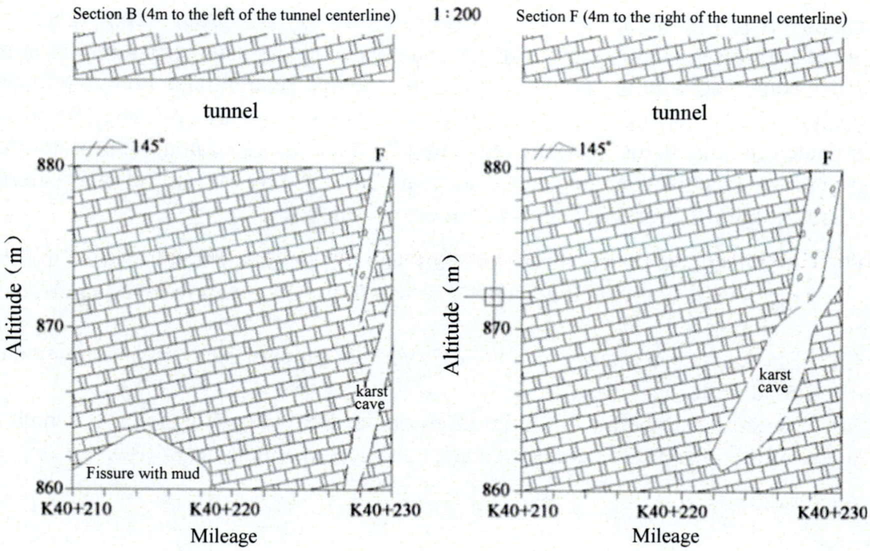

Fig. 7.57 Presumed section of ground penetrating radar

The comprehensive plan result map of this survey is shown in Fig. 7.58, which basically shows the shape and scale of the karst cave, as well as the extension and development of the karst cave to the lower and front of the tunnel. The shape and

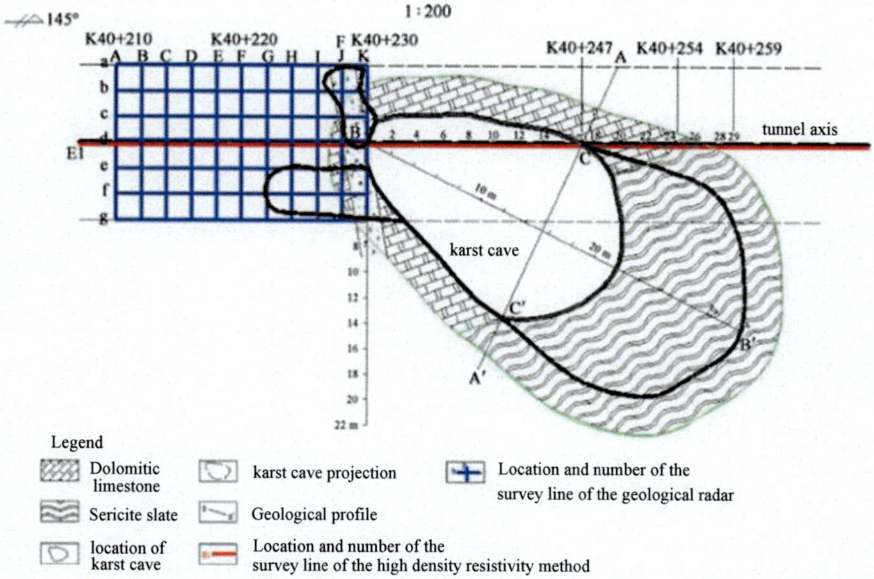

Fig. 7.58 Comprehensive section profile of K40+230 karst cave survey on the left of the tunnel

characteristics of the karst cave are as described above. Through comprehensive analysis of the geological survey inside and outside the cave and ground penetrating radar combined with high-density resistivity method section data along the tunnel central axis, in general, the formation of karst caves is mainly controlled by stratigraphic lithology to the lithology of the middle-thick layered dolomitic limestone of the Lower Triassic Yongningzhen Formation, followed by the control of structures (faults, joints). The specific conclusions are as follows:

(1) The length of the karst cave is 33 m, and the depth is greater than 36 m. The depth is greater than the length. The width of the karst cave in the tunnel axis direction is 17 m.
(2) The karst cave extends to K40+222 (vertical depth of 20 m) in the direction of small mileage.
(3) Sericite slate below dolomite limestone at the bottom of the cave, which is inferred from the occurrence state, intersects the tunnel at K40+254 m.